Medicinal Plants of Asia and the Pacific

Medicinal Plants of Asia and the Pacific

Christophe Wiart, Pharm.D.
Ethnopharmacologist

Taylor & Francis
Taylor & Francis Group
Boca Raton London New York

CRC is an imprint of the Taylor & Francis Group,
an informa business

Published in 2006 by
CRC Press
Taylor & Francis Group
6000 Broken Sound Parkway NW, Suite 300
Boca Raton, FL 33487-2742

International Standard Book Number-10: 0-8493-7245-3 (Hardcover)
International Standard Book Number-13: 978-0-8493-7245-2 (Hardcover)
Library of Congress Card Number 2005036199

Library of Congress Cataloging-in-Publication Data

Wiart, Christophe.
  Medicinal plants of Asia and the Pacific / Christophe Wiart.
    p. ; cm.
  Includes bibliographical references and index.
  ISBN-13: 978-0-8493-7245-2 (hardcover : alk. paper)
  ISBN-10: 0-8493-7245-3 (hardcover : alk. paper)
  1. Medicinal plants--Asia. 2. Medicinal plants--Pacific Area. 3. Ethnopharmacology--Asia. 4. Ethnopharmacology--Pacific Area. 5. Traditional medicine--Asia. 6. Traditional medicine--Pacific Area. [DNLM: 1. Plants, Medicinal--Asia--Handbooks. 2. Plants, Medicinal--Pacific Islands--Handbooks. 3. Ethnopharmacology--Asia--Handbooks. 4. Ethnopharmacology--Pacific Islands--Handbooks. 5. Medicine, Oriental Traditional--Asia--Handbooks. 6. Medicine, Oriental Traditional--Pacific Islands--Handbooks. QV 735 W631m 2006] I. Title.

RS179.W53 2006
615'.321--dc22   2005036199

Taylor & Francis Group
is the Academic Division of Informa plc.

Visit the Taylor & Francis Web site at
http://www.taylorandfrancis.com

and the CRC Press Web site at
http://www.crcpress.com

Dedication

I owe a special thanks to my family for their generosity in creating and sustaining a domestic milieu conducive to my work.

Preface

When I began thinking about this book, I was guided by the wish to solve a dilemma. After 10 years of carefully conducted ethnopharmacological research, I could not help but conclude that the hundreds of molecules of clinical value awaiting discovery in the Pacific Rim might never be discovered while the global prevalence of cancers, cardiovascular diseases, and microbial infections continued to grow.

One possible reason for the slow rate of discovery of drugs from plants is the fact that there are a few researchers who master and rationally interconnect botany, pharmacology, traditional medicines, pharmacy, and modern medicine. One can perhaps envision the creation of a new discipline of science which would encompass all these disciplines.

For the time being most drugs that are discovered from plants result from enormous strikes of good luck. The idea to shed some light on the pharmacological potentials of medicinal flora of the Pacific Rim was thus born, and I undertook the laborious task of writing this extensive work on 36 families of medicinal plants of great topicality. Each of the 173 medicinal plants described in this book is of particular interest and should be viewed as a starting point for further research, which may result in the discovery of drugs. Each plant in this book is described as accurately as possible, which allows nonbotanists to recognize the samples, which are accompanied by personally made botanical plates. The traditional uses of each plant are provided and the rationality of these uses is described and explained using chemotaxonomy, pharmacology, and medicinal chemistry. In addition, detailed chemical structures and indications for further fruitful investigations are provided.

This book is written for all who are interested in participating in the task to find cures from the medicinal plants of the Pacific Rim. My hope is that the readers of this book will appreciate the wealth of knowledge and information that is available in the field of drug research from medicinal plants. First, this book will allow the active researcher to examine his or her own work in light of detailed accounts by scientists engaged in similar fields of research. Second, the researcher will profit from the hundreds of references to pertinent publications summarized and critically commented upon in this book. Third, a vast number of readers in the fields of pharmacology, medicine, biotechnology, veterinary medicine, and biochemistry, as well as nonscientists, will have the opportunity to undertake a pleasant and colorful journey through the medicinal flora of the Pacific Rim.

I am most indebted to the individuals who have contributed to the production of this book and who have done so much to guarantee its success.

Christophe Wiart
Kuala Lumpur, Malaysia

About the Author

Christophe Wiart was born in Saint Malo, France. He earned a Doctorate of Pharmacy from the University of Rennes in 1996 and is currently an associate professor of pharmacognosy at the University of Malaya, Kuala Lumpur, Malaysia. Dr. Wiart has been studying medicinal plants of the Pacific Rim for the last 10 years. His activities and accomplishments include patenting, conferences, plenary lectures, and the publication of several peer-reviewed research articles and academic books. Contact: christophe_wiart@yahoo.com

Contents

CHAPTER 1

Introduction

When writing this introduction I could not help but think of *Ethnotherapies in the Cycle of Life: Fading, Being and Becoming,* edited by Christine E. Gottschalk-Batschkus and Joy C. Green. The feeling I had after reading this beautiful book was somewhat uneasy as it prompted with some embarrassment the utopian idea that the eradication of human illnesses will only be achieved when shamanism, traditional medicines, and science work side by side. In other words, traditional medicines and shamanism supported by strict scientific research might give birth to a hybrid concept that could put an end to existing human diseases.

Shall we see professors of medicine and shamans working together? In all probability, "yes," because we have no alternative. The logic of biological systems never allows a complete victory over anything, including a victory of drugs against diseases. We all know that at this moment we are right in the middle of a furious battle for survival. Not so long ago, giving birth and coughing were often followed by death. Certainly, we cannot deny that antibiotics have greatly improved the treatment of bacterial infections. However, at the beginning of the 21st century, we have to admit that the war with bacteria is far from won because resistance is common. The same can be said for viruses, parasites, and cancer cells. Many people also need sleeping pills and antidepressants to get through the day or sleep at night because of our stressful lifestyles, and we are likely to be blighted further by the emergence of massive epidemics or new diseases, since Mother Nature is very creative.

What is left of traditional systems of medicines? With the daily depletion of acres of rain forests, not much is left, but there is still enough to cover the health needs of most of the world's population. The last 50 years were the theater for the first great pharmaceutical discoveries and, at the same time, saw the progressive disappearance of traditional knowledge. Shamans and other healers came to be regarded as charlatans and were abandoned even by their own peoples who preferred taking aspirin instead of drinking bitter decoctions of roots. This increasing lack of interest in natural remedies has to be accepted as inevitable given the potency of modern pharmacochemistry.

Does this mean an end for even the vestiges of shamanism, rituals, and traditional medicines? How can the past resist the continuing attack of modern medicine with its accusations of placebo effects, clinical disappointments, and lack of scientific evidence? Who can tell? But, based on past evidence, there is also the possibility of finding new plants that can "hit the jackpot" of therapeutic effectiveness. If the Amazon and to a lesser extent Africa have seen the disappearance of traditional medicine and medicinal flora, the Pacific Rim still boasts the richest pharmacopoeia of traditional medicines and medicinal plants; it can be regarded as the very last gift of Mother Nature in the cause of human health. The mass of bioactive molecules represented by the medicinal flora of the Pacific Rim is formidable indeed. In this book I have chosen to present 173 of these species. The plant choices were guided by the exciting fact that there have been few studies of these species for their pharmacological effect. Readers are invited to pursue further research with the possibility of drug discovery.

The 173 medicinal plants described in this book are classified by families, starting from the most primitive ones and moving onto more recent discoveries. A pharmacological or ethnological classification would have been possible, but I prefer the botanical one as it allows a broad logical view of the topic with chemotaxonomical connections. The medicinal plants presented in this book are classified according to their botanical properties in the philosophical tradition of de Candolle, Bentham, Hooker, Hallier, Bessey, Cronquist, Takhtajan, and Zimmerman, which is my favorite.

The approach used in this book is strictly scientific, given that I am a scientist and not a shaman. Perhaps shamanism and alternative practices will become included in the curricula of schools of medicine, but for the moment this is not the case. Plants are described here as accurately as possible, and I hope that their traditional uses are clearly presented. The pharmacotoxicological substantiation of these uses in the light of chemotaxonomy is also discussed. I have produced a carefully drawn figure for each plant and noted its geographic location, which allows for quick field recognition for further investigation. I have tried to use all the available data obtained from personal field collections, ethnopharmacological investigations, and available published pharmacochemical evidence. At the same time, I have attempted to provide some ideas and comments on possible research development. I hope that this book will contribute to the discovery of drugs from these plants.

The pharmacological study of medicinal plants of the Pacific Rim has only recently begun to be useful to researchers and drug manufacturers who see in it a source of new wealth. A field of more than 6000 species of flowering plants is awaiting pharmacological exploration. One reason for this lack of knowledge is the fact that most of these plants grow in rain forests, hence the difficulties in collecting them from remote areas where modern infrastructures are not available. Let us hope that the future will see more successful business and scientific ventures between developing countries and developed ones with fair distribution of benefits, including those to villagers and healers who may have helped in finding "jackpot" plants.

The first 24 species of medicinal plants described are part of the Magnoliidae, which are often confined to primary tropical rain forests. Their neurological profile is due to the fact that neuroactive alkaloids are evenly distributed throughout the subclasses: Annonaceae, Myristicaceae, Lauraceae, Piperaceae, Aristolochiaceae, and Menispermaceae. These are often trees or woody climbers that can provide remedies for the treatment of abdominal pains, spasms, putrefaction of wounds, and inflammation, as well as curares for arrow poisons and medical derivatives.

A commonplace but interesting feature of these plants is their ability to elaborate isoquinoline alkaloids (benzylisoquinolines or aporphines), phenylpropanoids and essential oils, piperidine alkaloids phenylpropanoids, and nitrophenanthrene alkaloids. Alkaloids are of particular interest here as they may hold some potential as sources of anticancer agents, antibiotics, antidepressants, and agents for treating Alzheimer's and Parkinson's diseases.

The evidence presented so far clearly demonstrates that members of the family Annonaceae elaborate a surprisingly broad array of secondary metabolites that inhibit cancerous cells, including acetogenins, styryl-lactones, and isoquinoline alkaloids. Aristolochiaceae have attracted much interest in the study of inflammation, given their content of aristolochic acid and derivatives that inhibit phospholipase A_2. Other antiinflammatory principles may be found in the Myristicaceae, which produce a series of unusual phenylacylphenols. The evidence in favor of dopaminergic, serotoninergic, and GABA (gamma-amino butyric acid)-ergic alkaloids in the Magnoliidae is strong and it seems likely that anxiolytic or antidepressant agents of clinical value might be characterized from this taxon. Alkaloids of the Magnoliidae are often planar and intercalate with DNA, hence their anticancer properties. The Annonaceae and Lauraceae families abound with aporphinoid alkaloid topoisomerase inhibitors.

The next 42 species are members of the Dilleniidae, Elaeocarpaceae, Bombacaceae, Flacourtiaceae, Ebenaceae, Myrsinaceae, Cucurbitaceae, Passifloraceae, and Capparaceae. Most of these are used as antiinflammatory, counterirritant, or antiseptic agents in gynecological disorders. In comparison to the former group, the medicinal plants here abound with saponins which are cytotoxic, antiseptic, antiinflammatory, diuretic, and mucolytic; they elaborate a broad array of chemicals — cytotoxic oligostilbenes, quinones (Nepenthales), isothiocyanates (Capparales), cucurbitacins (Malvales and Violales), and naphthylisoquinoline alkaloids (Violales). Myrsinaceae produce an unusual series of benzoquinones, which have displayed a surprising number of pharmacological activities, ranging from inhibition of pulmonary metastasis and tumor growth to inhibition of lipooxygenase. Ebenaceae, particularly the *Diospyros* species, have attracted a great deal of interest for their dimers and oligomers of naphthoquinones which are antibacterial, antiviral, monoamine oxidase inhibitors, and cytotoxic via direct binding of topoisomerase. Note that Polygonaceae, Myrinaceae, and Ebenaceae are quinone producing families. Myrinaceae, Ebenaceae, and Sapinaceae abound with saponins.

Elaeocarpaceae elaborate an interesting series of indolizidine alkaloids derived from ornithine and cucurbitacins. Cucurbitacins are oxygenated steroids with chemotherapeutic potential, which have so far been found in the Cucurbitaceae, Datiscaceae, and Begoniaceae families. Capparaceae use isothiocyanates (mustard oils) as a chemical defense; they can make a counterirritant remedy. Isothiocyanates are interesting because they are cytotoxic, antimicrobial, and irritating, hence the use of Capparales to make counterirritant remedies. Medicinal Flacourtiaceae accumulate a series of unusual cyclopentanic fatty acids with potent activity against *Mycobacterium leprae,* hence their use to treat leprosy.

There are 68 species of medicinal plants belonging to the Rosidae, of which the families Connaraceae, Rosaceae, Anisophylleaceae, Thymeleaceae, Melastomataceae, Rhizophoraceae, Olacaceae, Icacinaceae, Euphorbiaceae, Sapindaceae, Anacardiaceae, Simaroubaceae, Meliaceae, and Rutaceae are presented in this book. Rosidae are in general tanniferous and provide astringent remedies that are used to check bleeding, to stop diarrhea and dysentery, to heal and inhibit the formation of pus, to cool, and to lower blood pressure. Tannins, which are often removed in extraction processes since they provide false positive results in high-throughput screenings, hold enormous pharmacological potential. With regard to the antineoplastic potential of Euphorbiaceae, most of the evidence that has emerged from the last 30 years lends support to the fact that they represent a vast reservoir of cytotoxic agents; one may reasonably expect the isolation of original anticancer drugs from this family if enough work is done.

Other principles of interest are essential oils, and oxygenated triterpenes in the Simaroubaceae, Meliaceae, and Rutaceae. The latter is of particular interest as a source of agents for chemotherapy. Rutaceae have attracted a great deal of interest for their ability to elaborate a series of cytotoxic benzo[c]phenanthridine and acid in alkaloids, a number of derivatives of which are of value in the treatment of acute leukemia in adults and malignant lymphomas, refractory to conventional therapy.

The last group of medicinal plants described encompasses Loganiaceae, Gentianaceae, Apocynaceae, Asclepiadaceae, Solanaceae, and Verbenaceae, making a total of 37 medicinal plants that are often used as analgesics, antipyretics, antiinflammatories, and to make poisons. These are plants with tubular flowers grouped in the Asteridae. The chemical weapons found in this subclass are mostly monoterpenoid indole alkaloids, pyrrolizidine alkaloids, iridoid glycosides, phenylethanoid glycosides, cardiotoxic glycosides, naphthoquinones, diterpenes, and sesquiterpenes. The most common medicinal properties of these plants are those of alkaloids, saponins, and iridoids. Alkaloids of the Apocynaceae are historically of value in fighting cancer, but many other molecules await discovery.

CHAPTER 2

Medicinal Plants Classified in the Family Annonaceae

2.1 GENERAL CONCEPT

One of the most exciting families of medicinal plants to start with when prospecting the flora of the Asia–Pacific for drugs is the Annonaceae (A. L. de Jussieu, 1789 nom. conserv., the Custard Apple Family). Annonaceae are widespread in the tropical world as a broad variety of trees, climbers, or shrubs which are quite easily spotted by their flowers that have a pair of whorls of leathery petals and groups of club-shaped fruits containing several seeds in a row (Figure 2.1). The inner bark itself is often fragrant and the plant is free of latex or sap; another feature is that the leaves are simple, alternate and exstipulate. In the Asia–Pacific, approximately 50 species from this family are medicinal, but to date there is not one on the market for clinical uses, a surprising fact since some evidence has already been presented that members of this family have potential for the treatment of cancer, bacterial infection, hypertension, and brain dysfunctions. Basically, there are three major types of active principles in this family: acetogenins, which often confer insecticidal properties, and isoquinolines and diterpenes of the labdane type (Figure 2.2).

Figure 2.1 Fruits of Annonaceae with club-shaped ripe carpels. **(See color insert following page 168.)**

2.2 *FISSISTIGMA LANUGINOSUM* (HK. F. ET TH.) MERR.

[From: Latin *fiss* = cleave and Greek *stigma* = mark made by pointed instrument, and Latin *lanuginosum* = wooly.]

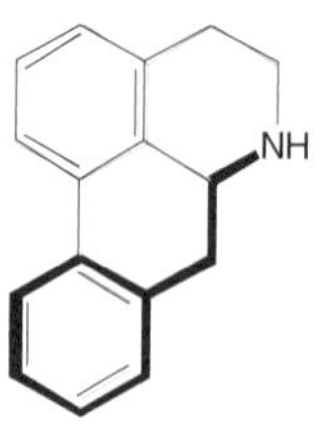

Isoquinoline

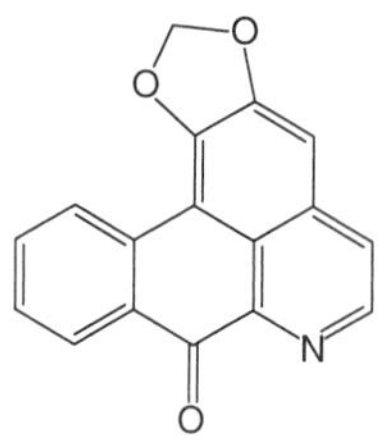

Liriodenine

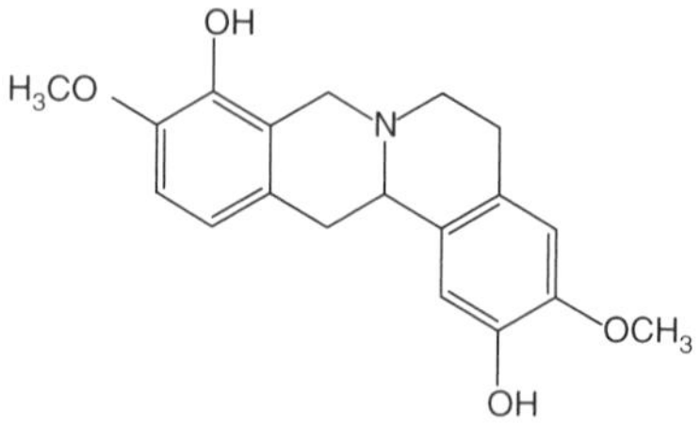

Discretamine

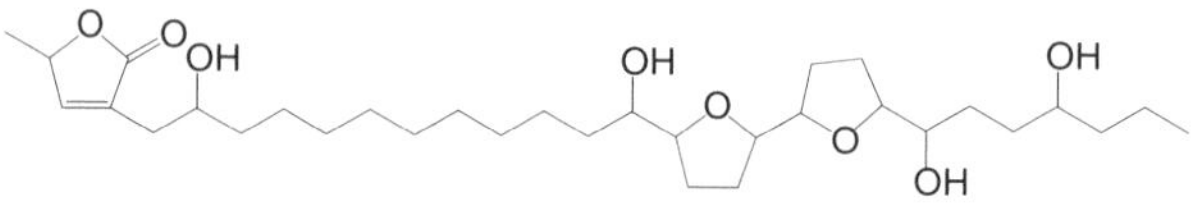

Acetogenine

Figure 2.2   Examples of bioactive natural products characteristic of the family Annonaceae.

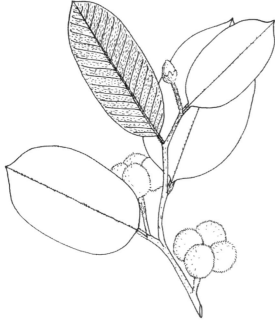

Figure 2.3   *Fissistigma lanuginosum* (Hk. f. et Th.) Merr. [From: Flora of Malaya. FRI No: 023427. Geographical localization: National Park, Pahang. Alt.: 700ft.]

2.2.1   Botany

Fissitigma lanuginosum (Hk. f. et Th.) Merr. (*Melodorum lanuginosum* Hk. et Th. and *Uvaria tomentosa* Wall.) is a climber which grows wild in the primary rain forests of Cambodia, Laos, Vietnam, and Peninsular Malaysia. The stems are rusty, tomentose, woody, and with numerous lenticels. The leaves are simple, alternate, exstipulate, dark green and glossy above, oblong or oblong–obovate. The midrib above is rusty and pubescent, and the entire lower surface is densely rufous. The blade is 9cm – 21cm × 4cm – 8cm. The petiole is 1.5cm long. The flowers are arranged in terminal cymes. The sepals are 1–1.5cm long and rufous. The petals are coriaceous, oblong–lanceolate, the outer petals are up to 3.5cm long, while the inner ones are smaller. The fruits are ripe carpels which are subglobose, 2cm in diameter and dark brown (Figure 2.3).

2.2.2 Ethnopharmacology

The Malays drink a decoction of roots as a postpartum remedy and to treat stomach troubles. Pedicin (2′,5′-dihydroxy-3′,4′,6′-trimethoxychalcone) from the plant inhibited tubulin assembly into microtubules with IC_{50} value of 300μM. Other chalcones, fissistin, and isofissistin are cytotoxic against KB cells.[1]

2.3 *FISSISTIGMA MANUBRIATUM* (HK. F. ET TH.)

[From: Latin *fiss* = cleave and Greek *stigma* = mark made by pointed instrument.]

2.3.1 Botany

Fissistigma manubriatum (Hk. f. et Th.) or *Melodorum manubriatum* Hk. f. et Th., *Uvaria manubriatum* Wall., *Melodorum bancanum* Scheff., *Melodorum korthalsii* Miq., is a stout climber which grows in the primary rain forest of Borneo. The plant can reach 30m long. The stems are rusty, tomentose when young. The leaves are 5cm – 13.5cm × 2 cm – 4.5cm, thinly coriaceous, dark green above, adpressed–tomentose beneath, oblong–lanceolate, acute or acuminate. The base of the blade is rounded. The blade shows 12–18 pairs of secondary nerves. The flowers have a faint melon fragrance and are attached to a 1cm-long, rusty, tomentose pedicel. The sepals are ovate–lanceolate, 3-nerved, tomentose and 7mm long. The petals are coriaceous, ovate–lanceolate, the outer 2–2.3cm long, the inner smaller. The fruits are ovoid globose ripe carpels which are tomentose and 2.5–3cm long (Figure 2.4).

Figure 2.4 *Fissistigma manubriatum* (Hk. f. et Th.). [From: Flora of Malay Peninsula. Field No: 6423. Field collector: H. I. Burkill, June 18, 1921. Botanical identification: J. Sinclair, Jan. 10, 1949.]

2.3.2 Ethnopharmacology

In Malaysia, a decoction of the roots is used as a drink to assuage stomachaches. It is most likely effective because of its content of isoquinoline alkaloids which are known to block the muscarinic receptors and therefore impede the secretion of gastric juices and the contraction of gastric smooth muscles.[2]

2.4 *PHAEANTHUS EBRACTEOLATUS* (PRESL.) MERR.

[From: Latin *ebracteolatus* = without bracteole.]

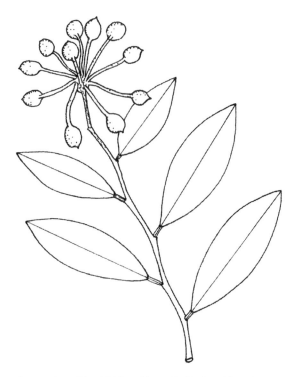

Figure 2.5   *Phaeanthus ebracteolatus* (Presl.) Merr. [From: Philippines Plant Inventory. Flora of the Philippines. Joint Project of the Philippine Natural Museum, Manila and B. P. Bishop Museum Honolulu. Supported by NSF/USAID. Field collectors: F. J. M. Garland et al. Geographical localization: Luzon, Llocos Norte Province, Mangoratao. Alt.: 60m. No: 9858. Date: Aug. 22, 1992.]

2.4.1 Botany

Phaeanthus ebracteolatus (Presl.) Merr. is a tree that grows to a height of 6m in the rain forests of the Philippines. The leaves are simple, alternate, exstipulate, and glossy dark green. The blade is lanceolate and 12cm – 9cm × 3.5cm – 5cm with discrete secondary nerves. The fruits are yellow to orange or red ripe carpels which are numerous, and are 1cm long on 4cm-long pedicels (Figure 2.5).

2.4.2 Ethnopharmacology

The plant contains a *bis*-benzylisoquinoline known as phaeantharine which has shown some potential as an insecticidal agent and exhibited some levels of antibacterial activity.[3,4] It would be interesting to learn whether this plant has any neuropharmacological potential.

REFERENCES

1. Alias, Y., Awang, K., Hadi, A. H., Thoison, O., and Pais, M. 1995. An antimitotic and cytotoxic chalcone from *Fissistigma lanuginosum*. *J. Nat. Prod.*, 58, 1160.
2. Lin, C. H., Chang, G. J., Su, M. J., Wu, Y. C., Teng, C. M., and Ko, F. N. 1994. Pharmacological characteristics of liriodenine, isolated from *Fissistigma glaucescens*, a novel muscarinic receptor antagonist in guinea-pigs. *Br. J. Pharmacol.*, 113, 275.
3. Knabe, J., Baldauf, J., and Hanke, B. 1988. Biological activities of phaeantharine chloride and some synthetic intermediates. *Arch. Pharm.*, 321, 35.
4. Van Beek, T. A., Verpoorte, R., Svendsen, A. B., Santos, A. C., and Olay, L. P. 1983. Revised structure of phaeantharine. *J. Nat. Prod.*, 46, 226.

CHAPTER 3

Medicinal Plants Classified in the Family Myristicaceae

3.1 GENERAL CONCEPT

The family Myristicaceae (R. Brown, 1810 nom. conserv., the Nutmeg Family) consists of approximately 16 genera and 380 species of tropical rain forest trees, which are in field collection, recognized easily by making a cut in the bark from which will exude a blood-like sap. Myristicaceae have attracted a great deal of interest since they produce indole alkaloids, which might hold potential for the treatment of depression and other central nervous system (CNS) diseases. *N,N*-dimethyl tryptamine, 5-methoxy-*N,N*-dimethyl tryptamine, 2-methyl-1,2,3,4-tetrahydro-β-carboline have been identified with *Virola sebifera*, which is used by South American shamans to cause hallucination (Figure 3.1). Other interesting principles from Myristicaceae are phenylacylphenols and phenylpropanoids. Examples of phenolic compounds of pharmacological value in Myristicaceae are kneracheline A and B, from *Knema furfuracea*, which inhibit the proliferation of bacteria cultured *in vitro*; also 3-undecylphenol and 3-(8Z-tridecenyl)-phenol from *Knema hookeriana*, which inhibit the proliferation of *Bursaphelechus xylophilus* cultured *in vitro* with a maximum effective dose of 4.5mg/cotton ball and 20mg/cotton ball, respectively.[1,2]

Note that phenolic compounds from the stem bark of *Knema glomerata* inhibit moderately the proliferation of human tumor cell lines cultured *in vitro*.[3] Phenylpropanoids are centrally active and myricetin and elemicin from nutmeg (*Myristica fragrans* Houtt.) are narcotic. In the Pacific Rim, approximately 20 species of plants classified within the family Myristicaceae are medicinal.

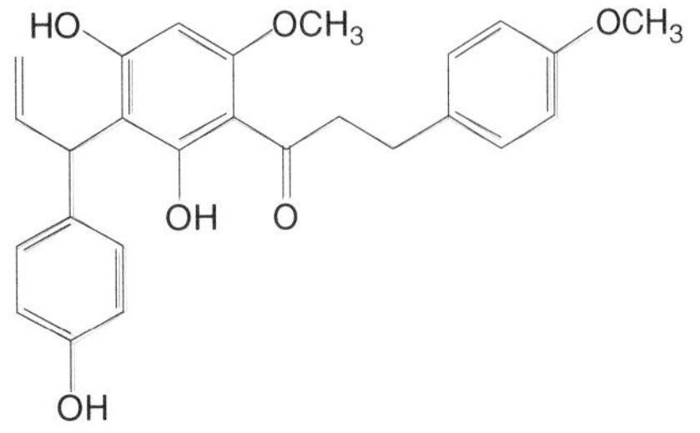

Iryantherin A

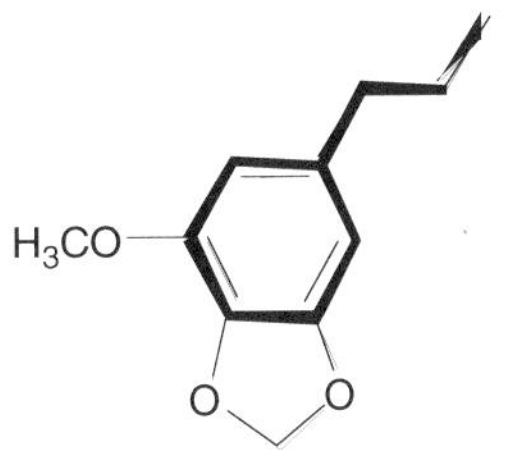

Myristicin

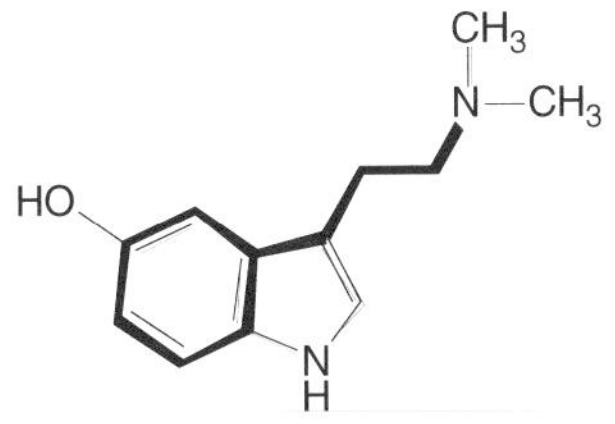

5-Hydroxy-*N, N*-dimethyl tryptamine

Figure 3.1   Examples of bioactive natural products from the family Myristicaceae.

Figure 3.2   *Knema glaucescens* Jack. [From: Flora of Borneo. Bukit Raya Expedition. Veldkamp No: 8522. Feb. 4, 1984. Geographical localization: Borneo, Batu Badinging, KCT, 47Km, 113°50′ E, 1°15′ S. 96Km, in primary dipterocarp forest.]

3.2 *KNEMA GLAUCESCENS* JACK

[From: Greek *knema* = internode and *glaucescens* = somewhat glaucous.]

3.2.1   Botany

Knema glaucescens Jack (*Knema palembanica* Warb.) is a tree that grows in the rain forest of Indonesia and Borneo to a height of 15m. The bark exudes a red sap after being incised. The stems are 4mm in diameter with a velvety apex. The leaves are simple, spiral, and exstipulate. The petiole is 7mm × 2mm, and velvety. The blade is lanceolate, shows 22 pairs of secondary nerves, and is 11.3cm × 3.2cm – 12.8cm × 2.6cm – 13.3cm × 3.6cm – 13cm × 3.2cm. The midrib is velvety above and the blade is glaucous below. The fruits are ovoid, and are 2.2cm × 1.7cm on an 8mm pedicel (Figure 3.2).

3.2.2   Ethnopharmacology

The plant is called *Kumpang* by the Iban tribes of Sarawak where a decoction of bark is used to treat abdominal discomforts. The pharmacological properties are unexplored. Are serotonin-like principles present here?

3.3 *KNEMA GLOBULARIA* (LAMK.) WARB.

[From: Greek *knema* = internode and Latin *globulus* = globe.]

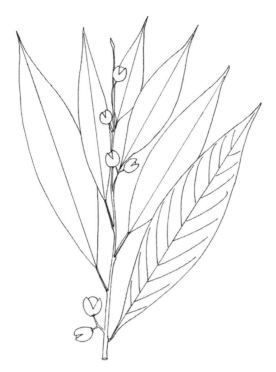

Figure 3.3   *Knema globularia* (Lamk.) Warb. [From: Distributed from the Botanic Gardens Singapore. Geographical localization: Duchong Island, Pahang. Sea level. Aug. 22, 1935. Botanical identification: J. Sinclair, Nov. 5, 1963. Field collector: E. J. Corner.]

3.3.1   Botany

Knema globularia (Lamk.) Warb. (*Myristica globularia* Lamk., *Myristica lanceolata* Wall., *Knema corticosa* Lour., *Knema corticosa* Lour. var. *tonkinensis* Warb., *Knema missionis* [Wall.] Warb., *Knema petelotii* Merr., *Knema sphaerula* [Hook. f.] Airy Shaw, *Knema wangii* Hu, *Myristica corticosa* [Lour.] Hook. et Thoms., *Myristica glaucescens* Hook., *Myristica sphaerula* Hook., and *Myristica missionis* Wall. ex King) is a tree that grows to a height of 15m with a girth of 25cm in the primary rain forests of China and Southeast Asia. The bark is grayish-brown, and exudes a red sap after incision. The stems are rusty tomentose at the apex. The leaves are simple, exstipulate, and spiral. The petiole is 1.5cm long. The blade is thin, oblong, lanceolate, 16cm × 3.9cm – 11cm × 2cm. The apex is acute or acuminate, the base is broadly cuneate to suborbicular, and shows 19 pairs of secondary nerves. The fruits are globose and yellow, 1.3cm × 1.2cm. The seeds are solitary and enveloped in a red aril (Figure 3.3).

3.3.2   Ethnopharmacology

The plant is known as Seashore Nutmeg, Small-Leaved Nutmeg, and *xiao ye hong guang shu* (Chinese). In Cambodia, Laos, and Vietnam, the seeds are used as an ingredient for an external preparation used to treat scabies. The therapeutic potential of *Knema globularia* (Lamk.) Warb. is unexplored. Knerachelimes with antibacterial potential are elaborated by this plant.

3.4 *MYRISTICA ARGENTEA* WARB.

[From: Greek *muron* = a sweet juice distilled from plants and Latin *argentea* = silvery.]

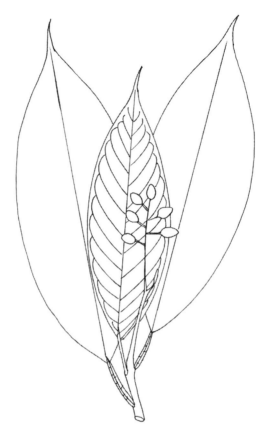

Figure 3.4   *Myristica argentea* Warb. [From: BOSWESEN. Nederlands Nieuw-Guine. Forestry Division Neth. New Guinea). Field collector: C. Kalkman. No: BW 6346. June 21, 1958. Botanical identification: J. Sinclair. Nov. 13, 1962. Geographical localization: Nederland's New Guinea, Fak–Fak, Agricultural Exp. Gard. Alt.: 75m.]

3.4.1   Botany

Myristica argentea Warb. is a tree that grows in the primary rain forests of Papua New Guinea. The leaves are simple and spiral. The petiole is stout, cracked transversally, channeled, and 2.8cm long. The blade is glossy, 20cm × 6.4cm – 13.5cm × 5.6cm – 19cm × 6cm, elliptic, acuminate at the apex in a tail, and shows 13–18 pairs of secondary nerves. The inflorescences are 4.5cm-long racemes. The fruits are globose and 6mm long (Figure 3.4).

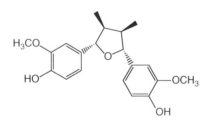

Nectandrin B

Figure 3.5

3.4.2   Ethnopharmacology

The plant is known as Macassar mace, female nutmeg, horse nutmeg, long nutmeg, Macassar nutmeg, New Guinea nutmeg, Papua mace, and Papua nutmeg. The fruits are used to treat diarrhea and to stimulate venereal appetite in Indonesia where it is called *pala negri, pala papoes*. The mace *Myristica argentea* Warb. abounds with a series of diaryldimethylbutane lignans of possible pharmacological value. Such lignans are erythro-austrobailignan-6 and meso-dihydroguaiaretic acid, myristargenol A, and myristargenol B from the aril of the seeds, and show some levels of activity against *Streptococcus mutans*.[4,5]

Erythro-austrobailignan-6, meso-dihydroguaiaretic acid, and nectandrin-B exert an antiproliferative effect on MCF-7 cells as well as antioxidant activity on the 1,1-diphenyl-2-picrylhydrazyl (DPPH) radical. In addition, Nectandrin-B (Figure 3.5) inhibits the enzymatic activity of 17β-hydroxysteroid dehydrogenase and antiaromatase activities.[6] Is the aphrodisiac property of the fruit linked to hormonal mechanisms?

3.5 *MYRISTICA ELLIPTICA* WALL. EX HOOK. F. THOMS.

[From: Greek *muron* = a sweet juice distilled from plants and Latin *elliptica* = elliptical, about twice as long as wide.]

3.5.1 Botany

Myristica elliptica Wall. ex Hook. f. Thoms. (*Myristica elliptica* var. *elliptica* J. Sinclair) is a large buttressed tree that grows to 10m in the primary rain forest of Southeast Asia in rain forest swamps and riverbanks. The bark exudes a sticky red sap after incision. The leaves are simple and exstipulate. The petiole is fissured, 2cm long, and channeled above. The blade is elliptic, 17cm × 6cm – 16cm × 5cm and shows 7–12 pairs of secondary nerves. The fruits are conspicuous, and up to 7cm × 5cm and attached to a 4mm-diameter pedicel (Figure 3.6).

3.5.2 Ethnopharmacology

In the Philippines, the seeds or a paste of bark is applied to itchy parts of the body. In Malaysia, the fruit is known as *buah penarahan* and known to be stupefying. The pharmacological potential of this plant is to date unexplored. One may, however, set the hypothesis that the stupefying property is owed to a series of phenylpropanoids.

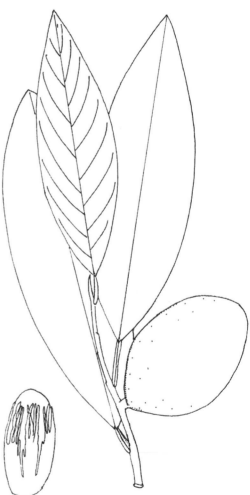

Figure 3.6   *Myristica elliptica* Wall. ex Hook. f. Thoms. [From: Phytochemical Survey of the Federation of Malaysia. KL No: 1530. June 3, 1959. Geographical localization: Ulu Langat, Selangor. Hill forest. Field collector: G. Umbai for A. N. Millard. Botanical identification: K. M. Kochummen]

REFERENCES

1. Alen, Y., Nakajima, S., Nitoda, T., Baba, N., Kanzaki, H., and Kawazu, K. 2000. Two antinematodal phenolics from *Knema hookeriana,* a Sumatran rain forest plant. *Z. Naturforsch.,* 55, 300.

2. Alen, Y., Nakajima, S., Nitoda, T., Baba, N., Kanzaki, H., and Kawazu, K. 2000. Antinematodal activity of some tropical rain forest plants against the pinewood nematode, *Bursaphelenchus xylophilus. Z. Naturforsch.,* 55, 295.

3. Zeng, L., Gu, Z. M., Fang, X. P., and McLaughlin, J. L. 1994. Kneglomeratanol, kneglomeratanones A and B, and related bioactive compounds from *Knema glomerata. J. Nat. Prod.,* 57, 76.

4. Filleur, F., Pouget, C., Allais, D. P., Kaouadji, M., and Chulia, A. J. 2002. Lignans and neolignans from *Myristica argentea* Warb. *Nat. Prod. Lett.,* 16, 1.

5. Nakatani, N., Ikeda, K., Kikuzaki, H., Kido, M., and Yamaguchi, Y. 1988. Diaryldimethylbutane lignans from *Myristica argentea* and their antimicrobial action against *Streptococcus mutans. Phytochemistry,* 27, 3127.

6. Filleur, F., Le Bail, J. C., Duroux, J. L., Simon, A., and Chulia, A. J. 2001. Antiproliferative, anti-aromatase, anti-17beta-HSD and antioxidant activities of lignans isolated from *Myristica argentea. Planta Med.,* 67, 700.

CHAPTER 4

Medicinal Plants Classified in the Family Lauraceae

4.1 GENERAL CONCEPT

The family Lauraceae (A. L. de Jussieu, 1789 nom. conserv., the Laurel Family) consists of 50 genera and 2000 species of trees and shrubs which are recognized in field collection by their aroma, the bark which is smooth and thick, the bay-like leaves, and their drupaceous fruits which are glossy and ovoid seated on a cupular vestigial perianth. *Laurus nobilis* L. (Sweet Bay Laurel, *Lauri fructus*; *Swiss Pharmacopoeia* 1934), *Cinnamomum zeylanicum* Nees (cinnamon), *Cinnamomum camphora* (L.) T. Nees & Eberm. (camphor), *Persea americana* Miller (avocado), *Sassafras albidum* (Nutt.) Nees (sassafras oil), *Umbellularia californica* (California Bay Laurel), *Persea nanmu* Oliv. (*nan-mu* wood), *Nectandra rodiaei* Schk. (green, heartwood), *Eusideroxylon zwageri* (ironwood), and *Ocotea bullata* E. Mey. are classical examples of Lauraceae. This family is interesting because the alkaloids it produces are cytotoxic and neuroactive (Figure 4.1). In the Asia–Pacific, there are approximately 150 species of plants classified within the family Lauraceae among which are *Cinnamomum sintoc*, *Beilschmiedia pahangensis*, *Beilschmiedia tonkinensis* Ridl., *Cryptocarya griffithiana*, *Litsea cubeba*, *Litsea odorifera*, and *Litsea umbellata*, which are discussed in this chapter.

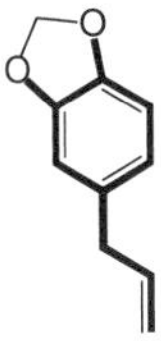

Safrole

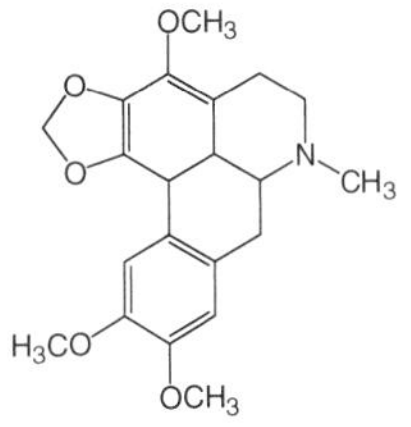

Ocoteine

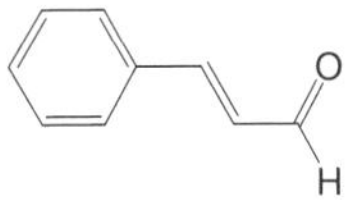

Trans-cinnamaldehyde

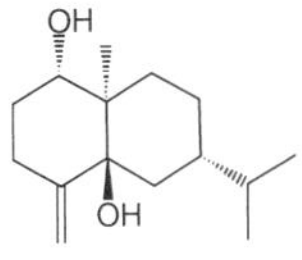

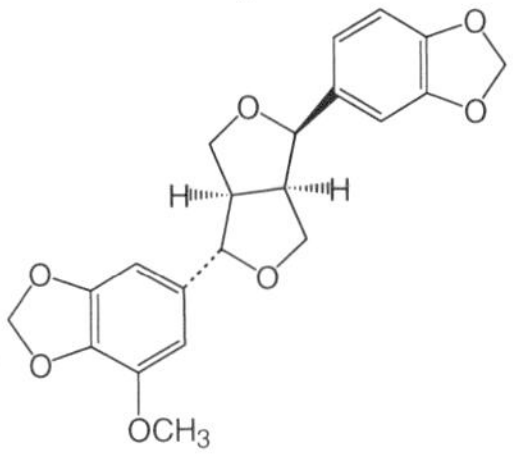

Verticillatol Demethoxyeoiexcelsin

Figure 4.1 Examples of bioactive natural products characteristic of the family Lauraceae.

4.2 *CINNAMOMUM SINTOC* BL.

[From: Greek *kinnamon* = cinnamon and Javanese *sintok* = vernacular name of *Cinnamomum sintoc* Bl.]

Figure 4.2   *Cinnamomum sintoc* Bl. [From: July, 28, 1998. Field collector: F. Mohd. Geographical localization: Larut Hill, Taiping. Alt.: 500m. FRI No: 42 939. Botanical identification: A. S. Mat.]

4.2.1 Botany

Cinnamomum sintoc Bl. (*Cinnamomum cinnereum* Gamb.) is a tall tree which grows to a height of 40m with a girth of 2.5m. The plant is quite common on the hill forests of Thailand, Indonesia, and Malaysia. The bark is gray–brown, smooth to shallow fissured. The inner bark is reddish with a strong aromatic smell. The sapwood is pale whitish. The leaves are simple, exstipulate, and subopposite. The petiole is 0.8–1.8cm long. The blade is leathery, ovate, lanceolate, 7cm – 22cm × 3cm – 8.5cm, and blunt at the apex. The margin of the leaves is characteristically wavy. The blade shows 3–4 pairs of secondary nerves. The inflorescences are axillary panicles that are up to 15cm long. The flowers are white to pale yellowish. The fruits are oblong, 1.8cm × 0.8cm seated on a cup-shaped entire rimmed perianth (Figure 4.2).

4.2.2 Ethnopharmacology

The plant is an esteemed remedy for chronic diarrhea and as an antispasmodic by the natives of the Malay coast of New Guinea where it is known as *sintok*. The pharmacological potential of *Cinnamomum iners* Reinw. ex Bl. would be worth studying, as interesting findings have been made in other *Cinnamomum* species such as the antidiabetic effect of *Cinnamomum cassia* and *Cinnamomum zeylanicum in vivo* and *in vitro*.[1]

4.3 *BEILSCHMIEDIA PAHANGENSIS* GAMB.

[After K. T. Beilschmied (1793–1848), pharmacist, and from Latin *Pahangensis* = from Pahang.]

4.3.1 Botany

Beilschmiedia pahangensis Gamb. is a tree which grows to a height of 15m and a girth of 90cm. The plant is quite common along the riverbanks in primary rain forests of South Thailand, Pahang, Kelantan, and Perak. The stems are slender and slightly flattened. The leaves are simple, alternate, and exstipulate. The petiole is 0.5–1cm long. The blade is elliptic to lanceolate, 7cm – 15cm × 2cm – 5.5 cm. The apex is blunt and the base is cuneate. The blade shows 5–10 pairs of secondary nerves. The flowers are arranged in axillary panicles. The fruits are ellipsoid–oblong, 3.5cm × 1.3cm, with a blunt apex and base (Figure 4.3).

Figure 4.3 *Beilschmiedia pahangensis* Gamb. [From: Flora of Malaya. Comm. Ex. Herb. Hort. Bot. Sing. Field collector: M. Shah. Nor. No: 2040. Geographical localization: Jeram Panjang, Pahang.]

4.3.2 Ethnopharmacology

In Peninsular Malaysia, a decoction of bark is used as a drink as a protective remedy after childbirth; it is also used to assuage stomach pains and to treat diarrhea. To date the pharmacological potential of this plant is unknown. Dehatrine *bis*-benzylisoquinoline alkaloid from the Indonesian medicinal plant, *Beilschmiedia madang* Bl. inhibits the survival of *Plasmodium falciparum* K1 strain (chloroquine resistant) cultured *in vitro* with similar activity to quinine.[2]

4.4 *BEILSCHMIEDIA TONKINENSIS* RIDL.

[After K. T. Bielschmied (1793–1848), pharmacist, and from Latin *tonkinensis* = from Tonkin in Indochina.]

4.4.1 Botany

Beilschmiedia tonkinensis Ridl. is a tree which grows to a height of 15m and a girth of 120cm in the rain forests of Vietnam, Cambodia, Laos, Thailand, and Malaysia. The stems are pale whitish. The leaves are aromatic, simple, alternate, and exstipulate. The petiole is 1–1.25cm long. The blade is leathery, elliptic, 7cm – 18cm × 3cm – 6cm. The apex is blunt and the base is cuneate. The blade

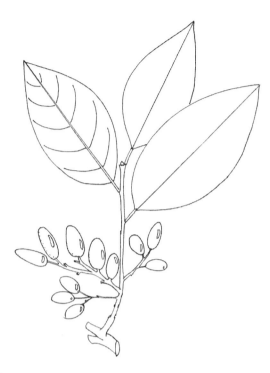

Figure 4.4 *Beilschmiedia tonkinensis* Ridl. [From: Flora of Malay Peninsula. Forest Department. Geographical localization: Kuala Rompin, Pahang. April 8, 1921. No: 4181. Botanical identification: I. H. Burkill.]

shows 6–9 pairs of secondary nerves, as well as tertiary nerves. The flowers are hairy and arranged in axillary panicles. The fruits are oblong, 2.5cm × 1.5cm with a slender 1cm-long stalk (Figure 4.4).

4.4.2   Ethnopharmacology

The leaves of the plant are used by Indonesians and Malays, who call it *medang pungok* or *medang serai*, to make poultices for application to broken bones. The pharmacology is unexplored.

4.5 *CRYPTOCARYA GRIFFITHIANA* WIGHT

[From: Greek *kryptos* = hidden and *karyon* = nut, and after W. Griffith (1810–1845), doctor and botanist of the East India Company.]

4.5.1   Botany

Cryptocarya griffithiana Wight is a tree that grows to a height of 20m and is 125cm in girth. The plant grows wild in the primary rain forests of Burma, Thailand, Malaysia, Indonesia, Borneo, and the Philippines. The bole is brownish and scaly. The inner bark is reddish-brown and granular. The sapwood is pale yellow. The stems are stout and covered with reddish-brown velvety hairs. The leaves are simple, exstipulate, and leathery. The petiole is 0.7–2.5cm long and velvety. The blade is elliptic to oblong, 12cm – 32cm × 8cm – 15cm. The upper surface is glabrous except for the midrib. The blade shows 5–8 pairs of secondary nerves. The lower surface is glaucous and densely velvety. The apex is rounded and the base is asymmetrical. The flowers are arranged in

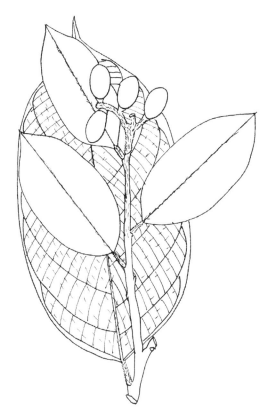

Figure 4.5 *Cryptocarya griffithiana* Wight. [From: Flora of Malaya. No: 13611. Geographical localization: Central Pahang, Tapah Hill, South Boundary Krau Game Reserve. Nov. 9, 1969. B. Everett.]

terminal and axillary reddish panicles. The fruits are greenish, oblong to ovate, and 2.5cm × 1.5cm (Figure 4.5).

4.5.2 Ethnopharmacology

The plant is not medicinal but has the reputation in Southeast Asia for being poisonous. African medicinal plants *Cryptocarya latifolia* Sonder, *Cryptocarya myrtifolia* Stapf., *Cryptocarya transvaalensis* Burtt Davy, *Cryptocarya woodii* Engl., and *Cryptocarya wyliei* Stapf., inhibit *in vitro* the enzymatic activity of COX-1 and COX-2.[3] What about *Cryptocarya griffithiana* and Southeast Asian congeners?

4.6 *CRYPTOCARYA TOMENTOSA* BL.

[From: Greek *kryptos* = hidden and *karyon* = nut, and from Latin *tomentosa* = densely covered with matted wool or short hair.]

4.6.1 Botany

Cryptocarya tomentosa Bl. is a medium-sized tree that grows to a height of 20m with a girth of 105cm in the primary rain forests of Thailand, Malaysia, Borneo, and Indonesia. It grows to

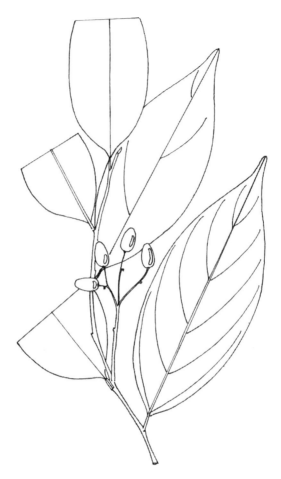

Figure 4.6 *Cryptocarya tomentosa* Bl. [From: Flora of Malaya. FRI No: 11592. Geographical localization: Lesong Botanical identification: F. S. P. Ng, Feb. 27, 2004.]

a height of 1000m in lowland to upper hill forests. The bark is reddish and finely scaly. The bole is buttressed. The inner bark is deep yellow, turning brown on exposure. The petiole is 0.5–1.5cm long, and finely hairy. The blade is elliptic, to oblong, to lanceolate, and 9.5cm – 20cm × 4cm – 9cm. The apex has a pointed base that is cuneate, and the lower surface is faintly glaucous and finely hairy. The midrib is sunken above and there are 6–8 pairs of secondary nerves. The tertiary nerves are scalariform. The fruits are ellipsoid, 2.8cm × 1.5cm, fleshy, and very black (Figure 4.6).

4.6.2 Ethnopharmacology

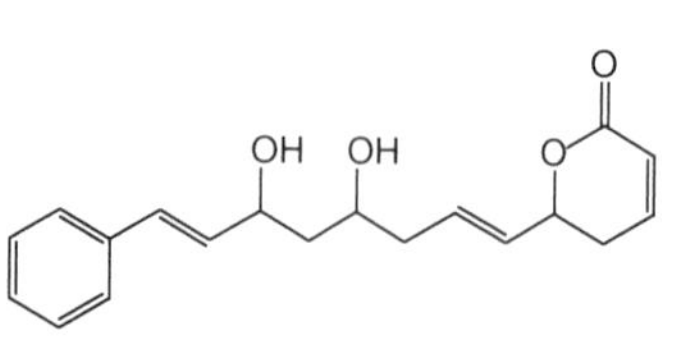

Cryptofolione

Figure 4.7

The bark of *Cryptocarya* has the reputation in Southeast Asia of being poisonous, probably due to substances of an isoquino-line-like nature. Note that *Cryptocarya* are interesting for the pyrone they elaborate, such as cryptofolione (Figure 4.7) and one might look into their potential as a source of anxiolytic agents[4] since cryptofolione has some chemical similitude with Kawain, the principle of kava (Piperaceae).

4.7 *LITSEA UMBELLATA* (LOUR.) MERR.

[From: Chinese *litse* = *Litsea* and from Latin *umbellata* = refers to the arrangement of the flowers which arise in a head from a central point, i.e., bearing an umbel.]

4.7.1 Botany

Litsea umbellata (Lour.) Merr. (*Litsea amara* Bl., *Litsea amara* var. *angusta* Meissn., and *Litsea amara* var. *attenuata* Gamb.) is a lowland forest tree that grows in India and Southeast Asia. The stems are petioles and the midrib is hairy. The leaves are simple, alternate, and exstipulate. The petiole is 5–7mm long. The apex of the blade is pointed or blunt. The base is rounded or cuneate. The midrib above is sunken and there are 9–13 pairs of secondary nerves. The tertiary nerves are scalariform. The blade is glaucous below. Axillary short racemes run off peduncled umbellules. Fruit is elliptic, up to 1cm long, black, glossy, and seated on a 4–6-lobed perianth (Figure 4.8).

4.7.2 Ethnopharmacology

The plant is known as *medang ayer* in Indo-Malaya; the leaves are used as a poultice to heal boils. The pharmacological potential of this plant is unknown. Both (+)-demethoxyepiexcelsin and verticillatol from *Litsea verticillata* have anti-HIV properties.[5] An interesting development would be the evaluation of Lauraceous lignans for antiviral properties.

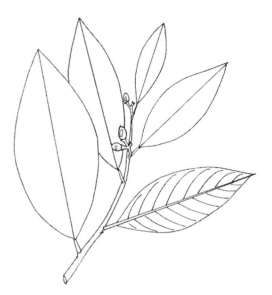

Figure 4.8 *Litsea umbellata* (Lour.) Merr. [From: Flora of Malay Peninsula. Forest Department. Geographical localization: Jalan Kulam ayer Raub, Pahang. Dec. 27, 1929. No: 20473. Field collector: Kalong. Botanical identification: J. G. H. Kostermans, Jan. 1, 1968.]

REFERENCES

1. Verspohl, E. J., Bauer, K., and Neddermann, E. 2005. Antidiabetic effect of *Cinnamomum cassia* and *Cinnamomum zeylanicum in vivo* and *in vitro*. *Phytother. Res.*, 19, 203–206.
2. Kitagawa, I., Minagawa, K., Zhang, R. S., Hori, K., Doi, M., Inoue, M., Ishida, T., Kimura, M., Uji, T., and Shibuya, H. 1993. Dehatrine, an antimalarial bisbenzylisoquinoline alkaloid from the Indonesian medicinal plant *Beilschmiedia madang*, isolated as a mixture of two rotational isomers. *Chem. Pharm. Bull. (Tokyo)*, 41, 997.
3. Zschocke, S. and van Staden, J. 2000. *Cryptocarya* species — substitute plants for *Ocotea bullata*? A pharmacological investigation in terms of cyclooxygenase-1 and -2 inhibition. *J. Ethnopharmacol.*, 71, 473.
4. Schmeda-Hirschmann, G., Astudillo, L., Bastida, J., Codina, C., Rojas De Arias, A., Ferreira, M. E., Inchaustti, A., and Yaluff, G. 2001. Cryptofolione derivatives from *Cryptocarya alba* fruits. *J. Pharm. Pharmacol.*, 53, 563.
5. Hoang, V. D., Tan, G. T., Zhang, H. J., Tamez, P. A., Hung, N. V., Cuong, N. M., Soejarto, D. D., Fong, H. H., and Pezzuto, J. M. 2002. Natural anti-HIV agents — part I: (+)-demethoxyepiexcelsin and verticillatol from *Litsea verticillata*. *Phytochemistry*, 59, 325.

CHAPTER 5

Medicinal Plants Classified in the Family Piperaceae

5.1 GENERAL CONCEPT

The family Piperaceae (C. A. Agardh, 1825 nom. conserv., the Pepper Family) consists of 10 genera and about 2000 species of tropical plants of which about 30 species are medicinal in Asia–Pacific. In field collection, Piperaceae can be recognized by three main features: articulate stems, asymmetrical or cordate leaves, and axillary spikes of little round berry-like fruits (Figure 5.1). Black Pepper (*British Pharmaceutical Codex*, 1949) and Long Pepper (*Indian Pharmaceutical Codex*, 1955), which consist of the dried unripe fruits of *Piper nigrum* L. and *Piper longum*, have been used since time immemorial in India. Black pepper at doses ranging from 300–600mg stimulates the tastebuds, produces a reflex increase in gastric secretion, reduces fever, and promotes urination. White pepper consists of dried unripe fruits of *Piper nigrum* L. deprived of the outer part of the pericarp. The taste of peppers is due to piperine, a piperdine alkaloid. The dried unripe fruit forms the condiment, cubebs. Cubebs (*British Pharmaceutical Codex*, 1934) consists of the dried unripe fully grown fruit of *Piper cubeba* L. f. It was formerly employed as a urinary antiseptic

Figure 5.1 Botanical hallmarks of Piperaceae: cordate leaf. **(See color insert following page 168.)**

(liquid extract: 1-in-1 dose 2–4 mL). Lozenges of cubebs have been used to treat bronchitis. Cubeb Oil (*British Pharmaceutical Codex*, 1949) is the oil obtained by distillation of cubebs. It has been used as an emulsion or in capsules as a urinary treatment. Other Piperaceae of relative pharmaceutical value are *Piper methysticum* Forst. (Kava, *British Pharmaceutical Codex*, 1934) and *Piper betle* (*British Pharmacopoeia*, 1934). A beverage prepared from the roots of *Piper methysticum* Forst. or kava has been used for centuries to calm and to promote sleep by a number of Polynesian

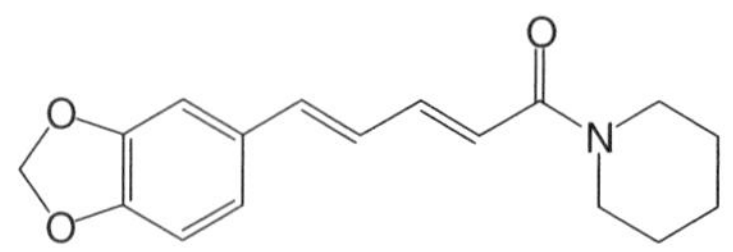

Piperine

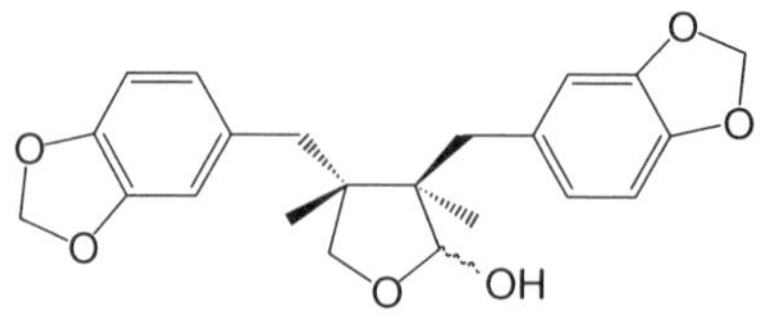

(-) – Cubebin

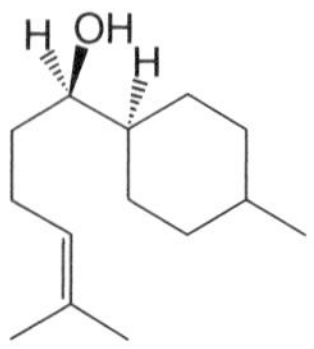

Bisabolol

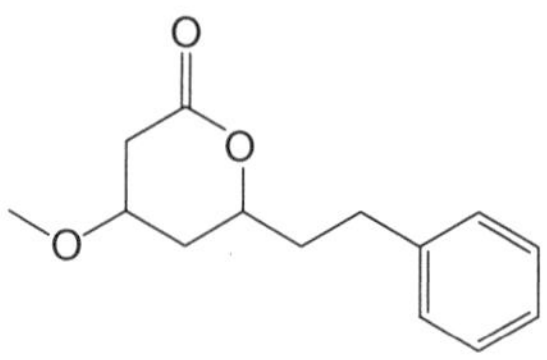

Kawain

Figure 5.2 Examples of bioactive natural products from the family Piperaceae.

people and, although toxic, has been marketed in Europe to treat sleep disorders and anxiety. Note that lignans of Piperaceae are of particular interest as a potential source of cytotoxic and antiviral agents (Figure 5.2).

5.2 *PIPER ABBREVIATUM* OPIZ

[From: Latin *piper* = pepper and *abbreviatum* = shortened or abbreviated in some fashion.]

5.2.1 Botany

Piper abbreviatum Opiz is a branching climber hugging trees with pendent lateral branches. The plant grows in Indonesia and the Philippines. The stems are fissured longitudinally, rooting, 3mm in diameter, and articulated. The leaves are simple, spiral, and exstipulate. The petiole is 8mm

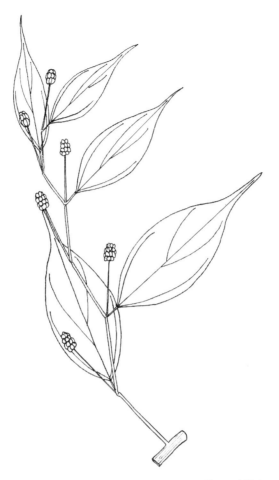

Figure 5.3   *Piper abbreviatum* Opiz. [From: Herbarium Bogoriense, Harvard University Herbarium. J. A. Mac-
Donald and R. Ismail. No. 4857, July 25, 1994. Plants of Indonesia. Geographical localization: Bali
Timur, Tabanan, Central Bali, Mt. Lesung, C.a. 6Km WNW of Bedugul, above Kebum Raya. Alt.:
1850m, 8°18′ S, 115°9′ E. Mountain forest, canopy 15m.]

long. The blade is elliptic, 8cm – 11cm × 2.2cm – 4cm, acuminate at the apex in a 2.2cm-long
tail, and shows two pairs of secondary nerves. The inflorescences are cream-colored spikes, which
are axillary, globose, and 1cm in diameter (Figure 5.3).

5.2.2   Ethnopharmacology

In the Philippines, a paste of leaves is used externally to treat splenomegaly. The pharmacological properties of *Piper abbreviatum* Opiz are unexplored. The medicinal property mentioned
above might be owed to counterirritant effects. The plant has not been studied for pharmacology.

5.3 *PIPER BETLE* L.

[From: Latin *piper* = pepper and from Malayalam *vettila* = *Piper betle*.]

5.3.1 Botany

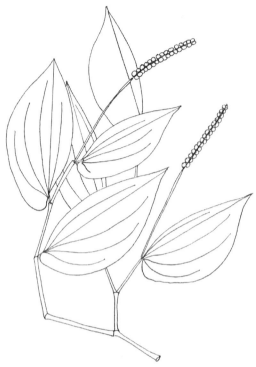

Piper betle L. (*Chavica betle*) is a climber that grows in India, Indonesia, Malaysia, the Philippines, Sri Lanka, Vietnam, and China. The stems are dichotomous, articulate, swollen, and rooted at nodes 3mm in diameter, woody, and with 4–2.5cm-long internodes. The leaves are simple, spiral, and exstipulate. The petiole is 5mm long, channeled, and pubescent. The blade is 10cm × 6cm – 9.5cm × 5cm, ovate to ovate–oblong, and light green below. The base of the blade is cordate and the apex is acuminate. The secondary nerves are in three pairs. The inflorescence is an axillary spike, which is 5.5cm long. The fruits are drupaceous, orange, and 3mm in diameter (Figure 5.4).

5.3.2 Ethnopharmacology

Figure 5.4 *Piper betle* L. [From: Forest Department. Field collector: M. Asri, March 12, 1982. Botanical identification: H. Kok, March 12, 1984. Geographical localization: Ulu Anap, Tatau, 4th Division, on logged-over forest, S. Kana.]

Betle (*British Pharmaceutical Codex*, 1934) consists of the dried leaves of *Piper betle*, which has been used as a stimulant, carminative, and antiseptic. It is used in India as a masticatory; in Malaysia, the leaves are usually mixed with lime and the scraping of *Areca* nuts. The plant is known as *lou ye* in China where the leaves are used as a condiment. The roots, leaves, and fruits are carminative, stimulant, corrective, and used for the treatment of malaria. In the *Pentsao*, an oil obtained from the leaves is used as a counterirritant in swellings, bruises, and sores. In Malaysia, the leaves are applied externally to the body after childbirth. They are also used to heal ulcers, boils, bruises, ulcerations of the nose, and as an antiseptic.

Hydroxychavicol is known to modulate benzo[a]pyrene-induced genotoxicity through the induction of dihydrodiol dehydrogenase, hence the increased potential of betle chewing and smoking in the development of oral squamous cell carcinoma (OSCC).[1] An aqueous extract of the leaves of *Piper betle* given orally during the initiation phase of 7,12-dimethylbenz[a]anthracene (DMBA)-induced mammary carcinogenesis in the rodent inhibited the emergence of tumors.[2] Note that a chloroform extract of *Piper betle* and *Piper chaba* showed some potential against *Giardia* cultured *in vitro*.[3]

The plant is known to produce phenylpropanoids such as hydroxychavicol and allylpyrocatechol, the latter being antibacterial against oral anaerobes responsible for halitosis.[4]

5.4 *PIPER OFFICINARUM* DC.

[From: Latin *piper* = pepper and *officinarum* = sold as an herb.]

5.4.1 Botany

Piper officinarum DC. (*Piper retrofractum* Vahl, *Chavica officinarum* Miq., and *Piper chaba* Hunt.) is a climber that grows in India, Indonesia, China, Malaysia, the Philippines, Thailand, and Vietnam. The stems are 2mm thick, terete, and striated with 4.2–1.8cm-long internodes. The leaves are simple, exstipulate, and spiral. The petiole is 9mm long. The blade is narrowly elliptic, ovate–oblong, or elliptical, 8.5cm – 16cm × 3.2cm – 7.5cm, papery, glaucous, and showing four pairs of secondary nerves and a few tertiary nerves. The inflorescences are 10cm × 4mm spikes attached to 1.5cm-long pedicels (Figure 5.5).

5.4.2 Ethnopharmacology

The plant is known as *kechundai* (Iban) and *jia bi ba* (Chinese). Pepper obtained from this species has been used as an adulterant for *Piper longum*. In Cambodia, Laos, and Vietnam, the plant is used to treat fever, jaundice, rheumatism, neuralgia, and boils. In the Philippines, the roots are chewed to promote digestion and externally used to heal wounds. The plant elaborates a series of unusual amides such as ridleyamide, brachystamide C, and retrofractamide C, the pharmacological potential of which would be worth assessing; one might set the hypothesis that such compounds mediate the antiinflammatory potential of the plant since retrofractamide B has significantly inhibited indomethacin-induced gastric lesions in the rodent.[5]

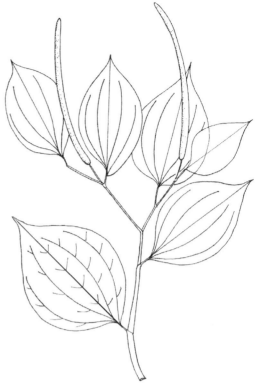

Figure 5.5 *Piper officinarum* DC. [From: Flora of Sabah, SAN No: 122498. Geographical localization: near Agricultural Station, District Tenom. Alt.: 500m. Field collector: W. Meijer, Oct. 15, 1987. In secondary forest along Pelalang Forest.]

5.5 *PIPER SARMENTOSUM* ROXB.

[From: Latin *piper* = pepper and *sarmentosum* = twiggy, with long, slender runners.]

5.5.1 Botany

Piper sarmentosum Roxb. (*Chavica hainana* DC., *Chavica sarmentosa* [Roxburgh] Miq., *Piper albispicum* DC., *Piper brevicaule* DC., *Piper gymnostachyum* DC., *Piper lolot* DC., *Piper pierrei* DC., and *Piper saigonense* DC.) is a shrub that grows to a height of 50cm in Cambodia, India, Burma, Thailand, Indonesia, Laos, Malaysia, the Philippines, Vietnam, and China. The leaves are simple, alternate, and exstipulate. The blade is lanceolate–elliptical, 10cm × 5cm – 14cm × 6cm – 9.9cm × 3.4cm – 12cm × 2cm, acuminate at the apex, rounded at the base, with two pairs of secondary nerves. The inflorescences are 5mm-long nerves. The fruits are green–red with 4mm × 3mm drupes (Figure 5.6).

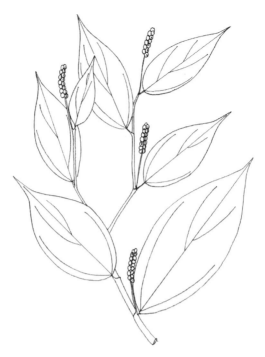

Figure 5.6 *Piper sarmentosum* Roxb. [From: Flora of Malaya. FRI No: 39468. Sept. 21, 1993. Field collector: M. Asri. Geographical localization: Perak, Ulu Perak, Halong River, ridge trail, Temenggor. In mixed diterocarp forest.]

5.5.2 Ethnopharmacology

The plant is known as *jia ju* in China where the leaves afford a treatment for fever and indigestion. The roots are used to assuage toothaches and to treat dermatomycoses. In Malaysia and South Thailand, the leaves are used externally to soothe headaches. In Indonesia, the roots are chewed for cough, asthma, and toothaches, and the leaves are used externally to mitigate chest pain. In Thailand the plant is called *chaplu*.

An aqueous extract of whole *Piper sarmentosum* Roxb. given orally at a dose of 0.125g/Kg for a week has lowered the glycemia of both streptozocin-induced diabetic rats and normal rats.[6]

Methanolic extract of leaves at concentrations of 3.2, 4.0, 4.8, and 6.4mg/mL exhibited an initially transient increase in twitch tension which was followed by a marked dose-related neurally evoked twitch depression with EC_{50} of 4.07mg/mL. This effect was antagonized by tetraethylammonium, suggesting neuromuscular blocking activity at the neuromuscular cholinergic junction.[7] A chloroform extract showed some levels of antiplasmodial activity.[8] The active principles involved in the antidiabetic properties of *Piper sarmentosum* are unknown. A remarkable advance in *Piper sarmentosum* pharmacological potential has been provided by Rukachaisirikul et al.[9] They isolated a series of amides including brachystamide B, sarmentine, brachyamide B, 1-piperettyl pyrrolidine, and lignans, and showed that sarmentine and 1-piperettyl pyrrolidine display antituberculosis and antiplasmodial activities. What is the pharmacological potential of amides from Piperaceae in diabetes?

REFERENCES

1. Ramji, N., Ramji, N., Iyer R., and Chandrasekaran, S. 2002. Phenolic antibacterials from *Piper betle* in the prevention of halitosis. *J. Ethnopharmacol.*, 83, 149.

2. Tang, D. W., Chang, K. W., Chi, C. W., and Liu, T. Y. 2004. Hydroxychavicol modulates benzo[a]pyrene-induced genotoxicity through induction of dihydrodiol dehydrogenase. *Toxicol Lett.*, 152, 235.

3. Sawangjaroen, N., Subhadhirasakul, S., Phongpaichit, S., Siripanth, C., Jamjaroen K., and Sawang-jaroen, K. 2005. The *in vitro* antigiardial activity of extracts from plants that are used for self-medication by AIDS patients in southern Thailand. *Parasitol Res.*, 95, 117.

4. Ramji, N., Ramji, N., Iyer, R., and Chandrasekaran, S. 2002. Phenolic antibacterials from *Piper betle* in the prevention of halitosis. *J. Ethnopharmacol.*, 83, 149.

5. Morikawa, T., Matsuda, H., Yamaguchi, I., Pongpiriyadacha, Y., and Yoshikawa, M. 2004. New amides and gastroprotective constituents from the fruit of *Piper chaba*. *Planta Med.*, 70, 152.

6. Peungvicha, P., Thirawarapan, S. S., Temsiririrkkul, R., Watanabe, H., Kumar Prasain, J., and Kadota, S. 1998. Hypoglycemic effect of the water extract of *Piper sarmentosum* in rats. *J. Ethnopharmacol.*, 60, 27.

7. Ridtitid, W., Rattanaprom, W., Thaina, P., Chittrakarn, S., and Sunbhanich, M. 1998. Neuromuscular blocking activity of methanolic extract of *Piper sarmentosum* leaves in the rat phrenic nerve-hemi-diaphragm preparation. *J. Ethnopharmacol.*, 61, 135.

8. Najib Nik, A., Rahman, N., Furuta, T., Kojima, S., Takane, K., and Ali Mohd, M. 1999. Antimalarial activity of extracts of Malaysian medicinal plants. *J. Ethnopharmacol.*, 64, 3, 249–254.

9. Rukachaisirikul, T., Siriwattanakit, P., Sukcharoenphol, K., Wongvein, C., Ruttanaweang, P., Wong-wattanavuch, P., and Suksamrarn, A. 2004. Chemical constituents and bioactivity of *Piper sarmento-sum*. *J. Ethnopharmacol.*, 93, 173.

CHAPTER 6

Medicinal Plants Classified in the Family Aristolochiaceae

6.1 GENERAL CONCEPT

The family Aristolochiaceae (A. L. de Jussieu, 1789 nom. conserv., the Birthwort Family) comprises approximately 5 genera and 300 species of poisonous climbers, which can be recognized in field collection by their corolla, often shaped like a little smoking pipe. Several species in this family have been used for medicinal purposes in the Western world: *Aristolochia reticulata* (serpentary, red river snakeroot, and Texan snakeroot), *Aristolochia serpentaria* (Virginian snakeroot), *Aristolochia clematis* (birthwort), and *Asarum europeaum*

Aristolochic acid   Isoboldine

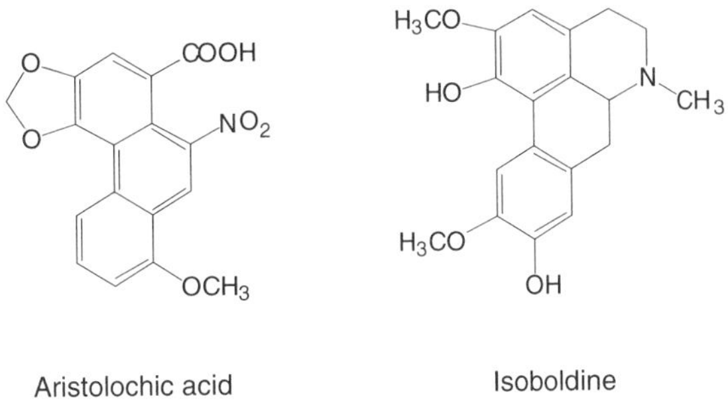

Figure 6.1   Examples of bioactive natural products derived from the Aristolochiaceae family.

(Asarabaca, *Spanish Pharmacopoeia* 1954). In China, *Aristolochia contorta, Aristolochia kaempferi*, and *Aristolochia recurvilabra* have been used in the traditional Chinese system of medicine since antiquity. Aristolochiaceae have the ability to elaborate a unique series of phenanthrene alkaloids, one of the best examples of which is aristolochic acid (Figure 6.1). The sodium salt of aristolochic acid has been tried as an antiinflammatory agent, but severe nephrotoxicity in humans and carcinogenicity in rodents aborted further developments. The Asia–Pacific region uses about 20 species of Aristolochiaceae for traditional medicine mainly to counteract snake-poisoning, to promote urination and menstruation, and to assuage stomachaches. It is also used to treat dropsy and skin diseases.

6.2 *ARISTOLOCHIA PHILIPPINENSIS* WARB.

[From: Greek *aristo* = best and *lochia* = delivery, and from Latin *philippinensis* = from the Philippines.]

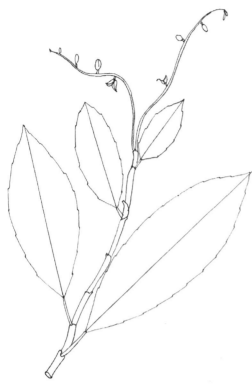

6.2.1 Botany

Aristolochia philippinensis Warb. is a climber that grows in the coastal forest of the Philippine Islands. The stems are slightly pubescent, terete, and articulate. The leaves are simple, exstipulate, and spiral. The blade is oblong–lanceolate, serrate, and 5cm – 18cm × 2cm × 5cm. The secondary nerves are inconspicuous. The flowers are arranged in terminal inflorescences. The fruits are capsular, up to 1cm long, and dehiscent (Figure 6.2).

6.2.2 Ethnopharmacology

In the Philippines, a decoction of the roots is used to assuage stomachache and to promote menses. Note that aristolochic acid and congeners share some similitude in chemical structure with our own steroidal hormones, hence their potency when acting as antiinflammatory and gynecological agents. Aristolochic acid inhibits *in vitro* and dose-dependently phospholipid hydrolysis by the human synovial fluid phospholipase A_2, snake venom phospholipase A_2, porcine pancreatic phospholipase A_2, and human platelet phospholipase A_2[1,2] which is a key enzyme in inflammation and possibly linked to the release of luteinizing and growth hormones from the anterior pituitary.[3]

Figure 6.2   *Aristolochia philippinensis* Warb. [From: Plants of the Philippines. Geographical localization: Pawalan Island. 9°17′N, 11°57′ E, Alt.: 0–5m. Coastal forest at base of limestone hill, above mangrove formation. D. D. Soejarto and D. A. Madulid, July 26, 1982. Botanical identification: June 1994.]

REFERENCES

1. Rosenthal, M. D., Vishwanath, B. S., and Franson, R. C. 1989. Effects of aristolochic acid on phospholipase A_2 activity and arachidonate metabolism of human neutrophils. *Biochim. Biophys. Acta.*, 1001.
2. Vishwanath, B. S., Fawzy A. A., and Franson, R. C. 1988. Edema-inducing activity of phospholipase A_2 purified from human synovial fluid and inhibition by aristolochic acid. *Inflammation*, 12, 549.
3. Thomson, F. J. and Mitchell, R. 1993. Differential involvement of phospholipase A_2 in phorbol ester-induced luteinizing hormone and growth hormone release from rat anterior pituitary tissue. *Mol. Cell. Endocrinol.*, 95, 75.

CHAPTER 7

Medicinal Plants Classified in the Family Nympheaceae

7.1 GENERAL CONCEPT

The family Nympheaceae consists of 5 genera and 50 species of aquatic rhizomatous herbs which are cosmopolitan in distribution. Nympheaceae are known to be an elaborate series of sesquiterpene alkaloids. The leaves arise directly from the rhizome, alternate, long petiolate, hastate, or peltate and floating. The flowers are solitary, axillary, and often showy. The fruits are spongy, often conical, and full of seeds, which are small and scanty. In the Asia–Pacific, *Brasenia schreberi* J.F. Gmel., *Euryale ferox* Salisb., *Nelumbo nucifera* Gaertn., *Nuphar japonicum* DC., *Nymphea sellata* Willd., and *Nymphea pubescens* Willd. are medicinal.

The evidence currently available suggests the family Nympheaceae to be an exciting source of cytotoxic, antiviral, and immunomodulating quinolizidine alkaloids; one can reasonably expect the discovery of clinical agents from this family in the relatively near future. Perhaps no other genus in this family has aroused more interest in the field of pharmacology than the genus *Nuphar*. Matsuda et al.[1] made the interesting observation that the rhizome of *Nuphar pumilum* contains dimeric sesquiterpene thioalkaloids, such as 6-hydroxythiobinupharidine, 6,6′-dihydroxythiobin-upharidine, and 6-hydroxythionuphlutine B, which inhibited the invasion of B16 melanoma cells across collagen-coated filters *in vitro* with IC_{50} 0.029, 0.087, and 0.36μM, respectively, indicating a clear antimetastatic potential (Figure 7.1). Using antisheep erythrocyte plaque-forming cell formation in mouse splenocytes assay, Matsuda et al. showed potent immunosuppressive activity of 6-hydroxythiobinupharidine, 6,6-dihydroxythiobinupharidine, 6-hydroxythionuphlutine B, and 6-hydroxythionuphlutine B, and observed that the 6- or 6-hydroxyl group at the quinolizidine ring of dimeric sesquiterpene thioalkaloids is essential for the immunosuppressive effect.

7.2 *NELUMBO NUCIFERA* GAERTN.

[From: Sri Lankan *nelumbu* = Nelumbo *nucifera* and Latin *nucifera* = nut-bearing.]

7.2.1 Botany

Nelumbo nucifera Gaertn. (*Nelumbium nelumbo* [L.] Druce, *Nelumbium speciosum* Willd., *Nelumbo komarovii* Gross., *Nelumbo nucifera* var. *macrorhizomata* Nak., and *Nymphaea nelumbo* Linn.) is an aquatic herb that grows in ponds, pools, rivers, and lakes in China, Bhutan, India,

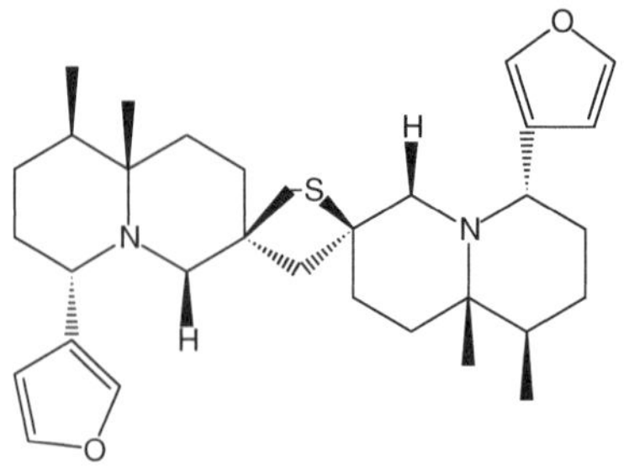

Thiobinupharidine

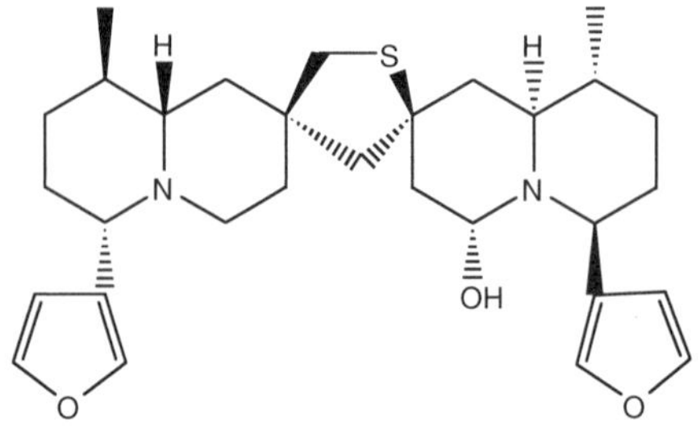

6-Hydroxythionuphlutine B

Figure 7.1   Examples of bioactive natural products derived from the family Nympheaceae.

Indonesia, Japan, Korea, Malaysia, Burma, Nepal, New Guinea, Pakistan, India, the Philippines, Sri Lanka, Thailand, Vietnam, Australia, and the Pacific Islands. The plant grows from a rhizome constricted at its nodes and is somewhat pinkish. The petiole is to 2m long, terete, fistulous, and glabrous. The blade is 25–90cm in diameter, round, thin, glabrous, and entire at the margin. The flowers are conspicuous, 10–25cm in diameter, pink or white, the petals oblong–elliptic to obovate, 5cm–11cm × 2.5cm–5cm. The fruits are conical, green, and up to 15cm long (Figure 7.2).

7.2.2   Ethnopharmacology

In Asia, the fruits of the lotus, or *lian*, *fu chu* (Chinese), and *teratai* (Malay), are sold in the market for the seeds, which are edible and medicinal. In China, the seeds are used to preserve the body's health and strength, to cool, to promote blood circulation, and to treat leucorrhea and gonorrhea. The rhizomes are edible and after cooking they form a sweet mucilaginous food that is taken to assuage a stomachache, to strengthen the body, to increase the mental faculties, and to quiet the spirit. The inflorescence is antihemorrhagic, and given as a postpartum remedy. The leaves are used to break fever, as an antihemorrhagic, to precipitate childbirth, and to treat skin diseases. The petiole is used to quiet the uterus. The flowers are spoken of in the *Pentsao* and believed to drive away old age and to give a fine complexion.

The plant is interesting since it elaborates antiviral isoquinolines (Figure 7.3): (+)-1(*R*)-coclaurine and 1(*S*)-norcoclaurine from the leaves of *Nelumbo nucifera* Gaertn., which inhibits the replication of HIV *in vitro* with EC_{50} values of 0.8 and <0.8g/μL, and therapeutic index values of >125 and >25, respectively. Liensinine and isoliensinine showed potent anti-Human Immunodeficiency Virus (HIV) activities with EC_{50} values of <0.8g/μL and Therapeutic Index values of >9.9 and >6.5g/μL. Nuciferine, an aporphine alkaloid, had an EC_{50} value of 0.8g/μL and a Therapeutic Index value of 36.[2] Isoliensinine exhibited a significant inhibitory effect on bleomycin-induced pulmonary fibrosis

Figure 7.2 *Nelumbo nucifera* Gaertn. [From: Field FRI No: 25712. Geographical localization: Seremban road-side. Oct. 19, 1976. Field collector and botanical identification: M. Asri.]

in male mice. The alkaloid lowered the hydroxyproline content and limited the lung histological injury induced by bleomycin, and also inhibited the overexpression of TNF-alpha and TGF-beta induced by bleomycin, and showed potent activity against bleomycin-induced pulmonary fibrosis.[3]

The antiinflammatory property of *Nelumbo nucifera* Gaertn. is confirmed *in vitro* and *in vivo*. A methanol extract of rhizomes given at doses of 200mg/kg and 400mg/Kg showed significant antiinflammatory activity in models of inflammation in rats as efficiently as phenylbutazone and dexamethasone.[4]

An extract inhibited proliferation in human peripheral blood mononuclear cells activated with phytohemagglutinin on account of NN-B-4 which is mediated, at least in part, through inhibition of early transcripts of interleukin (IL)-2, interferon (IFN)-γ, and cdk4, and arrest of cell cycle progression in the cells.[5] An ethanol extract of the petiole lowered normal body temperature in a dose of 200mg/Kg, while in yeast-induced fever it showed a dose-dependent lowering of body temperature as efficiently as paracetamol.[6]

Ethanol extract from seeds of *Nelumbo nucifera* showed antioxidant and hepatoprotective effects.[7] Oral administration of the ethanolic extract of rhizomes lowered the blood glucose levels of normal, glucose-fed hyperglycemic and streptozotocin-induced diabetic rats, and improved glucose tolerance, and also potentiated the action of exogenously injected insulin in normal rats.[8]

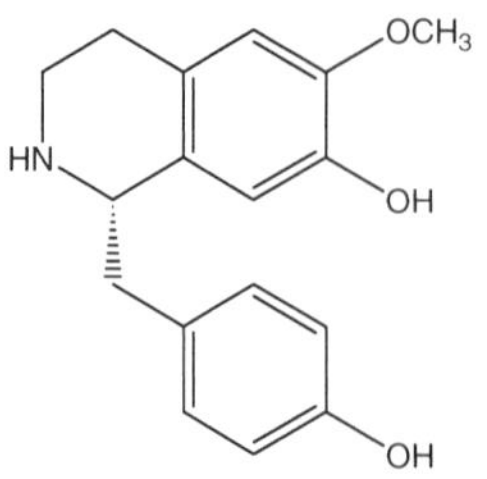

Coclaurine 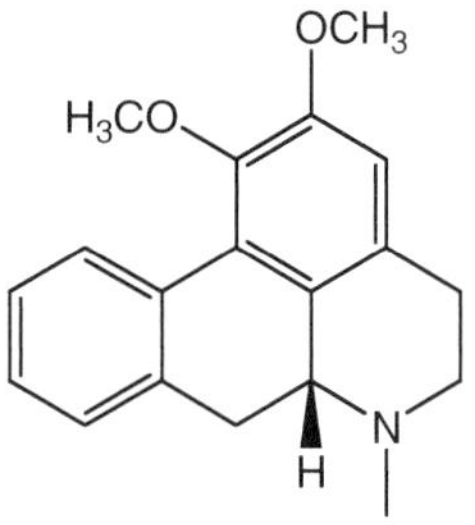   Nuciferine

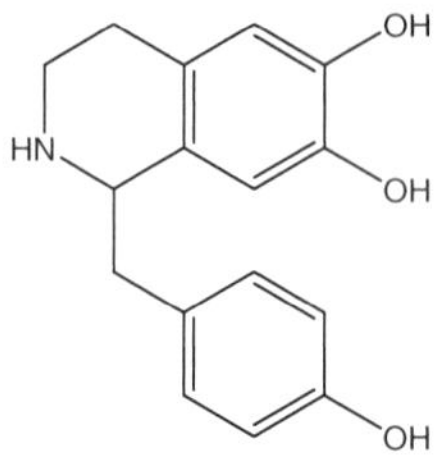

Norcoclaurine

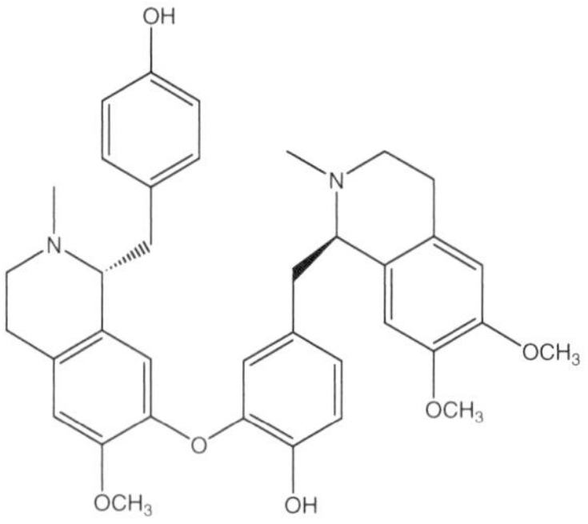

Liensinine

Figure 7.3   Antiviral alkaloids of *Nelumbo nucifera* Gaertn.

When compared with tolbutamide, the extract exhibited activity of 73% and 67% of tolbutamide in normal and diabetic rats, respectively.

Methanolic extract of rhizomes lowered spontaneous activity, decreased the exploratory behavioral pattern, and potentiated pentobarbitone-induced sleeping time in rats.[9]

7.3 *NYMPHEA PUBESCENS* WILLD.

[From: Greek *nymphaia*, referring to a water nymph, and from Latin *pubescens* = with soft, downy hair.]

7.3.1   Botany

Nymphea pubescens Willd. (*Nymphaea lotus* L. var. *pubescens* (Willdenow) J.D. Hook. & Thoms.) is an aquatic herb which grows from a 10cm rhizome. It is found in lakes, pools, and

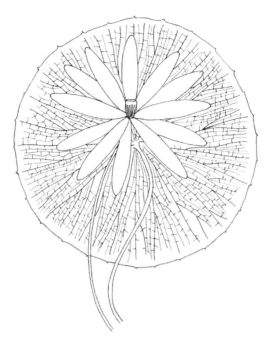

Figure 7.4 *Nymphea pubescens* Willd. [From: FRIM. Field collector: K. M. Kochummen. No: 67. *Fide Gards Bull. Sing.* Vol. 47 (1995), 390.]

rivers in a geographical area spanning Queensland, Papua New Guinea, India, and Southeast Asia. The petiole is up to 2m long. The blade is elliptic to orbicular, up to 45cm in diameter, hairy below; dentate at the margin. The flowers are showy and white above water. The calyx comprises 4–5cm, up to 9cm-long, obtuse sepals. The corolla has 19 petals that are oblanceolate and pinkish. The androecium consists of 60 flattened, thickened stamens with 2cm anthers. The gynoecium consists of 11 to 20 carpels united to form a plurilocular ovary (Figure 7.4).

7.3.2 Ethnopharmacology

Water lilies or *rou mao chi ye shui lian* (Chinese) are used in the Philippines to treat gonorrhea. It is also rubbed on the forehead to induce sleeping. The pharmacological property of this plant is unexplored. Are alkaloids with hypnotic properties present in this plant?

7.4 *NYMPHEA STELLATA* WILLD.

[From: Greek *nymphaia*, referring to a water nymph, and from Latin *stellata* = star-like shaped.]

7.4.1 Botany

Nymphea stellata Willd. (*Nymphaea nouchali* Burm. f., *Nymphaea minima* F.M. Bailey) is an aquatic herb that grows from a tuberous rhizome to 5cm in diameter in Australia, Papua New Guinea, Southeast Asia, and India in swamps and pools. The blade is 12.5cm × 35cm and the petiole is 20cm. The flowers are white. The calyx consists of four sepals which are 3cm long, green outside with purplish penciling. The corolla comprises 10 petals which are lanceolate, blue, pink, or white. The androecium consists of 20 stamens which are yellow. The flower pedicel is 13.5cm × 2mm. The flower bud is 4.2cm × 2.2cm (Figure 7.5).

Figure 7.5   *Nymphea stellata* Willd. [From: FRI No: 27165. Road between Kuala Kraw and Kuala Lompat, Pahang. In waterlogged ditches by the road. Aquatic herb. F. S. P. Ng, Jan. 10, 1972.]

7.4.2   Ethnopharmacology

The plant is known as *talipok* in Borneo and *yan yao shui lian* in China. In Laos, Cambodia, and Vietnam, the leaves are used externally to break fever, while the rhizomes are used to treat gastrointestinal disturbances. In India, the plant is used to treat liver disorders. An extract of flowers has protected albino rats against carbon tetrachloride-induced hepatic damage when given orally for 10 days.[10] What are the principles involved here? Thioalkaloids!

REFERENCES

1. Matsuda, H., Morikawa, T., Oda, M., Asao, Y., and Yoshikawa, M. 2003. Potent anti-metastatic activity of dimeric sesquiterpene thioalkaloids from the rhizome of *Nuphar pumilum*. *Bioorg. Med. Chem. Lett.*, 13, 4445.
2. Kashiwada, Y., Aoshima, A., Ikeshiro, Y., Chen, Y. P., Furukawa, H., Itoigawa, M., Fujioka, T., Mihashi, K., Cosentino, L. M., Morris-Natschke, S., and Lee, K. H. 2005. Anti-HIV benzylisoquinoline alkaloids and flavonoids from the leaves of *Nelumbo nucifera*, and structure–activity correlations with related alkaloids. *Bioorg. Med. Chem.*, 13, 443.
3. Xiao, J. H., Zhang, J. H., Chen, H. L., Feng, X. L., and Wang, J. L. 2005. Inhibitory effects of isoliensinine on bleomycin-induced pulmonary fibrosis in mice. *Planta Med.*, 71, 225.
4. Mukherjee, P. K., Saha, K., Das, J., Pall, M., and Saha, B. P. 1997. Studies on the anti-inflammatory activity of rhizomes of *Nelumbo nucifera*. *Planta Med.*, 63, 367.

5. Liu, C. P., Tsai, W. J., Lin, Y. L., Liao, J. F., Chen, C. F., and Kuo, Y. C. 2004. The extracts from *Nelumbo nucifera* suppress cell cycle progression, cytokine genes expression, and cell proliferation in human peripheral blood mononuclear cells. *Life Sci.*, 75, 699.

6. Sinha, S., Mukherjee, P. K., Mukherjee, K., Pal, M., Mandal, S. C., and Saha, B. P. 2000. Evaluation of antipyretic potential of *Nelumbo nucifera* stalk extract. *Phytother. Res.*, 14, 272.

7. Sohn, D. H., Kim, Y. C., Oh, S. H., Park, E. J., Li, X., and Lee, B. H. 2003. Hepatoprotective and free radical scavenging effects of *Nelumbo nucifera*. *Phytomedicine*, 10, 165.

8. Mukherjee, P. K., Saha, K., Pal, M., and Saha, B. P. 1997. Effect of *Nelumbo nucifera* rhizome extract on blood sugar level in rats. *J. Ethnopharmacol.*, 58, 207.

9. Mukherjee, P. K., Saha, K., Balasubramanian, R., Pal, M., and Saha, B. P. 1996. Studies on psycho-pharmacological effects of *Nelumbo nucifera* Gaertn. rhizome extract. *J. Ethnopharmacol.*, 54, 63.

10. Bhandarkar, M. R. and Khan, A. 2004. Antihepatotoxic effect of *Nymphaea stellata* Willd., against carbon tetrachloride-induced hepatic damage in albino rats. *J. Ethnopharmacol.*, 91, 61.

CHAPTER 8

Medicinal Plants Classified in the Family Menispermaceae

8.1 GENERAL CONCEPT

There are approximately 40 species of plants classified within the family Menispermaceae (A. L. de Jussieu, 1789 nom. conserv., the Moonseed Family). They are used for medicinal purposes in the Asia–Pacific, particularly to mitigate fever, as a diuretic, emmenagogue, carminative, tonic, antiinflammatory, and analgesic. The family Menispermaceae consists of 70 genera and approximately 400 species of tropical climbers that have attracted a great deal of interest on account of their ability to elaborate a series of diterpenes, benzylisoquinoline, and aporphine alkaloids. When looking for Menispermaceae for field collection, it is suggested to look for climbers in which the transverse section of the stem shows a very characteristic bicycle wheel-like aspect and a bright, yellowish color. Other distinctive features are the slender petiole often twisted at the base, leaves with a few nerves, and particularly the seeds which are muricate and horseshoe-like.

With regard to the pharmaceutical potential of Menispermaceae, *Anamirta paniculata* Coleb. (Levant berries) has been used in Western medicine to promote appetite and digestion on account of its bitterness. The dried transverse slices of roots of *Jateorrhiza palmata* Miers (Calumba, *British Pharmaceutical Codex*, 1954) are said to be a remedy for atonic dyspepsia; the dried stems of *Tinospora cordifolia* (Tinospora, *Indian Pharmaceutical Codex* 1953) have been used to promote digestion and appetite in the form of an infusion.

Examples of isolates of pharmaceutical interest are picrotoxin and tubocurarine (Figure 8.1). Picrotoxin is found in the seeds of *Anamirta cocculus* (*Anamirta paniculata*) and has been used for the treatment of barbiturate poisoning in the form of injection 3–6mg, intravenously (Picrotoxin, *British Pharmacopoeia*, 1963). Picrotoxin consists of a mixture of picrotoxinin and picrotin, picrotoxinin being a sesquiterpene with specific $GABA_A$ receptor-blocking activity, which impedes the GABAergic presynaptic inhibition of excitatory transmission of primary afferent neurones of the spinal cord. Picrotoxin is toxic, and as little as 20mg induces epileptiform convulsions, myosis, and dyspnea with more or less prolonged apnea.

Several Amazonian tribes have been using Menispermaceae from the genera *Chondrodendron, Curarea, Sciadotenia, Abuta, Telitoxicum,* and *Cissamplelos* to make arrow poisons or curares which abound with *bis*-benzyltertrahydroquinoline alkaloids such as (+)-tubocurarine, (+)-isochondrodendrine, (−)-curine, and (+)-chondrocurine. These alkaloids are anticholinergic at the neuromuscular synapse and provoke drastic relaxation of the skeletal muscles; hence their uses in surgical anesthesia (Tubocurarine Chloride, *British Pharmacopoeia*, 1963).

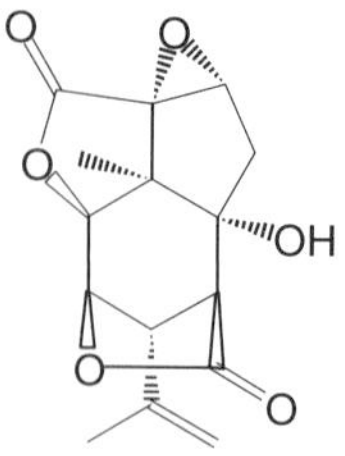

Picrotoxinin

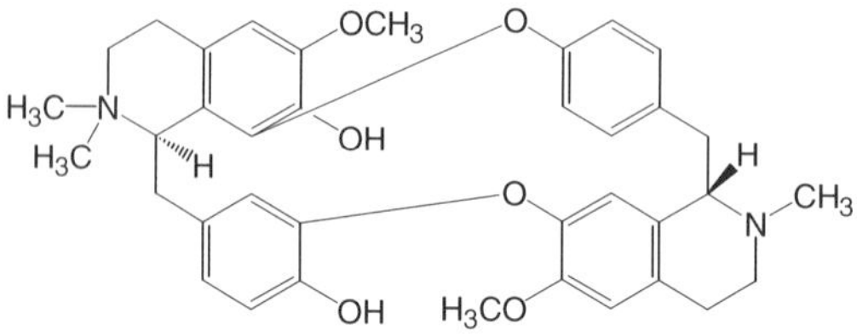

(+)-Tubocurarine

Figure 8.1   Examples of bioactive natural products from the family Menispermaceae.

8.2 *ARCANGELINA FLAVA* (L.) MERR.

[From: Latin *flavus* = pure yellow.]

8.2.1   Botany

Arcangelina flava (L.) Merr. (*Arcangelisia lemniscata* [Miers] Becc.) is a large climber that grows in the rain forests of Thailand, Malaysia, Indonesia, and the Philippines. The stems are 4mm large at the apex, smooth, and glabrous. The petiole is 3.5–9cm long. The leaves are simple, exstipulate, and spiral. The blade is 7cm × 4cm – 7.5cm × 13cm, elliptic, acuminate at the apex, thick, and recurved. The inflorescences are axillary and have 6cm-long panicles. The flowers are white. The fruits are globose, 3.5cm × 2.3cm – 2.5cm × 1.5cm, fleshy, yellow drupes (Figure 8.2).

8.2.2   Ethnopharmacology

The plant is known as *mengkunyit bukit* in Indonesia where a decoction of stems is used as a drink to treat jaundice, indigestion, and painful bowels. The wood is used to heal pox sores. In the Philippines a decoction of roots is used as a drink to reduce fever, to invigorate, to promote menses, and to abort; and a decoction of wood is used as an antiseptic for the skin. Cutting fresh stems of this climber reveals a bright yellowish-orange color which is accounted for by isoquinoline alkaloids, berberine, jatrorrhizine, dihydroberberine, and palmatine which abound in it (Figure 8.3). Berberine inhibits the growth of HepG2 cells by direct interaction with DNA in which it intercalates.[1] This intercalating property of berberine and congeners explains the broad range of antibacterial and

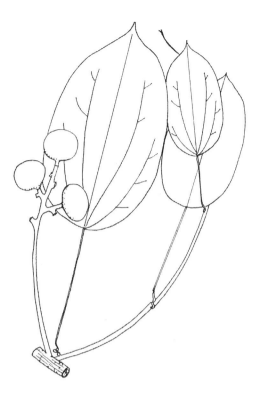

Figure 8.2 *Arcangelina flava* (L.) Merr. [From: Phytochemical Survey of the Federation of Malaya. KL No: 1615. July 21, 1959. Geographical localization: Ulu Langat, Selangor. Hill forest. Field collector: G. A. Umbai for A. H. Millard.]

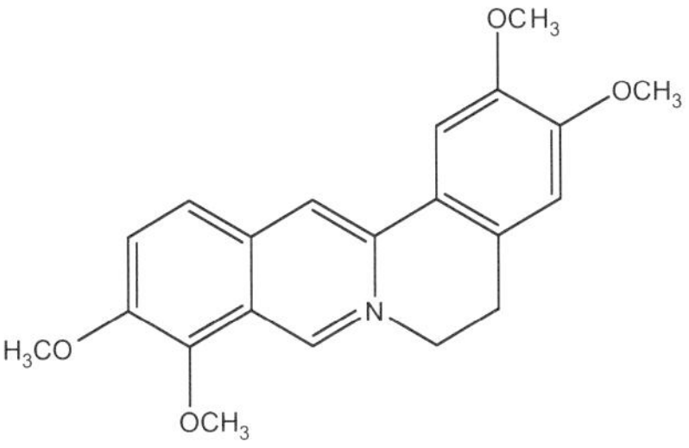

Palmatine

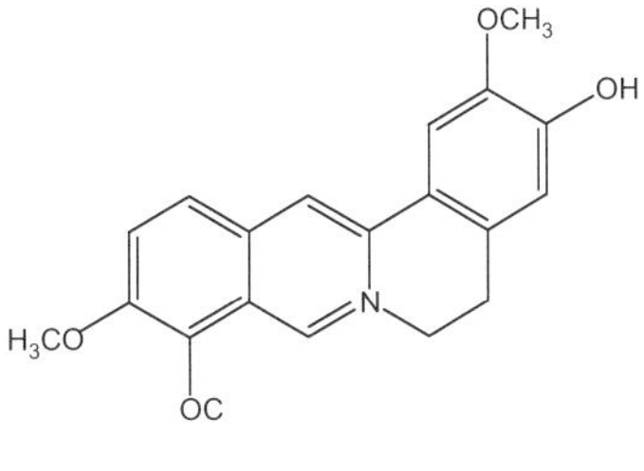

Jatrorrhizine

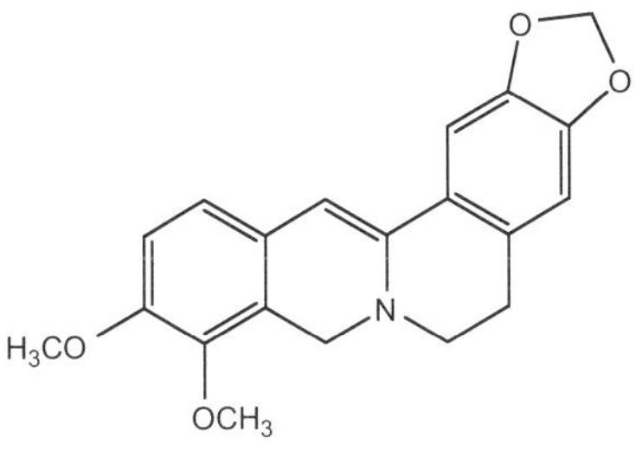

Dihydroberberine

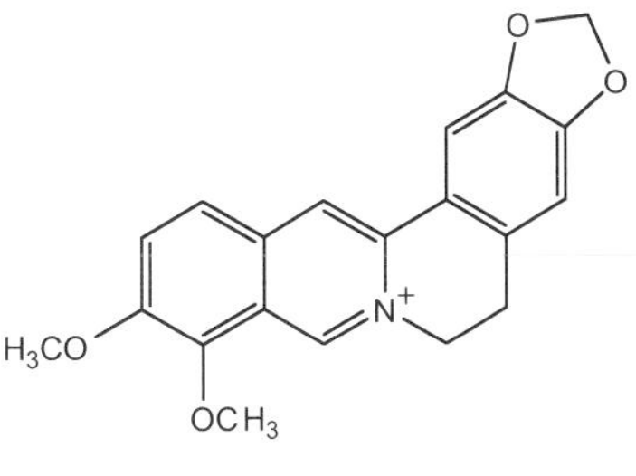

Berberine

Figure 8.3 Isoquinolines of *Arcangelina flava* (L.) Merr.

antiprotozoal effects elicited by the alkaloids and the medicinal properties mentioned above. Note also that berberine, extracted from *Arcangelisia flava* (L.) Merr., inhibits the enzymatic activity of *Plasmodium falciparum* telomerase dose-dependently at doses ranging from 30–300mM.[2] Palmatine, berberine, jatrorrhizine, and dihydroberberine inhibit the growth of *Babesia gibsoni* cultured *in vitro* at very small doses.[3]

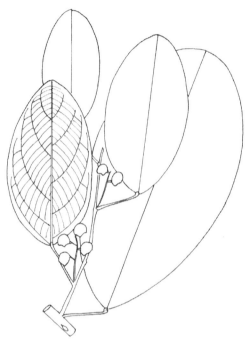

Figure 8.4 *Limacia oblonga* Hook. f. & Thoms. [From: Flora of Malaya. FRI No: 32543. Road to Fraser's Hill, ridge roadside. July 20, 1982. Field collector: K. M. Kochummen. Botanical identification: F. Jacques, October 2003.]

8.3 *LIMACIA OBLONGA* HOOK. F. & THOMS.

[From: Latin *oblonga* = oblong.]

8.3.1 Botany

Limacia oblonga Hook. f. & Thoms. is a climber that grows to a height of 3m in the rain forest of Malaysia. The leaves are simple, exstipulate, and spiral. The petiole is 3mm × 2mm. The blade is elliptic, 8.7cm × 4.4 cm – 11cm × 6cm – 14cm × 8cm – 15cm × 8cm – 19.5cm × 10cm and shows four to seven pairs of secondary nerves and scalariform tertiary nerves below. The inflorescence consists of axillary raceme. The flowers are greenish. The fruit is green and globose with 9mm berries (Figure 8.4).

8.3.2 Ethnopharmacology

The root plant is used externally by the Malays to heal sores. This property is probably mediated by berberine and congeners, which are known to occur in the plant.[4]

8.4 *STEPHANIA JAPONICA* (THUNB.) MIERS

[From: Greek *stephane* = wreath and Latin *japonica* = from Japan.]

8.4.1 Botany

Stephania japonica (Thunb.) Miers (*Stephania hernandifolia* Willd. Wap.) is a climber that is found in a geographical area ranging from India, South China, Taiwan, and Southeast Asia. The leaves are simple, exstipulate, and spiral. The petiole is 6.2cm long. The blade is broadly elliptic, 12cm × 2cm – 16cm × 11cm – 15cm × 8cm, acuminate at the apex, rounded at the base, and is attached to the petiole on its first half. The blade shows four pairs of secondary nerves, which are reddish. The flowers are minute and arranged in axillary cymes (Figure 8.5).

8.4.2 Ethnopharmacology

In Japan and Taiwan decoction of the plant is used as a drink to treat malaria and to invigorate. In Indonesia, the roots are used to assuage stomachaches, and a paste of the fruit is applied to

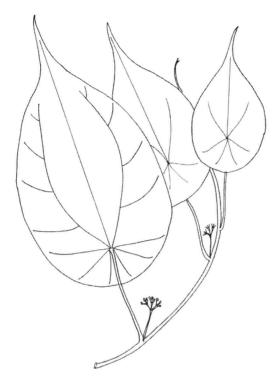

Figure 8.5 *Stephania japonica* (Thunb.) Miers. [From: Sarawak Forest Department. Field collector: M. Asri. No: S44854. Botanical identification: H. Kok, March 26, 1984. Geographical localization: Ulu Anap. 4th Division in secondary forest. Botanical identification: F. Jacques, October 2003.]

cancer of the breast. The antimalarial properties of the plant are very likely owed to the interesting array of isoquinolines, which abound in the plant, including homostephanoline, hasubanonine, prometaphanine, epistephanine, cyclanoline, hasubanol, isotrilobine, and trilobine.[5–13] Hall and Chang[14] made the interesting observation that isotrilobine in reverse doxorubicin resistance in human breast cancer cells might hold some potential for chemotherapy. Note that weight loss phytopharmaceuticals containing *Stephania tetrandra* S. Moore are banned from the European market because of their hazardous effect on the kidneys.

REFERENCES

1. Chi, C. W., Chang, Y. F., Chao, T. W., Chiang, S. H., P'eng, F. K., Lui, W. Y., and Liu, T. Y. 1994. Flow cytometric analysis of the effect of berberine on the expression of glucocorticoid receptors in human hepatoma HepG2 cells. *Life Sci.*, 54, 2099.
2. Sriwilaijareon, N., Petmitr, S., Mutirangura, A., Ponglikitmongkol, M., and Wilairat, P. 2002. Stage specificity of *Plasmodium falciparum* telomerase and its inhibition by berberine. *Parasitol Int.*, 51, 99.
3. Subeki, M. H., Matsuura, H., Takahashi, K., Yamasaki, M., Yamato, O., Maede, Y., Katakura, K., Suzuki, M., Trimurningsih, C., and Yoshihara, T. 2005. Antibabesial activity of protoberberine alkaloids and 20-hydroxyecdysone from *Arcangelisia flava* against *Babesia gibsoni* in culture. *J. Vet. Med. Sci.*, 67, 223.
4. Tomita, M., Juichi, M., and Furukawa, H. 1967. Studies on the alkaloids of menispermaceous plants. 248. Alkaloids of *Limacia oblonga* (Miers) Hook f. et. Thoms. *Yakugaku Zasshi*, 87, 1560.
5. Ibuka, T. and Kitano, M. 1967. Studies on the alkaloids of menispermaceous plants. CCXXXVII. Alkaloids of *Stephania japonica* Miers. (Supplement 17). Structure of homostephanoline. *Chem. Pharm. Bull. (Tokyo)*, 15, 1939.

6. Ibuka, T. and Kitano, M. 1967. Studies on the alkaloids of menispermaceous plants. Alkaloids of *Stephania japonica* Miers. 18. Structure of hasubanonine. *Chem. Pharm. Bull. (Tokyo),* 1809 (Supplement 1).

7. Ibuka, T., Kitano, M., Watanabe, Y., and Matsui, M. 1967. Studies on the alkaloids of menispermaceous plants. CCXXXVI. Alkaloids of *Stephania japonica* Miers. (Supplement 16). On the hofmann degradation of hasubanonine. *Yakugaku Zasshi,* 87, 1014.

8. Tomita, M., Inubushi, Y., and Ibuka, T. 1967. Studies on the alkaloids of menispermaceous plants. 230. Alkaloids of formosan *Stephania japonica* Miers. Structure of prometaphanine. *Yakugaku Zasshi,* 87, 381.

9. Watanabe, Y., Furukawa, H., and Kurita, M. 1966. Studies on the alkaloids of menispermaceous plants. 218. Alkaloids of *Stephania japonica* Miers. (Supplement 15). Dehydroepistephanine and its hydrogenation. *Yakugaku Zasshi,* 86, 257.

10. Furukawa, H. 1966. Studies on the alkaloids of menispermaceous plants. CCXVII. Alkaloids of *Stephania japonica* Miers. (Supplement 14). Hydrogenation of epistephanine. *Yakugaku Zasshi,* 86, 253.

11. Watanabe, Y., Matsui, M., and Ido, K. 1965. Studies on the alkaloids of menispermaceous plants. 213. Alkaloids of *Stephania japonica* Miers. (Supplement 11). Constitution of homostephanoline. *Yakugaku Zasshi,* 85, 584.

12. Ibuka, T. 1965. Studies on the alkaloids of menispermaceous plants. CCXII. Alkaloids of *Stephania japonica* Miers. (Supplement 10). Syntheses of phenanthrene derivatives related to degradative product of metaphanine. Syntheses of 1,5-diethoxy-2,6-di-methoxyphenanthrene, 1,6-dimethoxy-2,5-di-ethoxyphenanthrene, and 1,2-diethoxy-5,6-dimethoxyphenanthrene. *Yakugaku Zasshi,* 85, 579.

13. Tomita, M., Ibuka, T., Inubushi, Y., Watanabe, Y., and Matsui, M. 1965. Studies on the alkaloids of menispermaceous plants. CCX. Alkaloids of *Stephania japonica* Miers. (Supplement 9). Structure of hasubanonine and homostephanoline. *Chem. Pharm. Bull. (Tokyo),* 13, 538.

14. Hall, A. M. and Chang, C. J. 1997. Multidrug-resistance modulators from *Stephania japonica. J. Nat. Prod.,* 60, 1193.

CHAPTER 9

Medicinal Plants Classified in the Family Polygonaceae

9.1 GENERAL CONCEPT

The family Polygonaceae (A. L. de Jussieu, 1789 nom. conserv., the Buckwheat Family) consists of approximately 30 genera and 1000 species of bitter-tasting herbs, easily recognizable in the field by stems which are sourish, articulated, hollowed, and striated, and by their stipules that form some sort of membranous tubes at base of the leaves. The traditional system of medicines in the Pacific Rim uses about 30 plant species of Polygonaceae mainly for gastrointestinal disturbances, to expel intestinal worms, to allay fever, to invigorate, to regulate menses, to reduce liver discomfort, to treat skin infection, and to soothe inflammation. Classic examples of medicinal Polygonaceae used in Western medicine are *Rheum palmatum* L. var. *tanguticum* Maxim., and *Rheum officinale* H. Bn., which are laxatives; methoxystypandrone, a naphthoquinone; *Polygonum cuspidatum*, which has inhibited the enzymatic activity of HRV 3C-protease with an IC_{50} value of 4.6μM[1]; and two phenylpropanoid esters of sucrose: vanicoside B and lapathoside A, from the aerial part of *Polygonum lapathifolium*, which have inhibited the induction of Epstein–Barr virus early antigen (EBV-EA) by 12-*O*-tetradecanoylphorbol-13-acetate (TPA) and exhibited significant antitumor-promoting effects on mouse two-stage skin carcinogenesis.[2] Polygonaceae tend to elaborate resveratrol and congener, hence they have the potential as a source of chemotherapeutic agents (Figure 9.1).

9.2 *POLYGONUM CHINENSE* L.

[From: Greek *polus* = many and *gonos* = angled, and from Latin *chinense* = from China.]

9.2.1 Botany

Polygonum chinense L. is a perennial, rhizomatous herb that grows to a height of 1m in the wet valleys, mixed forests, thickets in valleys, and mountain grassy slopes of China, Taiwan, Himalaya, Japan, India, Malaysia, and the Philippines from sea level to 3000m. The stems are ligneous at the base, 3mm in diameter, red, striate, glabrous or hispid, and sour-tasting. Leaves: simple and alternate. The ochrea is tubular, 1.5–2.5cm long, membranous, glabrous, veined, and oblique at the apex. The petiole is 7mm long and auriculate at the base. The blade is ovate, elliptic, or lanceolate. The blade is 8cm × 3.5cm – 6cm × 2cm, and shows nine pairs of secondary nerves.

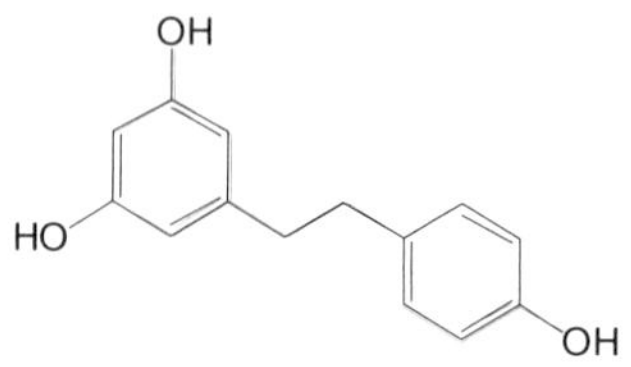

Resveratrol

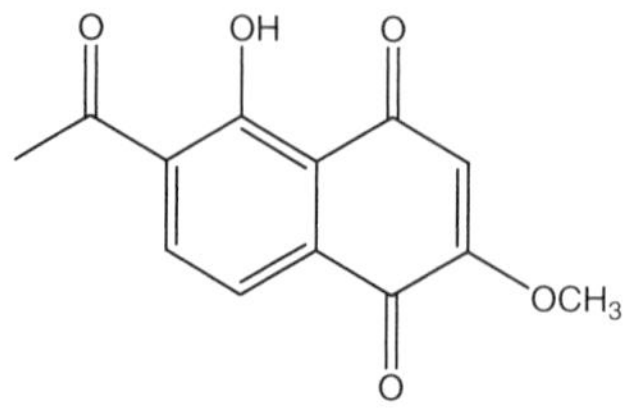

Methoxystypandrone

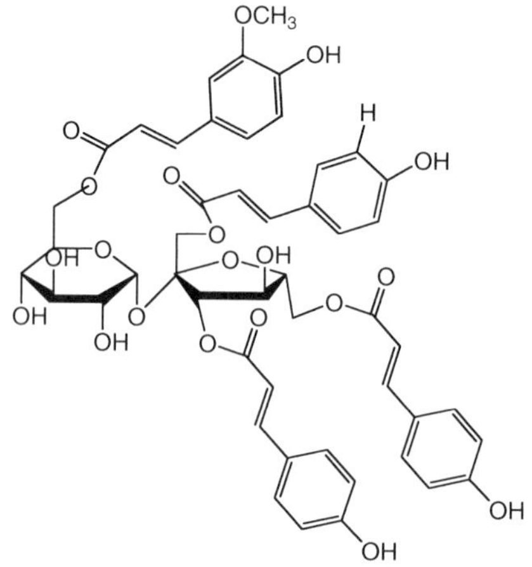

Vanicoside B

Figure 9.1 Examples of bioactive natural products from the family Polygonaceae.

The base of the blade is truncate or broadly cordate. The margin is entire, and the apex of the blade is shortly acuminate. The inflorescences are long and thin axillary clusters of very small flowers. The flowers are white or pinkish, and comprise a perianth made of five ovate lobes which are accrescent in fruits. The androecium consists of eight stamens with blue anthers. The gynaecium includes three styles, which are connate at the base. The fruits are broadly ovate, trigonous, and black achenes are included in the perianth (Figure 9.2).

9.2.2 Ethnopharmacology

Mountain knotweed, Chinese knotweed, or hill buckwheat are used medicinally in China where it is known as *huo tan mu, ch'ih ti li,* and *shan ch'iao mai* (Chinese). In China, the roots of *Polygonum chinense* L. are used to treat fluxes, to remove intestinal worms, and to counteract

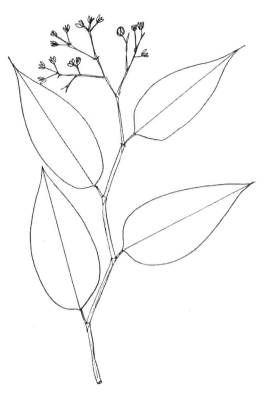

Figure 9.2 *Polygonum chinense* L. [From: Flora of Malay Peninsula. Geographical localization: T. Telan Pahang. Aug. 21, 1931. No: 25200.]

scorpion poisoning. The pharmacological property of this herb is unexplored. Note the presence of 25-R-spirost-4-ene-3,12-dione, stigmast-4-ene-3,6-ione, stigmastane-3,6-dione, hecogenin, and aurantiamide, which are antiinflammatory and antiallergic.[3]

9.3 *POLYGONUM MINUS* HUDS.

[From: Greek *polus* = many and *gonos* = angled, and from Latin *minus* = small.]

9.3.1 Botany

Polygonum minus Huds. is an annual herb which grows in roadsides, swamps, and ditches of Asia and the Pacific Islands. The stems are stoloniferous and decumbent. The ochrea show few short cilia. The leaves are lanceolate, membranous, 4cm × 9mm – 2cm × 5mm, and show a few secondary nerves. The flowers are white in spikes, which are 1–5cm long, linear-cylindrical, loosely but almost continuously flowered, and 3–5mm thick (Figure 9.3).

9.3.2 Ethnopharmacology

The plant is known as smartweed and in Sarawak is called *kasum* (Selakoh), *besanit* (Punan), and *rumput amak* (Iban) where it is used for sprains and body aches. They pound it with rice powder and make a paste which is rubbed or applied on the affected area. The plant is taken after childbirth and is also used as a remedy for indigestion. The pharmacological property of the plant

Figure 9.3   *Polygonum minus* Huds. [From: Plants of Indonesia. Bali Timur, Tabanan 2Km west of Candy Kuning in natural areas of Keban Raya, beyond the edge of Altingia Forest. Alt.: 1400m, 8°18′ S, 115°9′ E. Canopy, 15–20m tall. Common herbaceous.]

is unknown. 6,7-Methylenedioxy-5,3′,4′,5′-tetramethoxyflavone and 6,7-4′,5′ dimethylenedioxy-3,5,3′-trimethoxyflavone are known to occur in the plant.[4]

9.4 *POLYGONUM TOMENTOSUM* WILLD.

[From: Greek *polus* = many and *gonos* = angled, and from Latin *tomentosum* = densely covered with matted wool or short hair.]

9.4.1   Botany

Polygonum tomentosum Willd. (*Persicaria attenuata* subsp. *pulchra* [Bl.], K. L. Wilson *Polygonum pulchra* [Bl.], and *Polygonum tomentosum* Willd. non Schrank), is a perennial, rhizomatous floating creeper that grows in swamps and marshy areas in China, Taiwan, India, Indonesia, Malaysia, Burma, the Philippines, Sri Lanka, and Thailand. The rhizome is fibrous and the stems are erect to 1m tall, robust, pilose or glabrescent, and show fine reticulate roots at nodes and dry red cupper. The petiole is 1–2cm; the blade is 10–15cm × 1.5–3cm and broadly lanceolate. The inflorescence consists of terminal paniculate spikes, which are 4.5cm long. The perianth is green. The corolla is white, maturing orange. Seven or eight stamens are yellow and included. It has two free styles and the stigmas are capitate. Achenes are included in a persistent perianth, and are black, shiny, orbicular, biconvex, and 3–4mm in diameter.

9.4.2   Ethnopharmacology

In Burma, a decoction of roots is used to mitigate stomachaches in children. In Malaysia, the leaves are used to invigorate and to clean the blood. The Chinese call it *li liao*. The pharmacological potential of this herb is unexplored.

REFERENCES

1. Singh, S. B., Graham, P. L., Reamer, R. A., and Cordingley, M. G. 2001. Discovery, total synthesis, HRV 3C-protease inhibitory activity, and structure-activity relationships of 2-methoxystypandrone and its analogues. *Bioorg. Med. Chem. Lett.*, 11, 143.
2. Takasaki, M., Konoshima, T., Kuroki S., Tokuda, H., and Nishino, H. 2001. Cancer chemopreventive activity of phenylpropanoid esters of sucrose, vanicoside B and lapathoside A, from *Polygonum lapathifolium. Cancer Lett.*, 173, 133.
3. Tsai, P. L., Wang, J. P., Chang, C. W., Kuo, S. C., and Chao, V. 1998. Constituents and bioactive principles of *Polygonum chinensis. Phytochemistry*, 49, 1663.
4. Urones, J. G., Marcos, I. S., Pérez, B. G., and Barcala, P. B. 1990. Flavonoids from *Polygonum minus. Phytochemistry*, 29, 3687.

CHAPTER 10

Medicinal Plants Classified in the Family Myrsinaceae

10.1 GENERAL CONCEPT

The family Myrsinaceae consists of 30 genera and approximately 1000 species of tropical plants of which 40 species are medicinal in the Asia–Pacific, notably for the treatment of uterine disorders, inflamed throat, and as an analgesic. Myrsinaceae are recognized in the field by the presence of glands beneath the blade. The flowers are small, perfect, somewhat fleshy, and 5-merous. The leaves are simple, fleshy, elliptical with a peculiar green, and crenate. The fruits are often red berries, which are glossy.

A classic example of Myrsinaceae with pharmaceutical interest is *Embelia ribes* Burm. f., the seeds of which have been providing a treatment for intestinal worms (*Embelia, British Pharmaceutical Codex*, 1934) on account of benzoquinone: embelin (Figure 10.1). An interesting feature of Myrsinaceae family and *Ardisia* species is their ability to produce an unusual series of benzoquinones which have displayed a surprising number of pharmacological activities.[1] For instance, *Ardisia crispa* A. DC. produces 2-methoxy-6-tridecyl-1, 4-benzoquinone, which blocks platelet aggregation, B16-F10 melanoma cell adhesion to the extracellular matrix, and B16-F10 melanoma cell invasion; and inhibits pulmonary metastasis and tumor growth by blocking

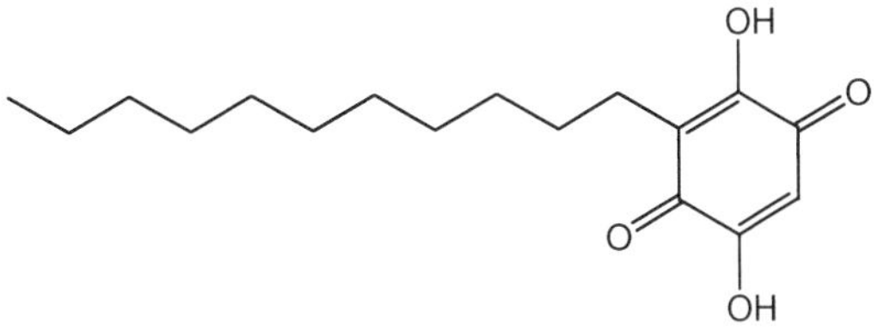

Embelin

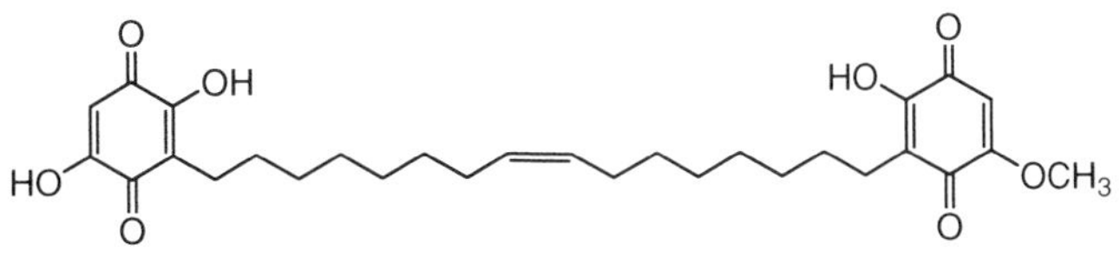

Ardisiaquinone A

Figure 10.1 Examples of bioactive benzoquinones characteristic of the family Myrsinaceae.

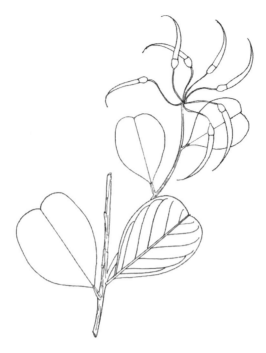

Figure 10.2 *Aegiceras corniculatum* Blco. [From: Flora of the Philippines, Bureau of Sciences. Agusan Sub-province, Mindanao. Field collector: A. Mallonga, April–July 1921. Forestry Bureau 28643.]

the integrin receptor.[2] Ardisiaquinones D, E, and F from *Ardisia sieboldii* inhibit the enzymatic activity of 5-lipo-oxygenase.[3,4]

10.2 *AEGICERAS CORNICULATUM* BLCO.

[From: Greek *aigos* = goat and *keras* = horn, and from Latin *corniculatum* = bearing little horns.]

10.2.1 Botany

Aegicera corniculatum Blco. (*Aegiceras majus* Gaertn.) is a tree that grows to a height of 4m in the mangroves of the tropical belt including Southeast Asia and the Pacific Islands. The stems are 3mm in diameter. The leaves are glossy above, simple, and cordate. The petiole is yellowish-orange and 8mm long. The blade is 6.5cm × 4.9cm – 5.2cm × 3.3cm, thick and the secondary nerves are inconspicuous in four to eight pairs. The fruits are horn-like, woody, and 4cm × 7mm attached to a 2.5cm-long pedicel (Figure 10.2).

10.2.2 Ethnopharmacology

In Vietnam, the plant is used to make a gargle. The plant is known to contain a series of oleanane triterpenes including 16α-hydroxy-l3, 28-epoxyoleanan-3-one 1, protoprimulagenin, aegicerin, as well as 2-methoxy-3-nonylresorcinol, 5-*O*-ethylembelin, 2-*O*-acetyl-5-*O*-methylembelin, 3,7-dihydroxy-2,5-diundecylnaphthoquinone, 2,7-dihydroxy-8-methoxy-3,6-diundecyldibenzofuran-1,4-dione, 2,8-dihydroxy-7-methoxy-3,9-diundecyldibenzofuran-1,4-dione, and 10-hydroxy-4-*O*-methyl-2,11-diundecylgomphilactone, 5-*O*-methylembelin, 3-undecylresorcinol, and 2-dehydroxy-5-*O*-methylem-belin, embelinone, and flavonoid glycosides.[5] Ardisiaquinones G, H, and I from *Ardisia teysmannia*

Figure 10.3 *Ardisia corolata* Roxb. [From: Distributed from the Botanic Gardens Singapore. Singapore Field No: 30192. Geographical localization: U. Bendong and B. Kajang, Nipah River, Kemaman. Alt.: 500ft. Nov. 3, 1935. Field collector and botanical identification: E. J. H. Corner.]

inhibit *in vitro* the first step of bacterial peptidoglycan synthesis with IC_{50} of 50µM, 26µM, and 16µM, respectively.[6] 5-*O*-ethylembelin is cytotoxic *in vitro* against HL-60, Bel (7402), U937, and Hela cell lines.[7] It will be interesting to learn whether or not more intensive future research on this plant discloses any molecules of therapeutic interest. It probably does.

10.3 *ARDISIA COROLATA* ROXB.

[From: Greek *ardis* = sharp.]

10.3.1 Botany

Ardisia corolata Roxb. (*Ardisia stylosa* Miq.) is a tree that grows to a height of 8m in lowland and hill forests in India, Malaysia, Thailand, and Indonesia. The stems are smooth, compressed, and ridged. The leaves are simple, exstipulate, and elliptic. The petiole is 1.3m long and channeled above. The blade is 16cm × 4.5cm – 23cm × 5.7cm – 11.5cm × 2.4cm, and shows 20 to 30 pairs of secondary nerves. The inflorescences are terminal pyramidal panicles up to 30cm long. The flowers are small, 5-merous and up to 3–5mm long, and are pink, white, or purple. The fruits are globose, deep red, and 6–6.5mm in diameter (Figure 10.3).

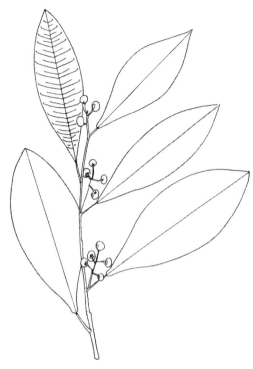

Figure 10.4 *Ardisia elliptica* Thunb. [From: Flora of Johor, Comm. Ex. Herb. Hort. Bot. Sing. Geographical localization: Jason Bay, Sedih, Johor. Date: June 16, 1972. Field collector: S. Ahmad. No: 22. Botanical identification: B. C. Stone, July 31, 1980.]

10.3.2 Ethnopharmacology

In Malaysia and Indonesia, the plant is known as *tinjau belukar*. The roots are used as a postpartum remedy. The fruits of the plant are known to contain ardisiphenols A–C, which scavenge 1,1-diphenyl-2-picrylhydrazyl (DPPH) free radicals and exhibit cytotoxicity against the murine breast cancer cell line, FM3A.[8]

10.4 *ARDISIA ELLIPTICA* THUNB.

[From: Greek *ardis* = sharp and from Latin *elliptica* = elliptic.]

10.4.1 Botany

Ardisia elliptica Thunb. (*Ardisia littoralis* Andr.) is a tree that grows to a height of 8m in lowland and hill forests in Southeast Asia and Hawaii. The stems are 3mm in diameter and finely fissured. The leaves are simple, exstipulate, and elliptic. The blade is 8cm × 3.5cm – 11cm × 4.4cm – 10cm × 3cm, 9.5cm × 3cm, and shows 15 pairs of secondary nerves. The inflorescences are axillary panicles up to 2.9cm long. The flowers are small, 5-merous, and pinkish. The fruits are globose, deep red, and the berries are 5mm in diameter (Figure 10.4).

10.4.2 Ethnopharmacology

Shoe Button *Ardisia* is used in Malaysia where a decoction of leaves is said to assuage retrosternal pains. The pharmacological potential of this plant is unexplored as of yet.

10.5 *ARDISIA FULIGINOSA* BL.

[From: Greek *ardis* = sharp.]

10.5.1 Botany

Ardisia fuliginosa Bl. is a treelet up to 3m in height that grows in Borneo. The leaves are simple, exstipulate, and elliptic. The blade is velvety below, and measures 16cm × 6.5cm – 15cm × 6cm, and shows 15 pairs of secondary nerves. The inflorescences are axillary panicles. The flowers are small, 5-merous, and pinkish. The fruits are globose, glossy, orange berries that are 8mm × 5mm. The fruit pedicel is 8mm long (Figure 10.5).

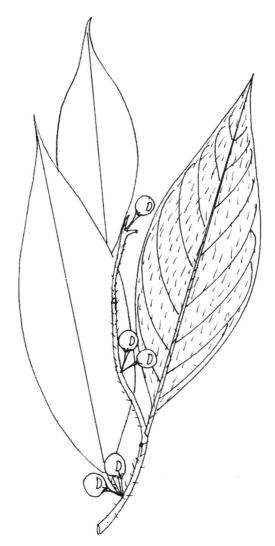

Figure 10.5 *Ardisia fuliginosa* Bl. [From Ex. Herb. Leiden, Herbarium Bogoriense. Flora of Borneo. Plants collected by J. P. Mogea and W. J. J. D. de Wilde during the Indonesian–Dutch Bukit Raya Expedition, 1982/1983. No: Mogea: 4401. Dec. 23, 1982. Geographical localization: Logging area, c. 8Km west of Central Base Camp. Alt.: 150m c. 65Km west of Badiding.]

10.5.2 Ethnopharmacology

Indonesians apply the sap squeezed from the stem to itchy parts of the skin. There is no evidence available on the pharmacological value of this plant. The plant is known in Borneo as *merjemah* (Sarawak).

10.6 *ARDISIA HUMILIS* VAHL.

[From: Greek *ardis* = sharp, and from Latin *humilis* = low-growing.]

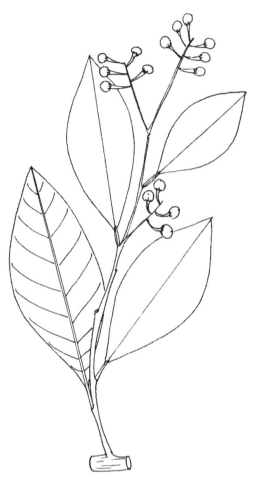

Figure 10.6 *Ardisia humilis* Vahl. [From: Sarawak Forest Department. Field collector: M. Asri. No: S52665. Geographical localization: Semunsan Wildlife Sanctuary, beach forest.]

10.6.1 Botany

Ardisia humilis Vahl. (*Ardisia hainanensis* Mez., *Ardisia pyrgina* Saint Lager, *Ardisia pyrgus* Roemer & Schultes, and *Tinus humilis* [Vahl.] Kuntze.) is a shrub that grows to a height of 2–5m tall in mixed forests, hillsides, and open fields; from sea level to 1100m in China, the Philippines, and Vietnam. The stems are glabrous, 5–7mm in diameter. The leaves are simple and exstipulate. The petiole is channeled above and 1cm long. The blade is obovate, elliptic, 4.9cm × 10.5cm × 2.2cm, leathery, glabrous, and inconspicuously pellucid punctuate. The apex of the blade is broadly acute to obtuse and the blade shows 12 pairs of secondary nerves. The inflorescences are terminal and axillary panicles are 4.2–20cm long. The flowers are fleshy, pink, or purplish red, 5–6mm on a 6–10mm-long pedicel. The fruits are dull red or purplish black, globose, 6mm in diameter and densely punctuate (Figure 10.6).

10.6.2 Ethnopharmacology

In Burma, the plant is used to treat menstrual disorders. Pharmacological properties are unexplored. It is called *Ai zi jin niu* in Chinese, *ati popa'a* in French Polynesia, and *merjemeh laut* in Malay.

10.7 *ARDISIA LANCEOLATA* ROXB.

[From: Greek *ardis* = sharp, and from Latin *lanceolata* = lance-shaped.]

10.7.1 Botany

Ardisia lanceolata is a tree that grows to a height of 8m in Malaysia, Singapore, Sumatra, Java, Borneo, and Celebes. The bark is grayish brown. The trunks are stout. The leaves are simple and exstipulate. The petiole is 1.5cm long. The blade is elliptical-oblong: 21cm × 7cm – 23cm × 8cm, 12.5cm × 4.5cm, chartaceous, rusty tomentose when young, and conspicuously pellucid punctuate. The apex of the blade is acute to obtuse and the blade shows 12–17 pairs of secondary nerves. The inflorescences are axillary panicles, which are short and minutely hairy. The pedicels are 8mm long. The flowers are purplish-pink, and the anthers are dark and glandular dotted. The gynaecium to the ovary is 5mm long. The flower buds are 7mm × 8mm. The fruits are dull red or purplish-black, globose, and 6mm in diameter (Figure 10.7).

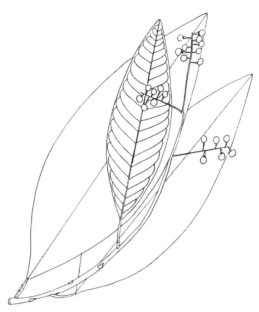

Figure 10.7 *Ardisia lanceolata* Roxb. [From: Distributed by The Botanic Gardens Singapore. Singapore Field No: 21311. Geographical localization: 5.5 miles from Kota Tinggi, Mawai Road from Johor. Feb. 2, 1935. Botanical identification: M. R. Henderson. Field collector: E. J. Corner. In swampy field.]

10.7.2 Ethnopharmacology

In Malaysia, the plant is used as a postpartum protective remedy. The pharmacological potential of this plant is unknown.

10.8 *ARDISIA ODONTOPHYLLA* WALL.

[From: Greek *ardis* = sharp, *odonto* = tooth, and *phullon* = leaf.]

10.8.1 Botany

Ardisia odontophylla Wall. is a shrub that is approximately 1m tall that grows in the rain forest of Malaysia. The stems are velvety. The leaves are simple, spiral, and exstipulate. The blade is obovate, velvety below, toothed, and shows 14–20 pairs of secondary nerves. The apex of the blade is acuminate to apiculate. The inflorescences are axillary panicles, which are up to 15cm long and hairy. The fruits are red (Figure 10.8).

10.8.2 Ethnopharmacology

A decoction of leaves is used to assuage stomachaches. The pharmacological potential of this plant is unexplored.

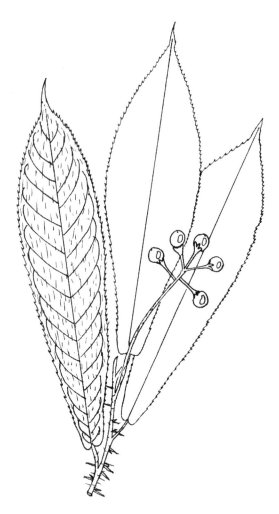

Figure 10.8 *Ardisia odontophylla* Wall. [From: Flora of Malaya. Comm. Ex Herb Hort Bot Sing. Geographical localization: Teku River, Pahang. Alt.: 200–300ft. Feb. 21, 1968. No: MS 1380. Botanical identification: B. C. Stone, April 1982.]

10.9 *ARDISIA OXYPHYLLA* WALL.

[From: Greek *ardis* = sharp.]

10.9.1 Botany

Ardisia oxyphylla Wall. is a treelet that grows in lowland and hill forests in Northeast India, Malaysia, Burma, Thailand, and Borneo. The bark is grayish-brown. The leaves are simple and exstipulate. The petiole is channeled and 2cm × 2mm. The blade is elliptic–oblong, 20mm × 5.5mm, villous, and the margin is cilate–pectinate. The apex of the blade is acuminate. The inflorescences are terminal panicles, which are up to 8cm long. The flowers are magenta with a densely tomentose ovary. The fruits are dull red or globose, and measure 5mm × 2mm (Figure 10.9).

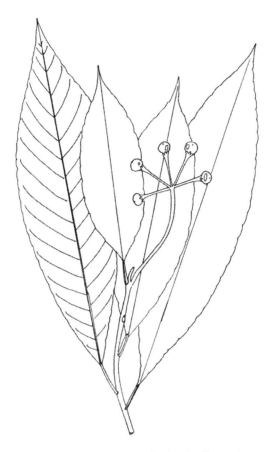

Figure 10.9 *Ardisia oxyphylla* Wall. [From: Flora of Malay Peninsula, Forest department. Geographical localization: Lalang River Reserve Kajang. March 28, 1930. No: 24078. Field collector: C. F. Symington.
Botanical identification: B. C. Stone, 1952.]

10.9.2   Ethnopharmacology

Medicinal uses and pharmacological properties: in Malaysia, a paste of leaves is used to heal
feet ulcers and cracks. The pharmacological potential of this plant is unexplored. One might set
the hypothesis that some benzoquinones with antilipoxygenase activity are responsible for the
traditional use of the plant.

10.10 *ARDISIA PYRAMIDALIS* (CAV.) PERS.

[From: Greek *ardis* = sharp.]

10.10.1   Botany

Ardisia pyramidalis (Cav.) Pers. is a treelet that is 7.5m in height with a girth of 7cm. The bark
is whitish. The inner bark is yellow. The leaves are simple and exstipulate. The petiole is long and
stout. The blade is elliptic–lanceolate, 16cm × 4cm – 26cm × 7 cm, 30cm × 6 cm, and denticulate
at the margin. The blade shows 21 pairs of secondary nerves below. The inflorescences are terminal
panicles of yellow flowers. The fruits are red and glossy (Figure 10.10).

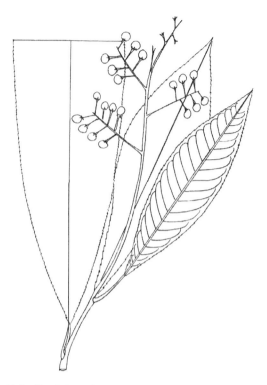

Figure 10.10 *Ardisia pyramidalis* (Cav.) Pers. [From: Flora of the Philippines. Philippine National Herbarium. No: 34223. Geographical localization: Mount Makiling, Laguna Province, Luzon. March 2, 1955. Field collector: M. Ebro.]

10.10.2   Ethnopharmacology

The plant is known as *gadong–gadon* in the Philippines where a decoction of roots is used as a drink to treat infection of the genitals, and to assuage toothaches. The leaves are used externally to mitigate headaches. The pharmacological potential of this plant is unexplored.

10.11   *ARDISIA RIDLEYI* KING & GAMBLE

[From: Greek *ardis* = sharp and after Ridley, British botanist of 19th century.]

10.11.1   Botany

Ardisia ridleyi King & Gamble is a treelet that grows wild in the rain forests of Thailand, Malaysia, and Sumatra. The stems are 2mm in diameter. The leaves are simple and exstipulate. The petiole is 9mm × 1mm. The blade is 8cm × 2.3cm – 14cm × 5.2cm, lanceolate, acuminate at the apex, and crenate at the margin. The inflorescences are terminal panicles, which are slender. The fruits are red and 8mm in diameter (Figure 10.11).

10.11.2   Ethnopharmacology

Malays call this plant *lutot hyam* and use it as a postpartum protective remedy. The pharmacological potential of this plant is unexplored.

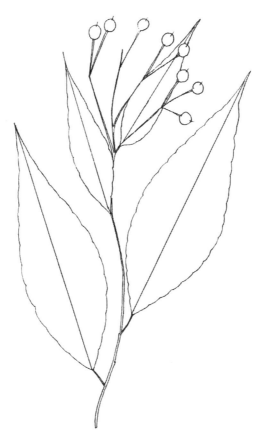

Figure 10.11 *Ardisia ridleyi* King & Gamble. [From: Flora of Malay Peninsula. Forest Department. Field collector: A. B. Yeob. Geographical localization: Maxwell Hill, Taiping. Alt.: 3200ft. Feb. 6, 1917. Botanical identification: H. N. Ridley.]

10.12 *ARDISIA SQUAMULOSA* PRESL.

[From: Greek *ardis* = sharp and from Latin *squamulosa* = squamulose.]

10.12.1   Botany

Ardisia squamulosa Presl. (*Ardisia boissieri* A. DC.) is a shrub that grows to a treelet height of 2.5m in the Philippines. The stem is 4mm in diameter and lenticelled. The leaves are simple and exstipulate. The petiole is pinkish. The blade is elliptic, glossy above, and 11.5xm × 6cm – 2 cm. The inflorescences are terminal panicles, which are 4.3cm long. The flowers are a pale, waxy pink. The stamens are grey with a yellow edge. The fruits are red, glossy globose, and 3mm in diameter (Figure 10.12).

10.12.2   Ethnopharmacology

In the Philippines, a paste of leaves is applied to a wound to promote healing. Chiang et al.[9] made the interesting observation that a water extract of *Ardisia squamulosa* inhibits the replication of adenovirus *in vitro*. Are bergerin and congeners involved here?

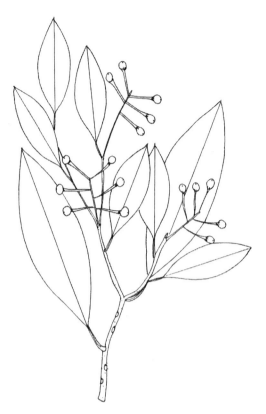

Figure 10.12 *Ardisia squamulosa* Presl. [From: Philippines Plant Inventory. USAID. PPI No: 6727. Field
collector: B. C. Stone et al. Geographical localization: Sibuyan Island, Province Romblon,
Geographical localization: Magdiwang, Barrio Hawasan, found along the Ating River, May 27,
1992.]

10.13 *MAESA CUMINGII* MEZ.

[From: Arabic *maas* = *Maesa lanceolata* Forrsk.]

10.13.1   Botany

Maesa cumingii Mez. is a shrub that grows wild in the Philippines. The stems are 2mm in
diameter. The leaves are simple and exstipulate. The petiole is 1.2cm × 1mm. The blade is broadly
elliptic, 2.5cm × 1cm – 4.5cm × 2.2cm, acute at the apex and shows 3–5 pairs of secondary nerves,
inflorescences, axillary, and is 1.5cm long. A pair of bracteoles subtends the base of the calyx
(Figure 10.13).

10.13.2   Ethnopharmacology

In the Philippines, the plant is known as *katiput* and provides a poison used for fishing.
The ichthyotoxic property is most likely owed to saponins that are known to abound in the
genus. Triterpenoid saponins, maesabalides I–VI, from *Maesa balansae,* destroy *Leishmania*
sp. Maesabalide III and IV destroy intracellular amastigotes with IC_{50} values of about 7 to

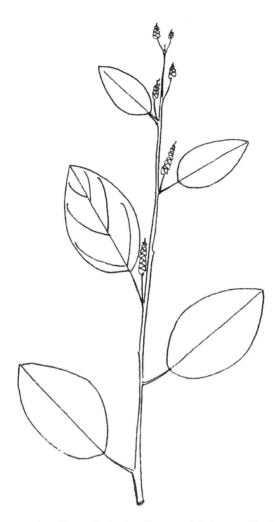

Figure 10.13 *Maesa cumingii* Mez. [From: Herbarium Bureau of Sciences, Manila, Philippines. Flora of the Philippines Islands. Plant of Luzon. Collected and presented by A. Loher. Geographical localization: Montalban, Rizal Province. January 1914. No: 13760.]

14mg/mL. A single subcutaneous dose at 0.2–0.4 mg/Kg has protected BALB/C mice against liver amastigote.[10]

10.14 *MAESA DENTICULLATA* MEZ.

[From: Arabic *maas* = *Maesa lanceolata* Forrsk. and from Latin *denticullata* = denticulate.]

10.14.1 Botany

Maesa denticullata Mez. is a shrub that grows wild in the Philippines. The stems are 2mm in diameter. The leaves are simple and exstipulate. The petiole is 2.5cm × 1mm. The blade is broadly elliptic, minutely serrate, 15cm × 7cm, acuminate at the apex and shows 3–5 pairs of secondary nerves. The inflorescences are axillary and 3.5cm long. A pair of bracteoles subtends the base of the calyx (Figure 10.14).

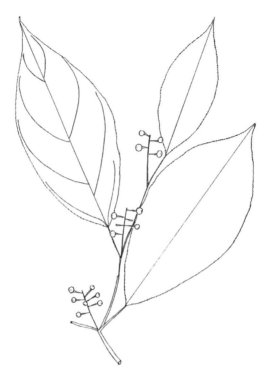

Figure 10.14 *Maesa denticullata* Mez. [From: Philippines Plants Inventory. USDAID. Field collectors: E. J. Reynoso et al. PPI No: 14486. Geographical localization: Northern Luzon, July 29, 1994. Ifugao Province, Brgy Pula, Mount Hagada, Banawie, 16°52.1′ N, 121°24′.7 E. In a submountain forest.]

10.14.2   Ethnopharmacology

In the Philippines the plant is used as fish poison. The pharmacological properties of this plant are unexplored as of yet. Is it antifungal?

10.15 *MAESA LAXA* MEZ.

[From: Arabic *maas* = *Maesa lanceolata* Forrsk., and from Latin *laxa* = lax.]

10.15.1   Botany

Maesa laxa Mez. is a shrub that grows in the Philippines. The stems are lenticelled. The leaves are simple and exstipulate. The petiole is slender and 3.2cm × 2mm. The blade is broadly elliptical, 6.5–12.5cm, acute at the apex and shows five pairs of secondary nerves. The margin is laxly toothed. The inflorescences are 10cm long racemes with 6mm pedicels. A pair of bracteoles subtends the base of the calyx. The fruits are 6mm in diameter berries (Figure 10.15).

10.15.2   Ethnopharmacology

Filipinos call the plant *tubing-aso* and use it to catch fish, in which it displays an ichthyotoxic tendency.

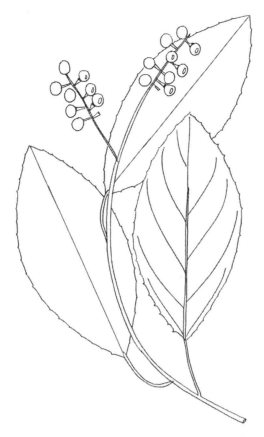

Figure 10.15 *Maesa laxa* Mez. [From: Flora of the Philippines Herbarium. Bureau of Sciences. No: 41553. Geographical localization: Cabalain, Leyte. Field collector: M. Ramos, December 1926.]

10.16 *MAESA PERLARIUS* (LOUR.) MERR.

[From: Arabic *maas* = *Maesa lanceolata* Forrsk.]

10.16.1 Botany

Maesa perlarius (Lour.) Merr. (*Dartus perlarius* Lour., *Maesa sinensis* A. DC., and *Maesa tonkinensis* Mez.) is a shrub that grows to a height of 3m in China, Taiwan, Thailand, and Vietnam. The stems are hirtellous and glandular granulose. The leaves are simple and exstipulate. The petiole is 10mm long and channeled. The blade is elliptical to broadly ovate, 7–11cm × 3–5cm, densely hirtellous when young. The base of the blade is acute, the margin is coarsely serrated, the apex is acute or acuminate; and the blade shows 7–9 pairs of secondary nerves. The inflorescences are axillary, paniculate, or rarely racemose, 2–4cm long, hirtellous, and glandular granulose. A pair of bracteoles are at the base of the calyx. The flowers are minute. The fruits are globose and 3mm in diameter (Figure 10.16).

10.16.2 Ethnopharmacology

In China, the plant known as *ii yu dan* is made into a paste of leaves which is applied to broken bones. In Cambodia, Laos, and Vietnam, the roots are used to promote digestion and urination,

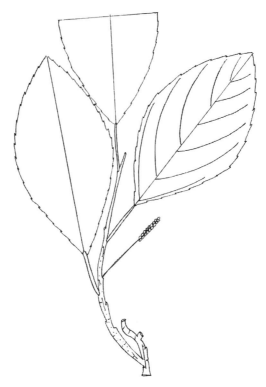

Figure 10.16 *Maesa perlarius* (Lour.) Merr. [From: Philippines Plant Inventory. Flora of the Philippines. FSN/USAID. Field collectors: E. J. Reynoso et al. PPI No: 7021. Geographical localization: Northern Luzon, Province: Ifugato, Mount Bunkung, Brgy Dalikan, Banawe. Found along the ridge in a secondary forest.]

and the leaves are used to treat measles. An infusion of leaves is used as a drink as a postpartum protective remedy. The pharmacological potential is unknown.

10.17 *MAESA RAMENTACEA* (ROXB.) A. DC.

[From: Arabic *maas* = *Maesa lanceolata* Forrsk. and from Latin *racemosa* = racemose].

10.17.1   Botany

Maesa ramentacea (Roxb.) A. DC. (*Baeobotrys ramentacea* Roxb.) is a shrub that grows to a height of 5m on mountain slopes, and in stream banks and shady places along jungle paths up to 1700m altitude in Bangladesh, Cambodia, India, Indonesia, Laos, Malaysia, Thailand, Burma, the Philippines, and Vietnam. The bark is brownish and the wood is yellow–red. The stems are angular, glabrous, minutely lenticelled, and 2mm thick. The leaves are simple and exstipulate. The petiole is 4mm × 1mm. The blade of the leaf is ovate to elliptic–lanceolate, 6.8cm × 3.5cm – 9.2cm × 3.3cm, and papery. The base is rounded, obtuse, to broadly cuneate, and the margin entire or undulate. The apex of the blade is acute or long acuminate. The blade shows six pairs of secondary nerves. Inflorescences are axillary or sometimes subterminal, paniculate, many-branched, and 16.5cm long. The fruit is yellowish-green, globose, 2–3mm, punctuate–lineate or veined (Figure 10.17).

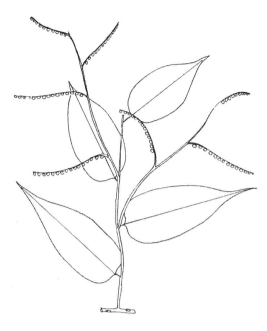

Figure 10.17 *Maesa ramentacea* (Roxb.) A. DC. [From: Flora of Thailand. No: 028.4. Geographical localization: Peninsular, Trang Chawng. Alt.: 100m. Oct. 10, 1948. Botanical identification: T. Smitinand.]

10.17.2   Ethnopharmacology

A paste of leaves is applied to scabies and other skin ailments. In Thailand, where the plant is called *ruai*, the leaves are mixed with rice and eaten to assuage retrosternal pains. The Chinese name for the plant is *cheng gan shu*.

The plant is known to elaborate an ichthyotoxic saponin known as saponin A.[11] An aqueous extract has inhibited growth of several species of a broad-spectrum fungus, probably because of its saponin content.[12]

10.18 *MAESA TETRANDRA* A. DC.

[From: Arabic *maas* = *Maesa lanceolata* Forrsk. and from Latin *tetrandra* = four anthers.]

10.18.1   Botany

Maesa tetrandra A. DC. is a shrub that grows in Indonesia. The stems are 2mm in diameter and the internode is 1.5cm long. The leaves are simple and exstipulate. The petiole is 7mm long and channeled. The blade is elliptic–lanceolate, 7.4cm × 2.5cm – 4.5cm × 1.6cm, velvety below, and has four pairs of secondary nerves which are inconspicuous. The margin is coarsely crenate, the apex is acute or acuminate. The inflorescences are axillary racemes which are 3–6cm long. A pair of bracteoles is at the base of the calyx. The fruits are globose (Figure 10.18).

10.18.2   Ethnopharmacology

In Indonesia, the roots are crushed and ingested as a remedy for fever, while a decoction of leaves and shoots affords a remedy for measles. The pharmacological potential of this plant is

Figure 10.18 *Maesa tetrandra* A. DC. [From: Herbarium Bogoriense, Harvard University Plants of Indonesia. Botanical identification: J. A. McDonald, 1993. Cat #: 3862. Geographical localization: Kabaena, Mountain Sabampolulu, 1Km SSE of Tangkeno, 5°16′ S, 121°56′ E. Alt.: 700–900m. Slope of rain forest.]

unexplored as of yet. One can reasonably expect quinones to be responsible for the antiviral property of the plant. Quinones abound in the family Ebenaceae, which is discussed in the next chapter.

REFERENCES

1. Kobayashi, H. and de Mejía, E. 2005. The genus *Ardisia*: a novel source of health-promoting compounds and phytopharmaceuticals. *J. Ethnopharmacol.*, 96, 347.
2. Kang, Y. H., Kim, W. H., Park, M. K., and Han, B. H. 2001. Antimetastatic and antitumor effects of benzoquinonoid AC7-1 from *Ardisia crispa*. *Int. J. Cancer*, 93, 736.
3. Fukuyama, Y., Kiriyama, Y., Kodama, M., and Iwaki, H. 1995. Naturally occurring 5-lipoxygenase inhibitors. VI. Structures of ardisiaquinones D, E, and F from *Ardisia sieboldii. Chem. Pharm. Bull. (Tokyo)*, 43, 1391.
4. Fukuyama, Y., Kiriyama, Y., Okino, J., Kodama, M., Iwaki, H., Hosozawa, S., and Matsui, K. 2001. Naturally occurring 5-lipoxygenase inhibitor. II. Structures and syntheses of ardisianones A and B, and maesanin, alkenyl-1,4-benzoquinones from the rhizome of *Ardisia japonica. Chem. Pharm. Bull. (Tokyo)*, 41, 561, 1993.
5. Yang, L. K., Khoo-Beattie, C., Goh, K. L., Chng, B. L., Yoganathan, K., Lai, Y. H., and Butler, M. S. 2001. Ardisiaquinones from *Ardisia teysmanniana. Phytochemistry*, 58, 1235.
6. Xu, M., Deng, Z., Li, M., Li, J., Fu, H., Proksch, P., and Lin, W. 2004. Chemical constituents from the mangrove plant *Aegiceras corniculatum. J. Nat. Prod.*, 67, 762.

7. Sumino, M., Sekine, T., Ruangrungsi, N., Igarashi, K., and Ikegami, F. 2002. Ardisiphenols and other antioxidant principles from the fruits of *Ardisia colorata*. *Chem. Pharm. Bull. (Tokyo)*, 50, 1484.

8. Sumino, M., Sekine, T., Ruangrungsi, N., and Ikegami, F. 2001. Ardisiphenols A–C, novel antioxidants from the fruits of *Ardisia colorata*. *Chem. Pharm. Bull. (Tokyo)*, 49, 1664.

9. Chiang, L. C., Cheng, H. Y., Liu, M. C., Chiang, W., and Lin, C. C. 2003. *In vitro* anti-herpes simplex viruses and antiadenoviruses activity of twelve traditionally used medicinal plants in Taiwan. *Biol. Pharm. Bull.*, 26, 1600.

10. Germonprez, N., Maes, L., Van Puyvelde, L., Van Tri, M., Tuan, D. A., and De Kimpe, N. 2005. *In vitro* and *in vivo* anti-leishmanial activity of triterpenoid saponins isolated from *Maesa balansae* and some chemical derivatives. *J. Med. Chem.*, 48, 32.

11. Tuntiwachwuttikul, P., Pancharoen, O., Mahabusarakam, W., Wiriyachitra, P., Taylor, W. C., Bubb, W. A., and Towers, G. H. 1997. A triterpenoid saponin from *Maesa ramentacea*. *Phytochemistry*, 44, 491.

12. Phongpaichit, S., Schneider, E.F., Picman, A. K., Tantiwachwuttikul, P., Wiriyachitra, P., and Arnason, J. T. 1995. Inhibition of fungal growth by an aqueous extract and saponins from leaves of *Maesa ramentacea* Wall. *Biochem. Sys. Ecol.*, 23, 17.

CHAPTER 11

Medicinal Plants Classified in the Family Ebenaceae

11.1 GENERAL CONCEPT

The family Ebenaceae (Gürke in Engler & Prantl, 1891), or Ebony Family, consists of five genera and approximately 450 species of trees known to elaborate a series of naphthoquinones and pentacyclic triterpenoid saponins (Figure 11.1). When searching for Ebenaceae, one might look into hillside primary rain forests. Ebenacea are recognized by their fruits, which appear like little persimmons, often brownish, and seated on a persistent calyx of significant hardness. The wood of Ebenaceae is dense, very hard, and blackens upon exposure to light. The principles responsible for the peculiar color of ebonies are naphthoquinones. With regard to the pharmacological potential of Ebenaceae, the evidence for the existence of possible therapeutic agents is strong and it seems quite likely that further studies will result in the isolation and identification of certain antibacterial, antiviral, cytotoxic, monoamine oxidase inhibitors, antioxidant monomer dimers, or oligomers of naphthoquinones.

As a matter of fact, the evidence currently available suggests that naphthoquinones, which are planar intercalate with DNA, interfere with the mitochondrial electron respiratory chain reaction because of its ketone moieties, which tend to generate noxious free radicals. Some naphthoquinones are marketed as a drug such as atovaquone, which is used to treat malaria and *Pneumocystis carinii* infection. An interesting development from naphthoquinones is their antiviral and central nervous system (CNS) properties.

Plumbagin, isodiospyrin, and 8′-hydroxyisodiospyrin inhibit significantly the proliferation of Hepa, KB, Colo-205, and HeLa cell lines cultured *in vitro* (Kuo et al., 1997).[1] Lemulinol A significantly inhibits the enzymatic activity of mouse liver monoamine oxidase (MAO).[3] The fruits of the *Diospyros* are astringent and often used to check bleeding and to treat diarrhea. The hydrolyzable tannins found in the fruits have displayed interesting pharmacological properties such as the lowering of blood pressure. Approximately 20 species of Ebenaceae are used for medicinal purposes in the Asia–Pacific, especially to expel intestinal worms and to treat viral infections. The seeds are often used for fishing, as these are ichthyotoxic.

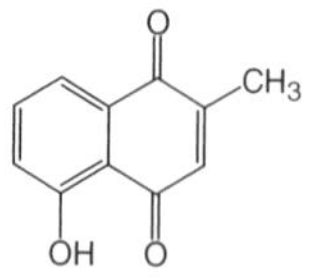

Plumbagin 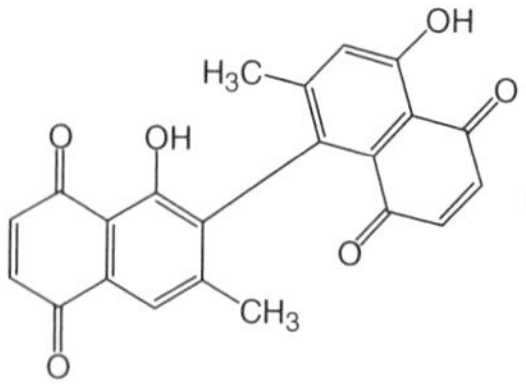   Isodiospyrin 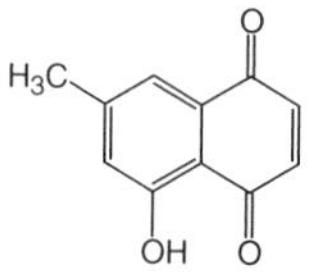   7-Methyljuglone

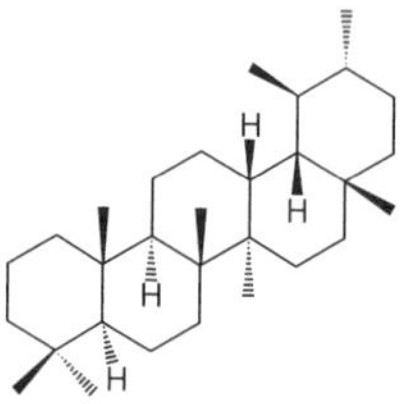

Ursane 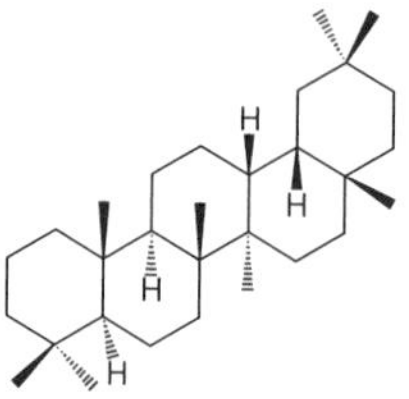   Oleanane

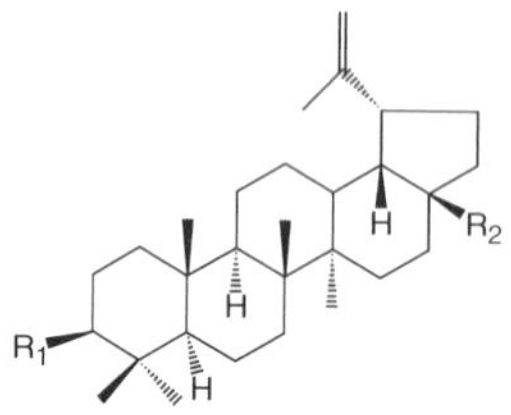

Lupeol R$_1$=β OH, R$_2$=CH$_3$ α Amyrin R$_1$=β OH, R$_2$=CH$_3$

Betulin R$_1$=β OH, R$_2$=CH$_2$OH   Ursolic acid R$_1$=β OH, R$_2$=COOH.

Betulinic acid R$_1$=β OH, R$_2$=COOH

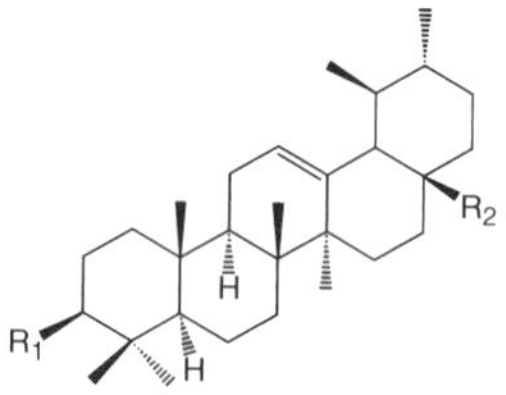

Figure 11.1   Examples of bioactive natural products from the Ebenaceae Family.

11.2 *DIOSPYROS LANCEIFOLIA* ROXB.

[From: Greek *diospyros* = persimmon fruit and from Latin *lanceifolia* = lanceolate leaves.]

11.2.1 Botany

Diospyros lanceifolia Roxb. is a timber that grows to a height of 20m with a girth of 70cm in lowland and hill rain forests to an altitude of 700m in India, Sumatra, Philippines, Malaysia, and Borneo. The bark is brown to black, smooth, or with fine cracks. The inner bark is bright yellow. The leaves are simple, oblong–elliptical to lanceolate, 4.5cm – 15cm × 2cm – 5cm with a base-pointed apex acuminate. The midrib is sunken above. It has up to nine pairs of secondary nerves. The petiole is 1cm long. The male inflorescence is a 3cm-long cyme. The male flowers are 4-merous, very small, and salver-shaped. The female flowers are solitary, small, and urseolate. The fruits are globose with a short apical beak, subglabrous, 2cm in diameter, and seated on a shallow 3–5-lobed calyx (Figure 11.2).

11.2.2 Ethnopharmacology

In Indonesia, the seeds are used as a fish poison. The principles involved here might be their content of naphthoquinone derivatives such as biplumbagin and chitranone, which are known to be ichthyotoxic, as has been shown with the seeds of *Diospyros maritima*.[4]

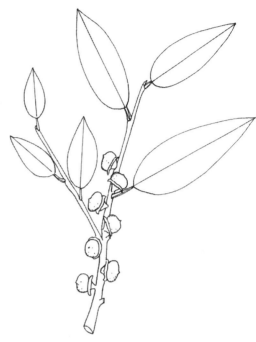

Figure 11.2 *Diospyros lanceifolia* Roxb. [From: Flora of Malaysia. FRI No: 17549. Geographical localization: Gunong Pulai Forest Reserve, Southwest Johor. Hillside. Alt.: 1600ft.]

11.3 *DIOSPYROS MALABARICA* (DESR.) KOSTEL.

[From: Greek *diospyros* = persimmon fruit and from Latin *malabarica* = from Malabar, India.]

11.3.1 Botany

Diospyros malabarica (Desr.) Kostel. (*Diospyros embryopteris* Pers., *Diospyros glutinosa* Koening, and *Diospyros siamensis* Ridl. [non Hoch.]) is a magnificent timber tree that grows to a height of 37m with a girth of 2m. The plant is found in a geographical area that spans India, Thailand, and North Malaysia. The bole is straight and the wood is hard and dense. The bark is black, smooth, and the inner bark turns bluish on exposure to sunlight. The leaves are simple and exstipulate. The blade is elliptic or ovate, pointed or rounded at the base, and shows a midrib sunken above. There are 4–8 pairs of secondary nerves curving upward to form several series of loops near the margin. Tertiary nerves are reticulate. The male flowers are formed in 3–5 flowered cymes axillary. Stamens

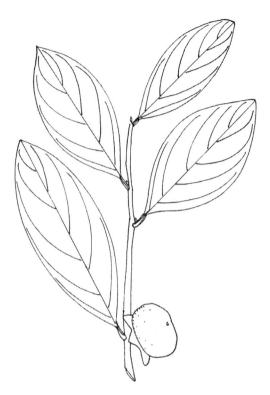

Figure 11.3 *Diospyros malabarica* (Desr.) Kostel. [From: Flora of Malaya. FRI No: 14133. Botanical identification: F. S. P. Ng, July 20, 1991. Geographical localization: East summit of Mountain Bongsu, Bongsu Forest Reserve. South Kedah, steep hillside on previously logged forest.]

vary in number between 26 and 47, mostly in antero posterior pairs. The female flowers are solitary, 4-merous, with four styles, and an 8-celled ovary. The fruits are globose, up to 3.5cm in diameter, and seated on a persistent calyx. The fruit is green, tinted red (Figure 11.3).

11.3.2 Ethnopharmacology

In China, the plant is known as *pei shih* (Chinese). The fruits are used to treat diarrhea and dysentery. The medicinal properties ascribed to it by the Chinese are somewhat remarkable. It is said to break fever, to be an antidote for snake poisoning, and to be demulcent. An extract of the fruit has been used as a vaginal injection in gonorrhea. A dark oil prepared from the fruit makes an excellent varnish for paper umbrellas and fans. In Cambodia, Laos, and Vietnam, the juice of the fruit is used to heal sores and wounds. The medicinal properties are most likely owed to tannins. Note that Choudhary et al. (1990)[5] made the observation that ethanolic leaf extracts of *Diospyros embryopteris* completely inhibited the libido of male rats when an oral dose of 100mg/Kg was given daily for 21 days.

11.4 *DIOSPYROS MULTIFLORA* BLCO.

[From: Greek *diospyros* = persimmon fruit and from Latin *multiflora* = numerous flowers.]

11.4.1 Botany

Diospyros multiflora Blco. is a timber tree that grows to a height of 12m in the rain forests of the Philippines. Leaves are simple, their base acute to acuminate, the apex is pointed. The petiole is grooved, woody, and up to 1cm long. The blade is lanceolate to elliptic, 5cm – 12cm × 2.5cm – 6.5cm and shows 9–12 pairs of secondary nerves. The midrib is sunken above. There are no tertiary nerves. The fruits are up to 2cm in diameter on a cup-shaped persistent calyx (Figure 11.4).

11.4.2 Ethnopharmacology

In the Philippines, the bark and leaves are used to treat herpetic eruptions. The antiviral property of the plant is unexplored as of yet. Some evidence has already been presented that indicates that naphthoquinones have antiherpes properties. In a recent study, Tandon et al.[6] synthesized and evaluated a series of naphthoquinone derivatives for antifungal, antibacterial, antiviral, and anticancer activities by using the standard assay and showed *in vitro* antiviral activity with the herpes simplex virus. The anti-

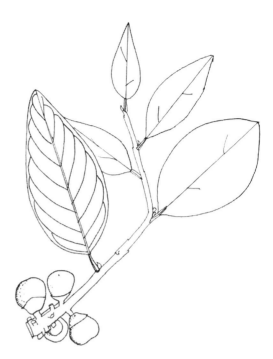

Figure 11.4   *Diospyros multiflora* Blco. [From: Flora of the Philippines. Philippine National Bureau. Corn. Juan Luna and Isaac Peral. Ermita, Manila, Luzon. Field collector: G. E. Adano, July 21, 1956, Philippines.]

viral property of naphthoquinones of the *Diospyros* species could be mediated by the inhibition of protease as reported by Matsumoto et al. (2001).[7]

11.5 *DIOSPYROS PILOSANTHERA* BLCO.

[From: Greek *diospyros* = persimmon fruit and from Latin *pilosanthera* = pilose anthers.]

11.5.1 Botany

Diospyros pilosanthera Blco. is a tree that grows to a height of 27m in the rain forests of Burma, Cambodia, Laos, Vietnam, Malaysia, and Indonesia. The bark is blackish, cracked, and fissured. The leaves are elliptical, 6cm – 14cm × 2.5cm – 6cm. The apex is acuminate, the base is pointed, and the midrib is sunken above. There are 8–16 pairs of secondary nerves. The male flowers are 5-merous, in 3–5 flowered cymes with 17 anthers. The female flowers are solitary, on 2mm peduncles. The fruits are depressed and ovoid, 2cm × 3cm, with a large woody calyx (Figure 11.5).

11.5.2 Ethnopharmacology

In the Philippines, Indonesia, and Malaysia, the plant is known as *kumu*. Filipinos drink a decoction of the bark to treat a cough. The pharmacological potential of this plant is unexplored

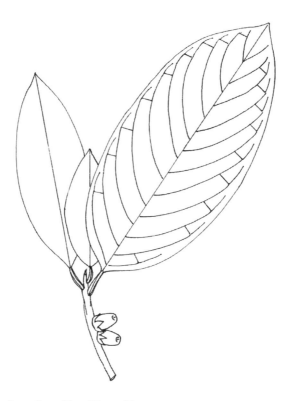

Figure 11.5   *Diospyros pilosanthera* Blco. [From: Singapore Field No: 37728. Distributed from The Botanic Gardens, Singapore. Botanical identification: F. S. P. Ng, March 17, 2000. Geographical localization: Mandai Road, Singapore. Field collector: M. S. Kiah, Aug. 1, 1940.]

as of yet. The bechic property might be due to saponins, which are known to abound in members of the Ebenaceae family.[8]

11.6 *DIOSPYROS SUMATRANA* MIQ.

[From: Greek *diospyros* = persimmon fruit and from Latin *sumatrana* = from Sumatra.]

11.6.1 Botany

Diospyros sumatrana Miq. (*Diospyros flavicans* [Wall.] Hiern, *Diospyros dumosa* King & Gamble, *Diospyros decipiens* King & Gamble, *Diospyros tubicalyx* Ridl., *Diospyros vestita* Bakh., *Diospyros velutinosa* Bakh., and *Diospyros hendersoni*) is a tree that grows to a height of 30m with a girth of 100cm in Indonesia, Thailand, and Malaysia (Borneo) in lowland rain forests to an altitude of 1500m above sea level. The trunks are hairy when young. The leaves are simple, elliptic, ovate, oblong, 3.5cm – 20cm × 1.2cm – 6.5cm, the apex is acuminate, the base is pointed, and the midrib sunken above. There are 3–11 pairs of secondary nerves. The male flowers are 5-merous, in 3–10 flowered subsessile cymes with 16 anthers. The female flowers are 4-merous, salver-shaped, with a 4-locular ovary. The fruits are globose, 1.2cm × 2.4cm, with a 2.5cm-diameter calyx (Figure 11.6).

Figure 11.6 *Diospyros sumatrana* Miq. [From: Oxford University Department of Forestry. Forest Herbarium. T. D. Pennington, Sept. 11, 1963. No: 7807. From FRIM Kepong. No: 94504.]

11.6.2 Ethnopharmacology

In Malaysia, the plant is known as *arang* or *kayu arang*. The seeds are poisonous and used to catch fish. The plant has not been studied for its pharmacological potential. The ichthyotoxic property could involve some naphthoquinones and/or saponins.[8]

11.7 *DIOSPYROS RUFA* KING & GAMBLE

[From: Greek *diospyros* = persimmon fruit and from Latin *rufa* = wrinkled.]

11.7.1 Botany

Diospyros rufa King & Gamble is a timber tree that grows to a height of 27m and a girth of 130cm in the lowland rain forest of Malaysia up to an altitude of 1300m. The bark is black to brown and smooth. The leaves are simple, oblong, oblong–obovate, 12–22cm × 4.5–8cm, the apex is acuminate, the base is pointed, and the midrib is sunken above. There are 13 pairs of secondary nerves. The petiole is 1–2.5cm long.

The male flowers are 4-merous, in 16 flowered subsessile clusters and show 16 anthers. The female flowers are 4-merous and show an 8-locular ovary. The fruits are round, brown with red hairs, and flattened; the apex is 3.5cm wide, in a 4-lobed, saucer-shaped, with a 2.5cm-diameter calyx (Figure 11.7).

11.7.2 Ethnopharmacology

In Malaysia, the seeds are poisonous and used to catch fish. The plant has not been studied for its pharmacological potential. The ichthyotoxic mechanism could involve some naphthoquinones or saponins.[7]

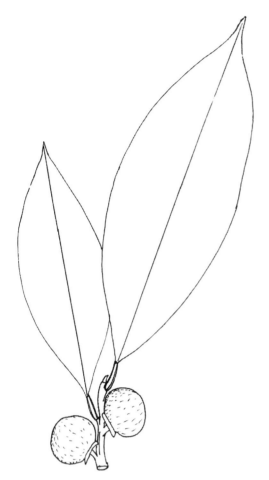

Figure 11.7 *Diospyros rufa* King & Gamble. [From: Flora of Malaya. FRI No: 35981. Geographical localization: River Durian Forest Reserve. Foot of hill near a stream.]

11.8 *DIOSPYROS TOPOSIOIDES* KING & GAMBLE

[From: Greek *diospyros* = persimmon fruit and from Indian *toposi* = *Diospyros toposia*.]

11.8.1 Botany

Diospyros toposioides King & Gamble is a timber tree that grows to a height of 13m and with a girth of 60cm in the lowland rain forests of Malaysia and Indonesia. The leaves are simple, oblong, oblong–ovate, 16cm – 33cm × 4cm – 14cm, the apex is acuminate, the base is rounded, and the midrib is sunken above. The secondary nerves are inconspicuous and loping at the margin. The male flowers are arranged in three flowered axillary cymes and show 35–96 anthers. The female flowers are 4-merous and show an 8-locular hairy ovary. The fruits are globose, up to 5cm in diameter, and are seated on a 3cm-wide calyx (Figure 11.8).

11.8.2 Ethnopharmacology

In Malaysia, the plant is known as *arang*, or *kayu arang*. The seeds are poisonous and used to catch fish. The plant has not been studied for its pharmacological potential. The ichthyotoxic property could involve some naphthoquinones or saponins.[8]

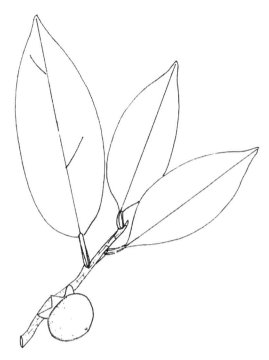

Figure 11.8 *Diospyros toposioides* King & Gamble. [From: Flora of Malaysia. Field No: 2417. Geographical localization: Chior, Perak. Field collector: K. M. Kochummen, July 11, 1967.]

REFERENCES

1. Kuo, Y. H., Chang, C. I., Li, S. Y., Chou, C. J., Chen, C. F., Kuo, Y. H., and Lee, K. H. 1997. Cytotoxic constituents from the stems of *Diospyros maritima*. *Planta Med.*, 63, 363.
2. Ting, C. Y., Hsu, C. T., Hsu, H. T., Su, J. S., Chen, T. Y., Tarn, W. Y., Kuo, Y. H., Whang-Peng, J., Liu, L. F., and Hwang, J. 2003. Isodiospyrin as a novel human DNA topoisomerase I inhibitor. *Biochem. Pharmacol.*, 66, 1981.
3. Okuyama, E., Homma, M., Satoh, Y., Fujimoto, H., Ishibashi, M., Yamazaki, M., Satake, M., and Ghazali, A. B. 1999. Monoamine oxidase inhibitory naphthoquinone and/or naphthalene dimers from *Lemuni hitam* (*Diospyros* sp.), a Malaysian herbal medicine. *Chem. Pharm. Bull. (Tokyo)*, 47, 1473.
4. Higa, M., Noha, N., Yokaryo, H., Ogihara, K., and Yogi, S. 2002. Three new naphthoquinone derivatives from *Diospyros maritima* Blume. *Chem. Pharm. Bull. (Tokyo)*, 590.
5. Choudhary, D. N., Singh, J. N., Verma, S. K., and Singh, B. P. 1990. Antifertility effects of leaf extracts of some plants in male rats. *Indian J. Exp. Biol.*, 28, 714.
6. Tandon, V. K., Singh, R. V., and Yadav, D. B. 2004. Synthesis and evaluation of novel 1,4-naphthoquinone derivatives as antiviral, antifungal and anticancer agents. *Bioorg. Med. Chem. Lett.*, 14, 2901.
7. Matsumoto, M., Misawa, S., Chiba, N., Takaku, N., and Hayashi, N. 2001. Selective nonpeptidic inhibitors of herpes simplex virus type 1 and human cytomegalovirus proteases. *Biol. Pharm. Bull.*, 24, 36.
8. Mallavadhani, U. V., Panda, A. K., and Rao, Y. R. 1998. Pharmacology and chemotaxonomy of *Diospyros*. *Phytochemistry*, 49, 901.

CHAPTER 12

Medicinal Plants Classified in the Family Bombacaceae

12.1 GENERAL CONCEPT

The family Bombacaceae (Kunth, 1822) consists of approximately 25 genera and 200 species of soft-wooded trees, which are widespread in tropical countries, especially in tropical America. They can be recognized by the fleshy or swollen aspect of the trunk and by the fruits which are often massive and capsular, loculicidal and often contain arillate seeds (Figure 12.1).

To date there is not much evidence to suggest whether Bombacaceae hold any pharmaceutical potential, but one should investigate this family thoroughly for pharmacology. The medicinal flora of Asia and the Pacific include a few species of Bombacaceae: *Bombax ceiba* L., *Bombax malabaricum* DC., *Gossampinus heptaphylla* (Houtt.) Bakh., *Gossampinus malabarica* (DC.) Merr., *Ceiba pentandra* (L.) Gaertn., *Eriodendron anfractuosum* DC., *Gossampinus rumphii* Schott, *Durio zibethinus* Murr., *Durio oxleyanus* Griff., and *Neesa altisima* Bl. These are often used for the treatment of inflammatory conditions and as diuretics.

Figure 12.1 Botanical hallmarks of Bombacaceae. **(See color insert following page 168.)**

12.2 *CEIBA PENTANDRA* (L.) GAERTN.

[From: *Latin pentandra* = 5 anthers.]

12.2.1 Botany

Ceiba pentandra (L.) Gaertn. (*Eriodendron anfractuosum* DC., and *Gossampinus rumphii* Schott) is a tree that grows to a height of 40m with a girth of 3m. The plant is native to Central America and has been introduced in tropical Africa and Asia. The wood is whitish and soft. The bark is smooth and greenish, and produces a few thorns, which are conical. The leaves are palmate and up to 40cm in diameter and consist of 7–8 folioles, which are 5cm × 1.5cm. The flowers are tubular, whitish to pink, in axillary fascicles. The fruits are green, fleshy, fusiform capsules 8cm – 14cm × 4.5cm – 7cm, containing up to 175 seeds 4–8mm long, which are minute and comose (Figure 12.2).

12.2.2 Ethnopharmacology

The Kapok tree is also known as *fromager* (French), *ceibo* (Mexican), and *kapok* (Malay). In Burma, the roots are used to invigorate and the leaves are used to treat gonorrhea. In Cambodia, the root is used to reduce fever. The bark is used to promote urination, to treat gonorrhea, to reduce fever, and to treat diarrhea. In Malaysia, the bark is used to treat asthma. In Indonesia, a decoction is used as a drink to treat gravels (small kidney calculi), and a decoction of leaves is used to treat syphilis. The juice squeezed from the leaves is used to treat asthma and coughs. In the Philippines, it is used to reduce fever and to promote libido, and the gummy exudate of the plant is eaten to treat dysentery, menorrhagia, and diabetes. Some evidence has already been presented, which lends support to the argument for its antidiabetic and antiinflammatory properties. Using streptozotocin-induced diabetes mellitus in experimental rats, Ladeji et al.[1] made a careful study of the antidiabetic properties of an aqueous bark extract given orally to rats for 28 days. There was, they report, a statistically significant reduction in plasma glucose levels. 5-Hydroxy-7,4',5'-trimethoxyisoflavone,3'-*O*-β-D-glucoside, and its aglycone, vavain, isolated from the bark of *Ceiba pentandra*, inhibited the enzymatic activity of cyclooxygenase-2 with IC_{50} values of 381, 97, and 80µM, respectively.[2]

Figure 12.2 *Ceiba pentandra* (L.) Gaertn. [From: Plants of Borneo. Geographical localization: Sabah, Village of Melangkap Tomis. Field collector: L. Lugas, April 18, 1993.]

Another possible pharmacologically interesting feature of *Ceiba pentandra* (L.) Gaertn. could be the production of sesquiterpenes and triterpenes inducing cell death or apoptosis *in vitro*. Hibasami et al.[3] have recently reported the presence of 2-*O*-methylisohemigossylic acid lactone, a sesquiterpene lactone from the roots of *Bombax ceiba* L., which induces cell death and morphological change indicative of apoptotic chromatin condensation in human promyelotic leukemia HL-60 cells. 2-*O*-methylisohemigossylic acid lactone affected the survival of human promyelotic leukemia HL-60 cells cultured *in vitro* accompanied with chromatin condensation, fragmentation of DNA to oligonucleosomal-sized fragments, which are characteristic of

Lupeol

Figure 12.3

apoptosis, an effect disciplined by inhibitors of caspases and proteolytic enzymes. Lupeol from *Gossampinus malabarica* (L.) Merr. induced the formation of apoptotic bodies in HL-60 cells cultured *in vitro* with an increase in hypodiploid nuclei up to 70.9% after a 3-day treatment with 150μM (Figure 12.3).[4]

12.3 *NEESIA ALTISSIMA* BL.

[After Nees von Esenbeck, 1787–1837, and from Latin *altissima* = tallest.]

12.3.1 Botany

Neesia altissima Bl. is a large tree that grows to a height of 40m with a girth of 4m, in the primary rain forests of Malaysia, Sumatra, Java, and Borneo. The bark is dark brown. The stems are glabrous, stout, and lenticelled. The leaves are simple and stipulate. The stipules are caducous, linear lanceolate, and 2–4 cm long. The petiole is up to 10cm long, woody, and shaped like a knee. The blade is obovate, thick, 30cm × 10cm, and shows 15–20 pairs of secondary nerves. The flowers are 1.5cm long, axillary, and show 25 stamens united at the base. The fruits are massive, purple, 5-angled, and 15–20cm × 10–15cm (Figure 12.4).

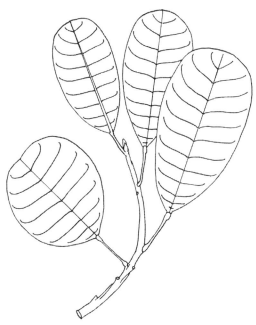

12.3.2 Ethnopharmacology

The plant is used medicinally in Indonesia where the empty capsules are burned and the charcoal obtained is used to make a remedy for the treatment of gonorrhea. The pharmacological potential of *Neesia altissima* Bl. and the genus *Neesia* in general is as of yet unexplored.

Figure 12.4 *Neesia altissima* Bl. [From: Herbarium Bogoriense. Field collector: A. J. Unesco and G. H. Kostermans. Geographical localization: Southwest Java, Udjung, Kulon Reserve.]

REFERENCES

1. Ladeji, O., Omekarah, I., and Solomon, M. 2003. Hypoglycemic properties of aqueous bark extract of *Ceiba pentandra* in streptozotocin-induced diabetic rats. *J. Ethnopharmacol.,* 84, 139.
2. Noreen, Y., el-Seedi, H., Perera, P., and Bohlin, L. 1998. Two new isoflavones from *Ceiba pentandra* and their effect on cyclooxygenase-catalyzed prostaglandin biosynthesis. *J. Nat. Prod.,* 61, 8.
3. Hibasami, H., Saitoh, K., Katsuzaki, H., Imai, K., Aratanechemuge, Y., and Komiya, T. 2004. 2-O-methylisohemigossylic acid lactone, a sesquiterpene, isolated from roots of mokumen (*Gossampinus malabarica*) induces cell death and morphological change indicative of apoptotic chromatin condensation in human promyelotic leukemia HL-60 cells. *Int. J. Mol. Med.,* 4, 1029.
4. Aratanechemuge, Y., Hibasami, H., Sanpin, K., Katsuzaki, H., Imai, K., and Komiya, T. 2004. Induction of apoptosis by lupeol isolated from mokumen in human promyelotic leukemia HL-60 cells. *Oncol. Rep.,* 11, 289.

CHAPTER 13

Medicinal Plants Classified in the Family Elaeocarpaceae

13.1 GENERAL CONCEPT

In field collection, trees of the family Elaeo-carpaceae (A. P. de Candolle, 1824 nom. con-serv., the Elaeocarpus Family) are recognized by three botanical features: the blade, the petiole, and the inflorescences. The blades are dark green, serrate, elliptical, and glossy. The petioles are woody, straight, slender, and kneeled at the apex. The inflorescences are terminal or pseudotermi-nal rectilinear racemes of bell-shaped, 5-merous flowers, which are white and ripen into olive-like glossy drupes (Figure 13.1). A noteworthy chem-ical feature of Elaeocarpaceae, Cucurbitaceae, and Begoniaceae is their ability to elaborate a series of oxygenated steroids or cucurbitacins, which have so far been found in the Cucurbita-ceae and Begoniaceae Families. The evidence available indicates that cucurbitacins, which abound in this family, hold some potential as a source of cytotoxic agents.[1-3] A remarkable advance in the chemotherapeutic evaluation of Elaeocarpaceae has been provided by Ito et al.[3] Using a bioassay-guided investigation of the bark of *Elaeocarpus mastersii*, they isolated cucurbi-tacin D and cucurbitacin F as cytotoxic principles instead of using KB cells cultured *in vitro*, together with ellagic acid derivatives.

Figure 13.1 Botanical hallmarks of Elaeocar-paceae with racemes of bell-shaped flowers.

Other principles of interest in Elaeocarpaceae are indolizidine alkaloids, which have attracted a great deal of interest on account of their ability to inhibit the enzymatic activity of glucosidases because of a structural similarity with glucose; hence there is some potential to explore it further in the treatment of Human Immunodeficiency Virus (HIV), diabetes, and cancer (Figure 13.2).

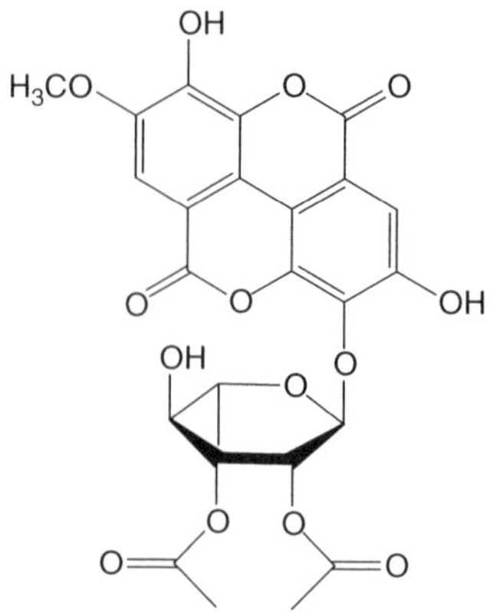

4'-*O*-methyllelagic acid 3-(2″,3″–di-*O*-acetyl)-α-L-rhamnoside

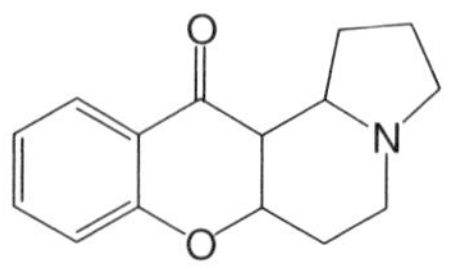

Indolizidine

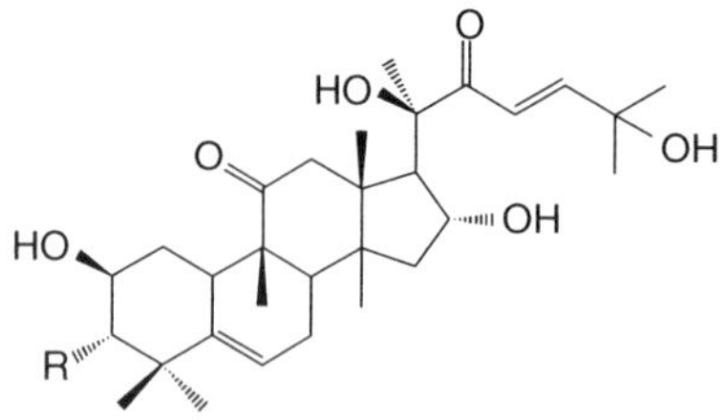

Cucurbitacin

Figure 13.2 Examples of bioactive natural products in the family Elaeocarpaceae.

Note that a small number of indolizidine alkaloids have emerged that have therapeutic indices favoring their introduction into clinical practice. Carroll et al.[4] made the interesting observation that grandisine A, B, and (−)-isoelaeocarpiline isolated from *Elaeocarpus grandis* from the Australian rain forest are able to bind to the delta type of opioid brain receptors. All these pharmacological events together lend considerable support to the view that Elaeocarpaceae with its 10 genera and 400 species of tropical trees would be worth screening thoroughly for cytotoxic agents. One can reasonably expect the discovery of molecules of chemotherapeutic value in this large family. An exciting approach would be to start with *Aceratium insulare* A.C., *Elaeocarpus grandiflorus* Smith., *Elaeocarpus madopetalus* Pierre, *Elaeocarpus petiolatus* Wall., *Elaeocarpus floribundus* Bl., *Elaeocarpus sphaericus* (Gaertn.) K. Sch., and *Mutingia calaruba*, which are used medicinally in the Asia–Pacific.

13.2 *ELAEOCARPUS FLORIBUNDUS* BL.

[From: Greek *elaion* = oil and *karpos* = fruit, and from Latin *floribundus* = profuse flowering.]

13.2.1 Botany

Elaeocarpus floribundus Bl. is a tree that grows up to 30m high with a girth of 2.4m, and is common in the lowland hills and mountains up to an altitude of 1500m in India, Burma, Thailand, Vietnam, Cambodia, Laos, Malaysia, Indonesia, and the Philippines. The bole is columnar. The bark is brown, rough, and lenticelled. The inner bark is pale yellow. The leaves, buds, flowers, and fruits are often dotted with small scabby or pimply pustules. The blade is ovate elliptic, thinly leathery, and 6.5cm × 3cm – 19cm × 8.5 cm. The apex is acuminate, the base is pointed, and the margin is toothed. The blade shows 5–7 pairs of secondary nerves. The petiole is 1–5.5cm long and kneed. The inflorescences are axillary racemes, which are 2.5–12cm long. The flowers are 5-merous, 0.5cm long, comprise 25 stamens, and have a hairy ovary. The flower pedicels are 0.4–1.2cm long. The fruits are smooth, glabrous, ellipsoid, narrowed at basal end, and contain a rugose of 3cm × 2.5cm stone (Figure 13.3).

Figure 13.3 *Elaeocarpus floribundus* Bl. [From: Flora of Malaya. FRI No: 17930. Geographical localization: Jerteh Kuala Terengganu, in forest reserve. Alt.: 1000ft.]

13.2.2 Ethnopharmacology

In Sumatra, an infusion of bark and leaves is used as a mouthwash for inflamed gums. The pharmacological activity of this plant is unexplored as of yet.

13.3 *ELAEOCARPUS OBTUSUS* BL. *SENSU* KING

[From: Greek *elaion* = oil, *karpos* = fruit, and from Latin *obtusus* = obtuse.]

13.3.1 Botany

Elaeocarpus obtusus Bl. *sensu* King (*Elaeocarpus macrocerus* [Turcz.] Merr. and *Elaeocarpus macrocerus litoralis* Kurz) is a large tree that grows to a height of 45m with a girth of 2.7m in the tidal swamps and riverbanks of Burma, Thailand, Malaysia, and Indonesia. The bole is buttressed, shows stilt roots, and pneumatophores shaped like an inverted Y. The bark is light brown, smooth, cracked, or lenticelled. The inner bark is pale orange and granular-crumbly. The stems are 5mm thick. The blade is obovate, leathery, 5cm × 2.5cm – 16cm × 6.5cm, subglaucous below, rounded at the apex, and tapered at the base. The margin is toothed. The blade shows 4–9 pairs of secondary nerves. The petiole is 9mm – 3.2cm long. The inflorescences consist of racemes which are 1–5cm long, and axillary. The flowers are 5-merous, 1–4cm long, and comprise more than 40 stamens and a hairy ovary. The fruits are smooth and glabrous, 3cm × 2.5cm, and contain a rugose stone (Figure 13.4).

Figure 13.4 *Elaeocarpus obtusus* Bl. *sensu* King. [From: Ex. Herb. Leiden. Geographical localization: Taytay municipality, 15Km north Embarcadero, area near Pncol, Philippines. Rain forest hill slope. Alt.: 140m. Field collector: A. C. Podzorski. SMHI No: 984. April 21, 1984.]

13.3.2 Ethnopharmacology

The Malays apply a paste of leaves to insect bites. The antiinflammatory property of *Elaeocarpus obtusus* Bl. *sensu* King is not confirmed as of yet. Note that the petroleum ether, benzene, chloroform, acetone, and ethanol extracts of the fruits of *Elaeocarpus sphaericus* stabilize the activity of mast cells.[5] The extracts at 50–200mg/Kg intraperitoneally or 200mg/Kg orally showed significant antiinflammatory action against both acute and subacute models, analgesic action, barbituratehypnosis potentiation, and antiulcerogenic activities in rats, and protected guinea pigs against asphyxia induced by histamine and acetylcholine aerosols.[6] Are indolizides involved here?

REFERENCES

1. Fang, X., Phoebe, Jr., C. H., Pezzuto, J. M., Fong, H. H., Farnsworth, N. R., Yellin, B., and Hecht, S. M. 1984 (November–December). Plant anticancer agents, XXXIV. Cucurbitacins from *Elaeocarpus dolichostylus. J. Nat. Prod.*, 47, 6, 988–993.
2. Rodriguez, N., Vasquez, Y., Hussein, A. A., Coley, P. D., Solis, P. N., and Gupta, M. P. 2003. Cytotoxic cucurbitacin constituents from *Sloanea zuliaensis. J. Nat. Prod.*, 66, 1515.
3. Ito, A., Chai, H. B., Lee, D., Kardono, L. B. S., Riswan, S., Farnsworth, N. R., Cordell, G. A., Pezzuto, J. M., and Kinghorn, A. D. 2002. Ellagic acid derivatives and cytotoxic cucurbitacins from *Elaeocarpus mastersii. Phytochemistry*, 61, 171.
4. Carroll, A. R., Arumugan, G., Quinn, R. J., Redburn, J., Guymer, G., and Grimshaw, P. 2005. Grandisine A and B, novel indolizidine alkaloids with human delta-opioid receptor binding affinity from the leaves of the Australian rain forest tree *Elaeocarpus grandis. J. Org. Chem.*, 70, 1889.
5. Singh, R. K., Bhattacharya, S. K., and Acharya, S. B. 2000. Studies on extracts of *Elaeocarpus sphaericus* fruits on *in vitro* rat mast cells. *Phytomedicine*, 7, 205.
6. Singh, R. K., Acharya, S. B., and Bhattacharya, S. K. 2000. Pharmacological activity of *Elaeocarpus sphaericus. Phytother. Res.*, 14, 36.

CHAPTER 14

Medicinal Plants Classified in the Family Capparaceae

14.1 GENERAL CONCEPT

The family Capparaceae (A. L. de Jussieu, 1789 nom. conserv., the Caper Family or *Capparidaceae*) family consists of approximately 45 genera and about 800 species of pungent treelets, shrubs, or herbs, which have the ability to elaborate a series of isothiocyanates (mustard oils), flavonoids, and occasionally pyrrolidine alkaloids. When looking for Capparaceae, search for plants with elongated receptacles called gynophores or androgynophores, a showy protruding androecium, and by the aspect of the corolla (Figure 14.1).

With regard to the pharmaceutical potential of Capparaceae, Shi et al.[1] showed that 5,3′-dihydroxy-3,6,7,8,4′-pentamethoxyflavone from *Polanisia dodecandra* inhibits a broad panel of cancer cells: central nervous system cancer (SF-268, SF-539, SNB-75, U-251), non-small-cell lung cancer (HOP-62, NCI-H266, NCI-H460, NCI-H522), small-cell lung cancer (DMS-114), ovarian cancer (OVCAR-3, SK-OV-3), colon cancer (HCT-116), renal cancer (UO-31), a melanoma cell line (SK-MEL-5), and leukemia cell lines (HL-60, SR), cultured *in vitro*. The cellular mechanism by which this flavone is toxic is based on inhibition of the

Figure 14.1 Botanical hallmarks of Capparaceae showing protruding androecium. **(See color insert following page 168.)**

polymerization of tubulin into the mitotic spindle ($IC_{50} = 0.83\mu M$). Other interesting cytotoxic natural products from Capparaceae are triterpenes of the dammarane type such as polacandrin, which inhibits the proliferation of KB (ED_{50}: 0.6μg/mL), the P388 (ED_{50}: 0.9μg/mL), and RPMI-7951 (ED_{50}: 0.62μg/mL) cell lines, as well as 17α-hydroxycabraleahydroxylactone ($IC_{50} = 3.1\mu g/mL$), 12β-acetoxycleocarpone ($C_{50} = 8.9\mu g/mL$), 3-*O*-acetyl-12β-acetoxy-25-*O*-ethylcleo-

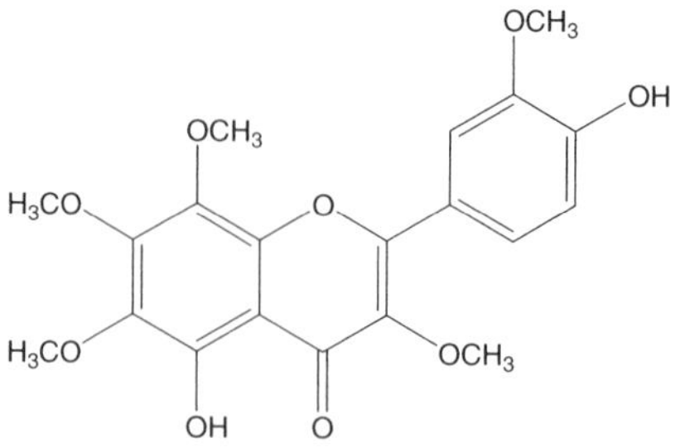

5,4'-Dihydroxy-3,6,7,8,3'-pentamethoxyflavone

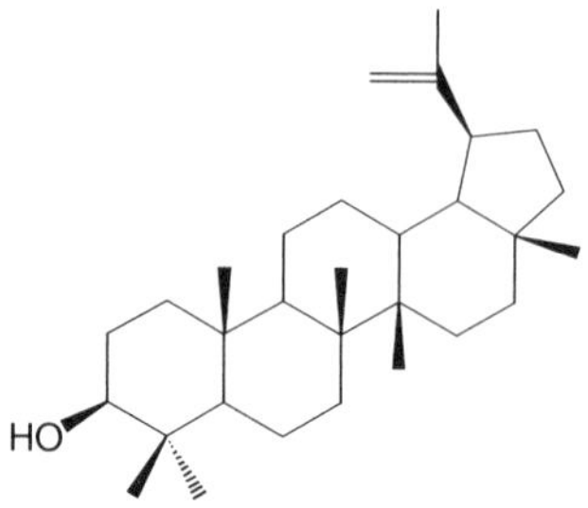

Lupeol

Figure 14.2 Examples of bioactive natural products from the family Capparaceae.

carpanol (IC_{50} = 3.9μg/mL), 1(12), 22(23)-tetradehydrocabralealactone (IC_{50} = 4.1μg/mL), Δ1,2-dehydrocabralealactone (IC_{50} = 1.9μg/mL), α-hydroxycabraleahydroxylactone (IC_{50} = 3.1μg/mL), and 12β-acetoxycleocarpone (IC_{50} = 8.9μg/mL).[2,3]

It will be interesting to learn whether or not more intensive research on Capparaceae discloses any secondary metabolites of chemotherapeutic interest (Figure 14.2). Approximately 20 species of plants classified within the family Capparaceae are used medicinally in the Pacific Rim. These are often used as counterirritant remedies. The pharmacological potential in *Capparis micrantha* DC. (*Capparis myrioneura* Hall. F.) and *Crateva religiosa* Forst. are described next.

14.2 *CAPPARIS MICRANTHA* DC.

[From: Greek *kapparis* = caper and from Latin *micrantha* = very small anthers.]

14.2.1 Botany

Capparis micrantha DC. (synonym: *Capparis myrioneura* Hall. f., *Capparis bariensis* Pierre ex Gagnep., *Capparis donnaiensis* Pierre ex Gagnep., and *Capparis roydsiaefolia* Kurz) is a climber from the rain forests of Southeast Asia. The leaves are simple, spiral, and exstipulate. The petiole is 5–8mm long. The blade is lanceolate, 7–15cm × 3–5cm, and shows 7–10 pairs of secondary nerves. The apex is acute acuminate and the base is obtuse. The flowers are up to 5cm long on 2.5cm pedicels (Figure 14.3).

Figure 14.3 *Capparis micrantha* DC. [From: Flora of Malaya. FRI No: 3689. Geographical localization: Southeast
Pahang Aur Forest Reserve at River Aur. Climber at riverside. Nov. 5, 1967. T. C. Whitmore.]

14.2.2 Ethnopharmacology

Caper Thorn is used in Cambodia, Laos, and Vietnam for medicinal purposes. The juice
squeezed from the roots is used as a drink to reduce fever, to promote urination, and as a remedy
for a cough. The wood is smoked to heal syphilitic ulceration of the nose and the seeds are used
to treat a cough. In the Philippines, a decoction of the root is used as a drink to assuage stomachache,
and to invigorate after childbirth. Indonesians use the wood to assuage stomachache and to treat
biliousness and dizziness. The pharmacological potency of *Capparis micrantha* DC. is to date
unexplored.

Note that the plant would be worth studying for its cytotoxic
properties, given that Wu et al.[4] have identified an unusual cytotoxic alkaloid, cappamansin A, from the roots of *Capparis sikkimensis* subsp. *formosana*. Cappamansin A has encouraged
potently the survival of ovarian (1A9), lung (A549), ileocecal
(HCT-8), breast (MCF-7), nasopharyngeal (KB), and vincristine-
resistant (KB-VIN) human tumor cell lines cultured *in vitro* (Figure 14.4). With regard to the antiinflammatory effect of the plant,
one might offer the hypothesis that the antiinflammatory principles involved could be of an isoprenic nature since al-Said et al.[5]
identified capparenol-13 from *Capparis spinosa*, as an inhibitor
of paw edema in rodents.

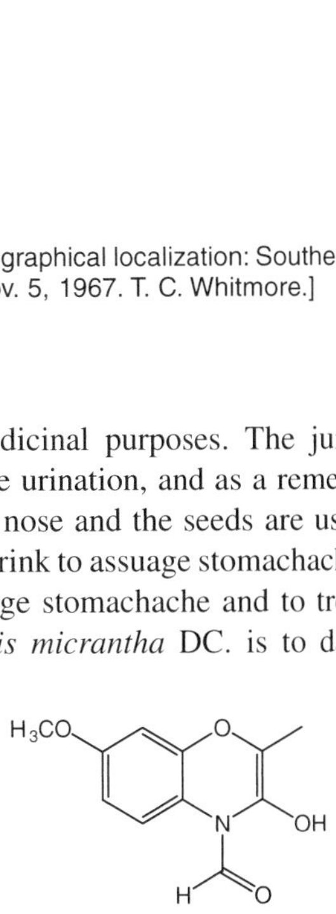

Cappamansin A

Figure 14.4 Cappamansin A, antitumor agent from the
Cappara species.

14.3 *CRATEVA RELIGIOSA* FORST.

[After *kratevas*, a Greek root gatherer of antiquity.]

Figure 14.5 *Crateva religiosa* Forst. [From: Herbarium of the Forestry Department (SAN), Sepilok, Sandakan Sabah. Geographical localization: Sabah, Kinabatangan, River Lokan Forest Reserve, riverside dipterocarp forest. Field collectors: K. Perreira, S. Diwol, and G. Pin, et al., May 10, 2000.]

14.3.1 Botany

Crateva religiosa Forst. (*Crateva macrocarpa* Kurz) is a treelet that grows in lowland forests in a geographical area ranging from India to Papua New Guinea including the Ryuku Islands. The leaves are trifoliate and exstipulate. The petiole is slender and 12cm long. The folioles are 6cm × 3cm – 13cm × 5cm, papery, and show 7–10 pairs of secondary nerves. The flowers are large and yellow and show a conspicuous androecium. The fruits are caper-like and up to 2.5cm long on top of 5–9cm-long pedicels (Figure 14.5).

14.3.2 Ethnopharmacology

The leaves of *Crateva religiosa* Forst. (or Spider Tree, Garlic Pear) are used in China to promote digestion. In the Solomon Islands, a liquid from the bark is used to empty the bowels, and the leaves are applied externally to assuage earache. In Taiwan, the leaves are used to assuage headache, stomachache, and dysentery. The medicinal properties are very probably owed to isothiocyanates, which cause irritation of the skin and are counterirritant.

REFERENCES

1. Shi, Q., Chen, K., Li, L., Chang, J. J., Autry, C., Kozuka, M., Konoshima, T., Estes, J. R., Lin, C. M., and Hamel, E. 1995. Antitumor agents, 154. Cytotoxic and antimitotic flavonols from *Polanisia dodecandra*. *J. Nat. Prod.*, 58, 1470.
2. Shi, Q., Chen, K., Fujioka, T., Kashiwada, Y., Chang, J. J., Kozuka, M., Estes, J. R., McPhai, A. T., McPhai, D. R., and Lee, K. H. 1992. Antitumor agents, 135. Structure and stereochemistry of polacandrin, a new cytotoxic triterpene from *Polanisia dodecandra*. *J. Nat. Prod.*, 55, 1488.
3. Nagaya, H., Tobita, Y., Nagae, T., Itokawa, H., Takeya, K., Halim, A. F., and Abdel-Halim, O. B. 1997. Cytotoxic triterpenes from *Cleome africana*. *Phytochemistry*, 44, 1115.
4. Wu, J. H., Chang, F. R., Hayashi, K., Shiraki, H., Liaw, C. C., Nakanishi, Y., Bastow, K. F., Yu, D., Chen, I. S., and Lee, K. H. 2003. Antitumor agents. Part 218: Cappamensin A, a new *in vitro* anticancer principle, from *Capparis sikkimensis*. *Bioorg. Med. Chem. Lett.*, 13, 2223.
5. al-Said, M. S., Abdelsattar, E. A., Khalifa, S. I., and el-Feraly, F. S. 1988. Isolation and identification of an anti-inflammatory principle from *Capparis spinosa*. *Pharmazie*, 43, 640.

CHAPTER 15

Medicinal Plants Classified in the Family Flacourtiaceae

15.1 GENERAL CONCEPT

Members of the family Flacourtiaceae (A. P. de Candolle, 1824 nom. conserv., the Flacourtia Family) are tropical rain forest trees that are mainly characterized by their fruits, which are globose, often large, and woody. In pharmaceutical interest, the little evidence currently available suggests that the members of this family, and especially the *Caesaria* species, might hold some potential as a source of antioxidant, cytotoxic, and antibacterial agents.[1] Examples of cytotoxic agents found in Flacourtiaceae are clerodane diterpene esters, corymbulosins A–C, which were isolated from an organic extract of the fruit of *Laetia corymbulosa*.[2] Of special interest in this family are a series of antiviral salicilin derivatives which affect the replication of Human Immunodeficiency Virus (HIV) and Herpes Simplex Virus (HSV) as tremulacin (Figure 15.1).[3]

Approximately 20 species of plants in the Flacourtiaceae family are medicinal in the Asia–Pacific, notably *Hydnocarpus alcalae* C. DC. (Philippines), *Hydnocarpus anthelminticus* Pierre (Cambodia, Laos, Vietnam), *Hydnocarpus wightiana* (India), *Oncoba echinata* (*gorli* oil), and *Hydnocarpus kurzii* (King) Warb. (Chaulmoogra oil), which were the only treatments available for leprosy until the discovery of dapsone around 1940, and they have been used externally for a very long time to treat leprosy. *Flacourtia jangomas* (Lour.) Raeusch., *Homalium tomentosum* (Vent.) Benth., are *Hydnocarpus kurzii* ssp. *australis* Sleumer are discussed next.

15.2 *FLACOURTIA JANGOMAS* (LOUR.) RAEUSCH.

[After E. de Flacourt (1607–1660), French Governor of Madagascar.]

15.2.1 Botany

Flacourtia jangomas (Lour.) Raeusch. (*Flacourtia cataphracta* Roxb. and *Stigmarota jangomas* Lour.) is a treelet native to Burma that is found cultivated in a geographical area ranging from India to Hawaii. The plant grows to a height of 10m. The stems are thorny. The bark is yellowish to brown, smooth, and lenticelled. The petiole is 6–8mm long. The blade is elliptic, serrate, 7cm – 11cm × 3.5cm – 4cm, papery, and show 3–6 pairs of secondary nerves. The inflorescences are axillary racemes of 1–2cm long. The flowers are white to greenish and comprise 4 or 5 ovate-triangular, 2mm-long sepals that are puberulous outside. The androecium consists of

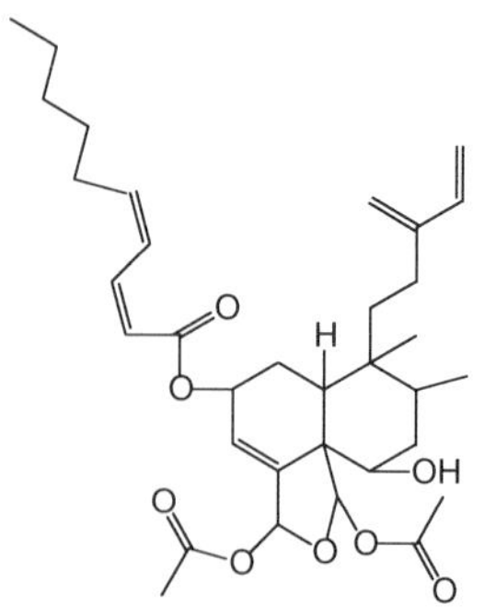

Corymbulosine

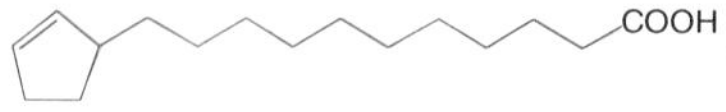

Hydnocarpic acid

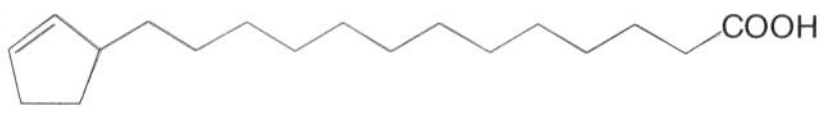

Chaulmoogric acid

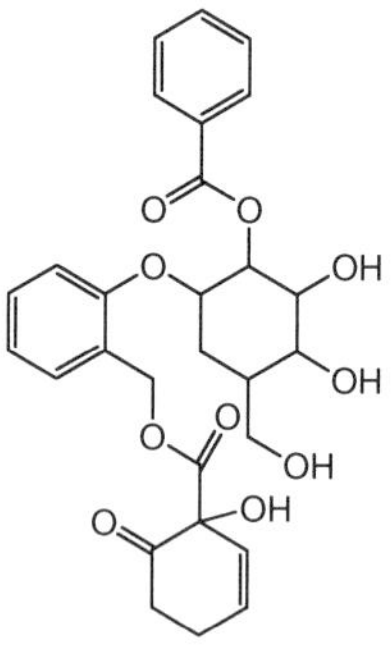

Tremulacin

Figure 15.1 Examples of bioactive natural products from the family Flacourtiaceae.

numerous anthers, which are ovate to suborbicular. The ovary is 4–6-celled, with two ovules per locule. The fruits are dark red, 1.5cm × 2.5cm, fleshy, with a persistent short style and contain 4–5 seeds (Figure 15.2).

15.2.2 Ethnopharmacology

The fruits of *Flacourtia jangomas* (Lour.) Raeusch. or Coffee Plum, *paniala*, Chinese Plum, *prunier d'Inde* (French), *ciruela forastera*, *venevene pāma* (Cook Islands), *palamu* (Niue), and *yun nan ci li mu* (Chinese) are eaten in Burma to promote digestion. The bark is used to promote digestion. In Malaysia, a decoction of leaves is used as a drink to treat diarrhea, to promote digestion, and the juice, squeezed from the roots, is used to treat herpes infection. In Cambodia, Laos, and Vietnam, a decoction of the leaves is used as a drink to abort, or the fruits are eaten for the same purpose. A paste of roots is applied to sores, ulcers, and to soothe an inflamed throat.

The pharmacological potential of *Flacourtia jangomas* (Lour.) Raeusch. is as of yet unexplored. It would be interesting to learn whether the antipyretic and antiinflammatory properties could be ascribed to a mechanism involving the inhibition of the enzymatic activity of phospholipase A$_2$,

Figure 15.2 *Flacourtia jangomas* (Lour.) Raeusch. [From: Federated Malay States. Geographical localization: Village of Pasir Bojak, Pangkor Island. Dec. 23, 1918. Field collector: M. S. Hamid. Botanical identification: H. Sleumer, October 1954.]

since some evidence has already been presented which indicates a phospholipase A_2 potential in a member of the closely related genus *Caesaria*. A crude extract of *Caesaria sylvestris* has inhibited phospholipase A_2 activity in snake venoms and mitigated the hemorrhagic and myotoxic activities caused by crude venoms.[4] Izidoro et al.[5] made a careful study of the antivenom property of *Casearia mariquitensis* against the South American pit viper (*Bothrops neuwiedi pauloensis*) and clearly demonstrated that an aqueous extract from the leaves annihilates the pulmonary hemorrhage induced by the venom with protection of a beta fibrinogen chain. An interesting development from this observation would be to identify the agent responsible for such activity. Are salicilin derivatives involved here?

15.3 *HOMALIUM TOMENTOSUM* (VENT.) BENTH.

[From: Greek *homalos* = uniform and from Latin *tomentosum* = tomentose.]

15.3.1 Botany

Homalium tomentosum (Vent.) Benth. is a deciduous, medium-sized tree that grows to a height of 40m and a diameter of 1m in the rain forests of Southeast Asia. The bole is buttressed. Leaves are simple and spiral. The blade is broadly obovate to obovate–oblong, 12.5cm × 5cm, crenate, obtuse to apiculate at the apex, dull and glabrescent above, tomentose below, and shows 15 pairs of secondary nerves. The inflorescences are 10cm spikes. The flowers are 5–6-merous, sessile, or woolly and minute (Figure 15.3).

15.3.2 Ethnopharmacology

Homalium tomentosum (Vent.) Benth. (*gerinseng* [Indonesian] and *chang phuek luang* [Thai] is medicinal. In Cambodia and Burma the plant is known as *phloeuo nien* and *myauk chaw*, respectively. The roots are used there as well as in Laos and Vietnam as an astringent. The pharmacological potential of this plant is unknown. Note that cochinchiside A and tremulacin from *Homalium* showed some level of activity against HSV-1 and HSV-2, while tremulacin is active against HIV-1.[3] What are the antiviral principles of *Homalium tomentosum*?

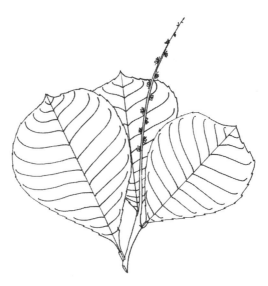

Figure 15.3 *Homalium tomentosum* (Vent.) Benth. [From: Burma Forest School Herbarium. Geographical localization: Ainggye, Pyinmana, Aug. 18, 1926. Botanical identification: H. Sleumer, November 1953.]

15.4 *HYDNOCARPUS KURZII* SSP. *AUSTRALIS* SLEUMER

[From: Greek *hudron* = truffle, *karpos* = fruit, and from Bengali *cul-mugr* = *Hydnocarpus kurzii*.]

15.4.1 Botany

Hydnocarpus kurzii ssp. *australis* Sleumer (*Taraktogenos kurzii* King and *Hydnocarpus castanea* Hook. f.) is a treelet that grows to a height of 2m with a girth of 7cm. The bark is smooth and pale yellowish. The bole is straight. The leaves are simple, oblong, 15cm × 4–32cm × 8cm with a tapering apex and wedge-shaped base. Fruits are round up to 10cm in diameter, light fawn rugose, and subglabrous. The walls with radial lines are on a 6mm-long pedicel. The seeds are irregular, grayish, angled with blunt ends, brownish-yellow, and 2–3cm long (Figure 15.4).

15.4.2 Ethnopharmacology

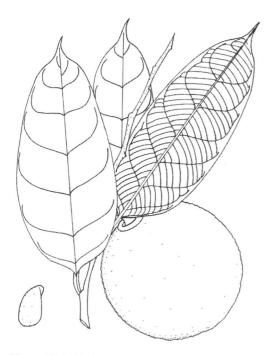

Figure 15.4 *Hydnocarpus kurzii* ssp. *australis* Sleumer. [From: Ex, Herb. Hort. Bot. Singaporense. No: 22417. Geographical localization: Kota Kelanggi, Pahang, Aug. 4, 1929. Det.: H. Sleumer, 1937.]

The plant is known as the Chaulmoogra oil tree. *Oleum hydnocarpi*, or chaulmoogra oil (*International Pharmacopoeia*, 1967), is the fixed oil obtained from the fresh ripe seeds of *Hydnocarpus wightiana, Hydnocarpus anthelminthica, Hydnocarpus heterophylla,* and *Hydnocarpus kurzii* ssp. *australis* Sleumer, which has been used in the treatment of leprosy, especially in endemic areas in the East where it has been administered by direct infiltration of the lesions and was effective in decreasing the size of nodules, anesthetic patches, and skin lesions. The active principles are unusual cyclopentenic fatty acids including chaulmoogric acid. The precise molecular mycobactericidal mode of action is still unknown. Note that an infusion of bark of the plant is used as a drink to improve general health and to treat skin diseases.

An exciting development from this plant is the observation that the oil of *Hydnocarpus* increases the strength of scar tissue and strengthens the collagen tissue in wounds inflicted on male Wistar rats; hence some cosmetological and or dermatological application may be possible.[6] In addition, one should probably look into the cytotoxic potential of this plant and the *Hydnocarpus* species in general, as Sharma and Hall[7] have isolated lignans hydnowightin, hydnocarpin, and neohydnocarpin, which are active against murine L-1210 leukemia growth of human KB, nasopharynx, colon adenocarcinoma, osteosarcoma, and HeLa-S3 uterine growth (Figure 15.5).

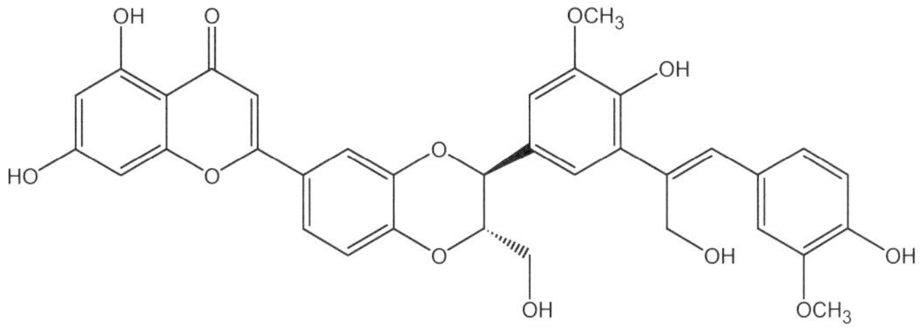

Hydnowightin

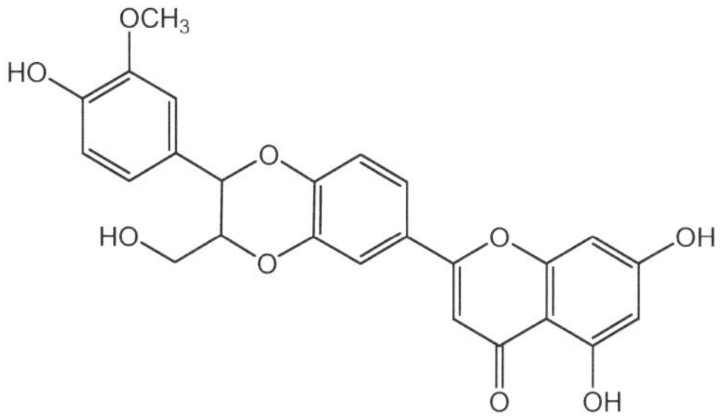

Hydnocarpin

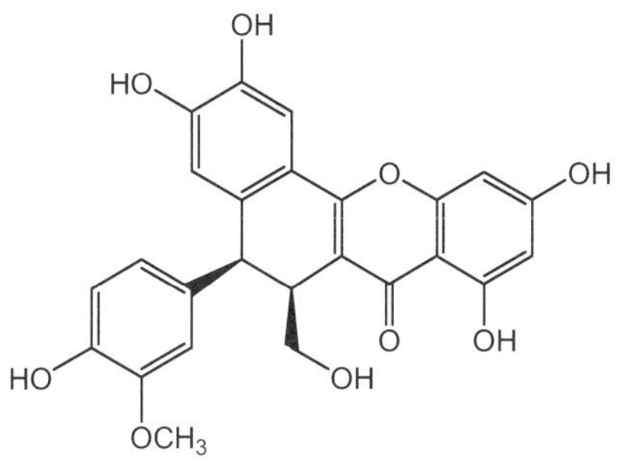

Neohydnocarpin

Figure 15.5 Cytotoxic lignans of the *Hydnocarpus* species.

REFERENCES

1. Mosaddik, M. A., Banbury, L., Forster, P., Booth, R., Markham, J., Leach, D., and Waterman, P. G. 2004. Screening of some Australian *Flacourtiaceae* species for *in vitro* antioxidant, cytotoxic and antimicrobial activity. *Phytomedicine,* 11, 461.

2. Beutler, J. A., McCall, K. L., Herbert, K., Johnson, T., Shoemaker, R. H., and Boyd, M. R. 2000. Cytotoxic clerodane diterpene esters from *Laetia corymbulosa. Phytochemistry,* 55, 233.

3. Ishikawa, T., Nishigaya, K., Takami, K., Uchikoshi, H., Chen, I. S., and Tsai, I. L. 2004. Isolation of salicin derivatives from *Homalium cochinchinensis* and their antiviral activities. *J. Nat. Prod.,* 67, 659.

4. Raslan, D. S., Jamal, C. M., Duarte, D. S., Borges, M. H., and De Lima, M. E. 2002. Anti-PLA2 action test of *Casearia sylvestris* Sw. *Boll. Chim. Farm.,* 141, 457.

5. Izidoro, L. F., Rodrigues, V. M., Rodrigues, R. S., Ferro, E. V., Hamaguchi, A., Giglio, J. R., and Homsi-Brandeburgo, M. I. 2003. Neutralization of some hematological and hemostatic alterations induced by neuwiedase, a metalloproteinase isolated from *Bothrops neuwiedi pauloensis* snake venom, by the aqueous extract from *Casearia mariquitensis* (Flacourtiaceae). *Biochimie,* 85(7), 669–675.

6. Oommen, S. T., Rao, M., and Raju, C. V. 1999. Effect of oil of *Hydnocarpus* on wound healing. *Int. J. Lepr. Other Mycobact. Dis.,* 67, 154.

7. Sharma, D. K. and Hall, I. H. 1991. Hypolipidemic, anti-inflammatory, and antineoplastic activity and cytotoxicity of flavonolignans isolated from *Hydnocarpus wightiana* seeds. *J. Nat. Prod.,* 54, 1298.

CHAPTER 16

Medicinal Plants Classified in the Family Passifloraceae

16.1 GENERAL CONCEPT

The family Passifloraceae consists of 16 genera and 650 species of climbers widespread in tropical regions and known to produce cyanogenic glycosides and a series of centrally active β-carboline alkaloids such as passiflorine. The botanical hallmark of Passifloraceae is the flower, which shows an elongated androgynophore (Figure 16.1). Examples of Passifloraceae of pharmaceutical interest are *Passiflora incarnata* (*Passiflora, French Pharmacopoeia*, 1965), the dried flowering and fruiting tops of which have been used as an antispasmodic and sedative. It has been used as a nerve sedative and for its anodyne properties in various neuralgias, as a liquid extract (1 in 1 dose: 0.5mL to 1mL) and as a tincture (1 in 5 dose: 0.5–2mL) from the leaves of *Passiflora alata* (*Folium Passiflorae, Brazilian Pharmacopoeia*, 1959).

Figure 16.1 Botanical hallmarks of Passifloraceae.

A remarkable advance in the neuropharmacological investigation of *Passiflora incarnata* Linn. has been provided by Speroni and Miughetti,[1] Soulimani et al.,[2] and Dhawan et al.[3,4] Lyophilized hydroalcoholic extracts of the aerial parts of *Passiflora incarnata* L. (*Passifloraceae*, Passion Flower) showed anxiolytic and sedative properties at 400mg/Kg. Dhawan et al.[5,6] showed that a fraction derived from the methanol extract of the plant exhibits significant anxiolytic activity at a dose of 10mg/Kg in mice, and suggested the active constituent to be a substituted flavone. The question arises as to whether or not the carboline alkaloids present in the plant are involved in its anxiolytic properties. Harmine and congeners have indole moieties which make them good potential candidates as serotoninergic agents (Figure 16.2). In the Pacific Rim, *Adenia cardiophylla* Engl., *Adenia populifolia* Engl., *Adenia acuminata* King, *Adenia cordifolia* Engl., *Adenia zuca* (Blco.) Merr., *Passiflora quadrangularis* L., *Passiflora foetida* L., and *Passiflora laurifolia* L. are of medicinal value.

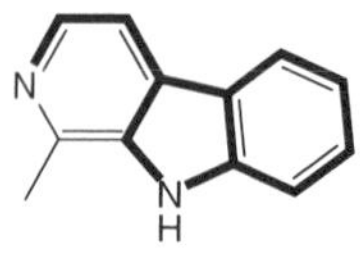

Harman 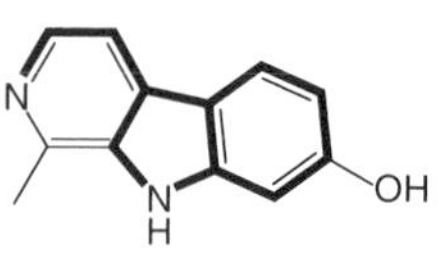   Harmol 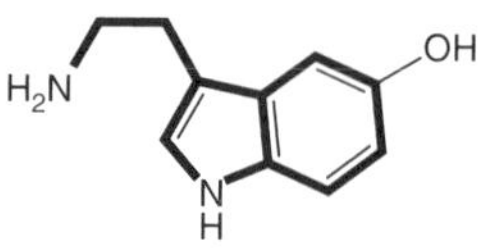   Serotonine

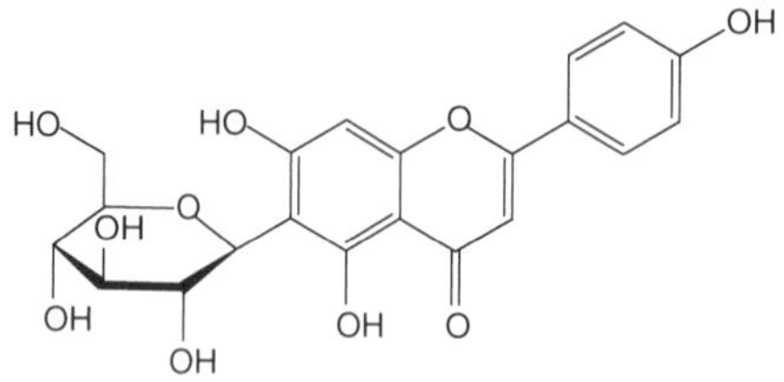

Isovitexin

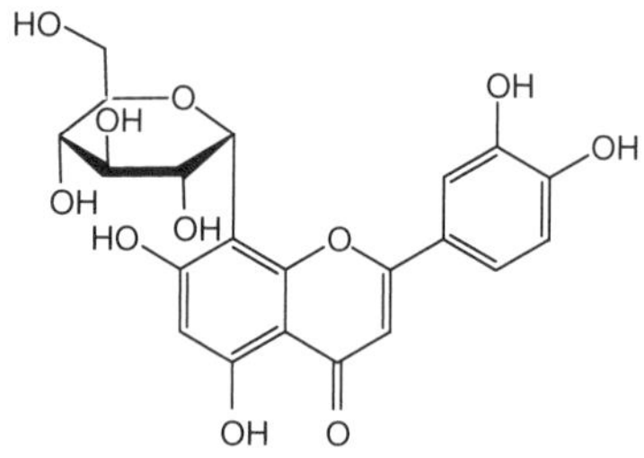

Orientin

Figure 16.2 Examples of bioactive natural products from the family Passifloraceae.

16.2 *ADENIA CORDIFOLIA* ENGL.

[From: Greek *aden* = a gland and Latin *cordifolia* = heart-shaped leaves.]

16.2.1 Botany

Adenia cordifolia Engl. is a climber of the peat swamp forests of Malaysia and Indonesia, and grows to a length of 10m. The stems are woody, 2mm in diameter. The tendrils are 5–7cm long. The leaves are simple. The petiole is 1.5–2.5cm long. The blade is membranaceous, 9cm × 7.5cm × 4cm, and shows five pairs of secondary nerves. The margin is crenate. A small, primitive leaf is present at the base of the blade. The fruits are bright red, capsular, and split into three valves (Figure 16.3).

16.2.2 Ethnopharmacology

In Indonesia, the sap of the stem is used to soothe inflamed eyes. The pharmacological properties of this plant are unexplored. The roots of *Adenia volkensii* Harms contain a volkensin, a type 2

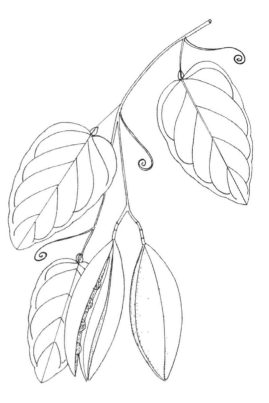

Figure 16.3 *Adenia cordifolia* Engl. [From: Flora of Malaya. Botanical identification: F. S. P. Ng, September 1975. Geographical localization: P. P. Telok. Field collector: K. M. Kochummen, Feb. 17, 1946.]

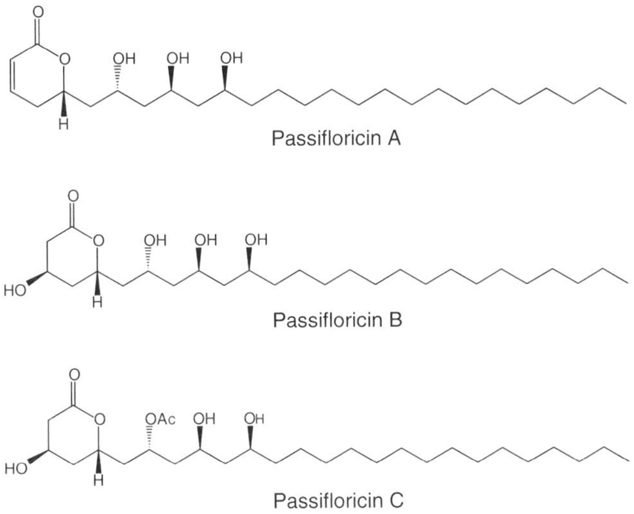

Passifloricin A

Passifloricin B

Passifloricin C

Figure 16.4 Passifloricins — unusual polyketide pyrones from *Passiflora foetida*. Note the similitude of these molecules with goniothalamin and kavain. Are passifloricins proapoptotic and anxiolytic?

ribosome-inactivating protein.[7] Gummiferol from the leaves of *Adenia gummifera* is a polyacetylenic diepoxide, which inhibits the growth of KB cells cultured *in vitro*.[8] Does it inhibit inflammation?

16.3 *PASSIFLORA FOETIDA* L.

[From: Latin *passio* = passion and *flos* = flower. The plant parts seemed to represent aspects of Christ: the corona was the crown of thorns, the five stamens were the five wounds, the three styles were the three nails, and the ten petal-like parts were the ten faithful apostles; and from Latin *foetida* = offensively malodorous.]

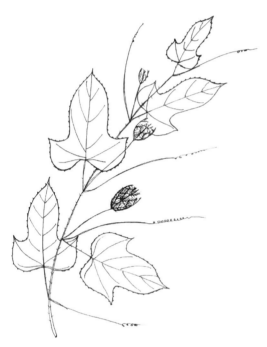

Figure 16.5 *Passiflora foetida* L. [From: Plants of Borneo. Geographical localization: Sabah, Village of Melangkap Tomis, 100m from Bali Raya, Sept. 12, 1992.]

16.3.1 Botany

Passiflora foetida L. is a perennial climber native to America and is pantropical in distribution. The stems are hirsute to pilose with 5cm-long tendrils. The petiole is 8cm long. The blade is hispid–hirsute, 9.5cm × 5cm, trilobed, and membranaceous. It shows four pairs of secondary nerves. The flowers are solitary in axils, 4–5cm wide, purple and white on pedicels that are 3–7cm long. The fruits are yellow to orange, subglobose, papery thin, and enclosed in finely laciniate sepals. The fruits are edible and the leaves are eaten as a vegetable (Figure 16.5).

16.3.2 Ethnopharmacology

Stinking Passion Flower, Love-in-a-Mist, also known as *tairis* (Subuk), *kanda kanda* (Sungei), *lapak lapak* (Tawau), *pasio vao* (Samoa), *qaranidila* (Fiji), *pohapoha* (Hawaii), *kudamono* (Palau), and *tomates* (Yap). The Malays use the plant to calm itching. In the Philippines, the leaves are applied to wounds to promote healing. Echeverri et al.[9] isolated a series of unusual polyketide pyrones from the plant (Figure 16.4). Decoctions of the fruits of *Passiflora foetida* var. *albiflora* inhibited the enzymatic activity of gelatinase matrix metalloproteinase (MMP)-2 and MMP-9, both metalloproteases involved in tumor invasion, metastasis, and angiogenesis.[10]

16.4 *PASSIFLORA QUADRANGULARIS* L.

[From: Latin *passio* = passion and *flos* = flower. The plant parts seemed to represent aspects of Christ: the corona was the crown of thorns, the five stamens were the five wounds, the three styles were the three nails, and the ten petal-like parts were the ten faithful apostles; and from *quadrangularis* = four-angled.]

16.4.1 Botany

Passiflora quadrangularis L. is a climber native to tropical America. It is cultivated for its fruits, which are palatable, and also used as an ornamental plant. The stems are strongly quadrangular, fleshy, and 3mm in diameter. The petiole is 3.7cm long. The blade shows 11 pairs of secondary nerves below, and a few tertiary nerves. The stipule is 3.2cm × 1.4cm. The blade is broadly lanceolate, 11.5cm × 10cm – 10.5cm × 12.5cm – 9cm × 7.5cm. The flowers are 7–10cm in diameter, purple, filamentous, and large. The fruits are oblong, with 20cm – 30cm × 10cm – 20cm berries (Figure 16.6).

16.4.2 Ethnopharmacology

The vernacular names for *Passiflora quadrangularis* L. include: Granadilla, Giant Granadilla; *barbadine* (French), *temu belanda* (Malay), *papatini*, *kkuma* (Cook Islands), *para pautini* (French Polynesia), *kudamono* (Palau), *pasio* (Samoa), and *pasione* (Tonga). In Burma, the roots are eaten to expel worms from the intestines. In Cambodia, Laos, and Vietnam, the fresh roots are known to induce narcosis. The plant contains quadrangulaside, which is a cycloartane triterpene.[11] An extract of the plant counteracted the hemorrhagic effect of the venom of *Bothrops atrox*.[12]

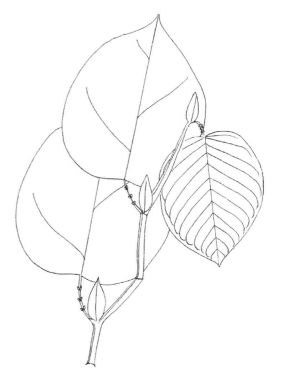

Figure 16.6 *Passiflora quadrangularis* L. [From: Herbarium Forest Department, North Borneo, Sandakan. Forest District: Elopura, Sandakan, Sepilok, Forest Reserve Compt 13. Feb. 6, 1947. Botanical identification: Singapore.]

16.5 *PASSIFLORA LAURIFOLIA* L.

[From: Latin *passio* = passion and *flos* = flower. The plant parts seemed to represent aspects of Christ: the corona was the crown of thorns, the five stamens were the five wounds, the three styles were the three nails, and the ten petal-like parts were the ten faithful apostles; and from *laurifolia* = laurel-leaved.]

16.5.1 Botany

Passiflora laurifolia L. is a climber that originates from South America. The stem is slender and produces 5–7cm-long tendrils. The leaves are simple, spiral, and stipulate. The stipules are linear–lanceolate, up to 1cm long. The blade is coriaceous, oblong–elliptical, minutely tipped at the apex, and shows about 12 pairs of secondary nerves. The flowers are solitary, 6–7cm in diameter. The sepals and petals are white or purplish. The fruits are cucumber-like yellow berries, ovate to globose, 3.5cm – 4.5cm × 2cm – 2.6cm, and edible (Figure 16.7).

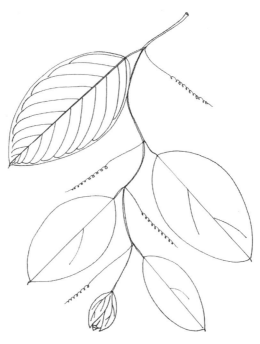

Figure 16.7 *Passiflora laurifolia* L. [From: Flora of North Borneo. Geographical localization: Sandakan. Field collector: H. G. Keith. No: 488177. Botanical identification: Singapore, July 21, 1950.]

16.5.2 Ethnopharmacology

Yellow Granadilla, Belle Apple, Yellow Water-Lemon, also known as *pomme-liane*, *pomme d'or* (French), *bua susu* (Malay), *pasio vao* (Samoa), and *vaine 'ae kuma* (Tonga), is known to be toxic in Cambodia, Laos, and Vietnam. The pharmacology of this plant is unexplored. Cyanogenetic glycosides are most likely responsible for the toxicity of the plant.

REFERENCES

1. Speroni, E. and Minghetti, A. 1988. Neuropharmacological activity of extracts from *Passiflora incarnata*. *Planta Med.*, 54, 488.
2. Soulimani, R., Younos, C., Jarmouni, S., Bousta, D., Misslin, R., and Mortier, F. 1997. Behavioral effects of *Passiflora incarnata* L. and its indole alkaloid and flavonoid derivatives and maltol in the mouse. *J. Ethnopharmacol.*, 57, 11.
3. Dhawan, K., Dhawan, S., and Sharma, A. 2004. *Passiflora*: a review update. *J. Ethnopharmacol.*, 94, 1.
4. Dhawan, K., Kumar, S., and Sharma, A. 2001. Anti-anxiety studies on extracts of *Passiflora incarnata* Linneaus. *J. Ethnopharmacol.*, 78, 165.
5. Dhawan, K., Kumar, S., and Sharma, A. 2002. Nicotine reversal effects of the benzoflavone moiety from *Passiflora incarnata* Linneaus in mice. *Addict. Biol.*, 7, 435.
6. Dhawan, K., Dhawan, S., and Chhabra, S. 2003. Attenuation of benzodiazepine dependence in mice by a tri-substituted benzoflavone moiety of *Passiflora incarnata* Linneaus: a non-habit forming anxiolytic. *J. Pharm. Sci.*, 6, 215.
7. Orsini, F., Pelizzoni, F., and Verotta, L. 2004. Volkensin from *Adenia volkensii* Harms (kilyambiti plant), a type 2 ribosome-inactivating protein. *Eur. J. Biochem.*, 271, 108
8. Fullas, F., Brown, D. M., Wani, M. C., Wall, M. E., Chagwedera, T. E., Farnsworth, N. R., Pezzuto, J. M., and Kinghorn, A. D. 1995. Gummiferol, a cytotoxic polyacetylene from the leaves of *Adenia gummifera*. *J. Nat. Prod.*, 58, 1625.

9. Echeverri, F., Arango, V., Quinones, W., Torres, F., Escobar, G., Rosero, Y., and Archbold, R. 2001. Passifloricins, polyketides-pyrones from *Passiflora foetida* resin. *Phytochemistry*, 56, 881.
10. Puricelli, L., Dell'Aica, I., Sartor, L., Garbisa, S., and Caniato, R. 2003. Preliminary evaluation of inhibition of matrix-metalloprotease MMP-2 and MMP-9 by *Passiflora edulis* and *P. foetida* aqueous extracts. *Fitoterapia*, 74, 302.
11. Orsini, F., Pelizzoni, F., and Verotta, L. 1985. Quadranguloside, a cycloartane triterpene glycoside from *Passiflora quadrangularis*. *Phytochemistry*, 25, 191.
12. Otero, R., Nunez, V., Barona, J., Fonnegra, R., Jimenez, S. L., Osorio, R. G., Saldarriaga, M., and Diaz, A. 2000. Snakebites and ethnobotany in the northwest region of Colombia. Part III: Neutralization of the hemorrhagic effect of *Bothrops atrox* venom. *J. Ethnopharmacol.*, 73, 233.

CHAPTER 17

Medicinal Plants Classified in the Family Cucurbitaceae

17.1 GENERAL CONCEPT

The family Cucurbitaceae (A. L. de Jussieu, 1789 nom. conserv., the Cucumber Family) consists of approximately 90 genera and 700 species of climbers of which about 50 are medicinal in the Pacific Rim. The cardinal botanical features to seek when collecting Cucurbitaceae are juicy and often hispid stems with tendrils and cucumber smell; palmate, rough, fleshy leaves; showy flowers which are membranaceous white or yellow; and cucumber-like berries (Figure 17.1). Fruits of the Cucurbitaceae family provide cucumber (*Cucumis sativus* L.), melon (*Cucumis melo* L.), pumpkin (*Cucurbita pepo* L.), and watermelon (*Citrulus lanatus* [Thunb.] Mansf.).

In the Western system of medicine several species from this family have been used as a laxative on account of a series of bitter and acrid oxygenated steroids known as cucurbitacin (Figure 17.2). Included, for instance, are *Bryonia cretica* L. subsp. *dioica* (Jacq.) Tutin (*bryony*), *Citrullus colocynthis* (L.) Schrad. (Bitter Gourd) (Colocynth, *British Pharmaceutical Codex*, 1963), and *Ecballium elaterium* (L.) A. Rich. (Wild Cucumber, Squirting Cucumber) (Elaterium, *British Pharmaceutical Codex*, 1934). Some evidence has already been pre-

Figure 17.1 Botanical hallmarks of Cucurbitaceae include membraneous flowers and cucumber-like berries. **(See color insert following page 168.)**

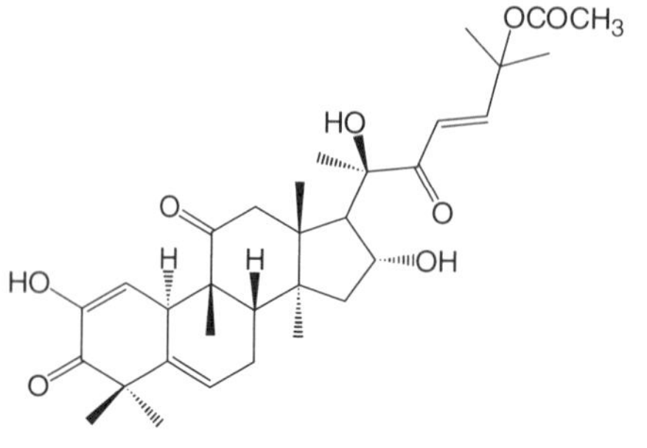

Cucurbitacin E (elaterin)

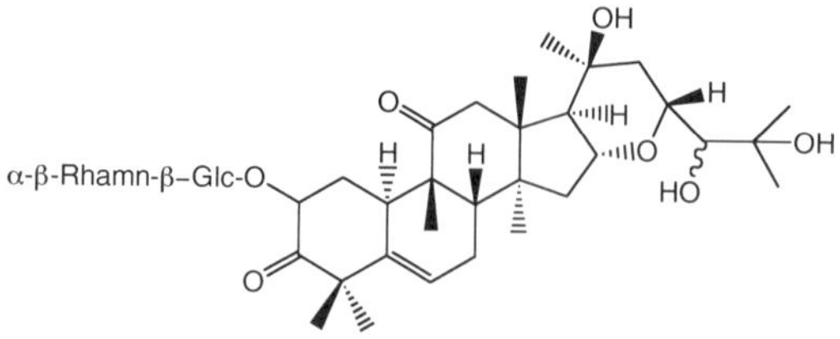

Cucurbitacin

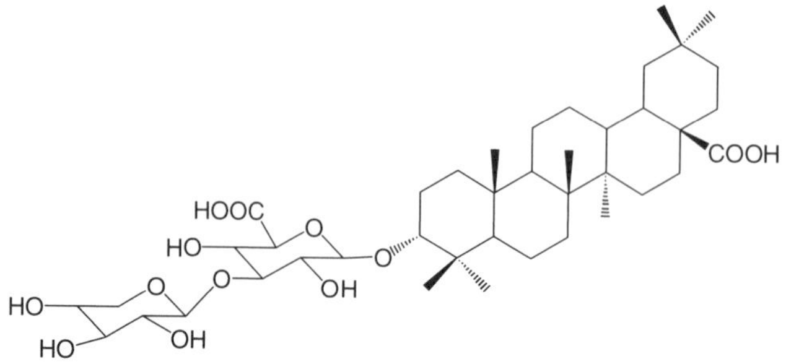

Momordin I

Figure 17.2 Examples of bioactive natural products from the family Cucurbitaceae.

sented which indicates that cucurbitacins are cytotoxic predominantly in renal tumor, brain tumor, and melanoma. They are very worthwhile exploring further.

In the Pacific Rim, the fruits are often used to promote urination, to allay fever, to soothe inflamed parts, to check hemorrhages, to counteract poisoning, to treat diabetes, jaundice, scabies, to expel intestinal worms, and to calm anxiety.

17.2 *GYMNOPETALUM COCHINCHINENSE* (LOUR.) KURZ

[From: Greek *gymnos* = naked and petalon = petal, and from Latin *cochinchinense* = from Cochinchina.]

17.2.1 Botany

Gymnopetalum cochinchinense (Lour.) Kurz (*Bryonia cochinchinensis* Lour., *Evonymus chinensis* Lour., *Gymnopetalum cochinchinense* [Lour.] Kurz, *Gymnopetalum quinquelobatum* Merr.) is a perennial climber that grows in forests and thickets on mountain slopes up to an altitude of 900m in India, Malaysia, Thailand, Vietnam, and China. The stems are hispid and develop axillary linear tendrils. The leaves are simple, exstipulate, and spiral. The petioles are 2–4cm long. The blade is membranaceous, cordate, 3–5-lobed, 4cm – 8cm × 4cm– 8cm, the apex is acuminate, the base is cordate, and both surfaces are scabrous. The flowers are male or female. The male flowers are solitary and grouped by 3–8 in a raceme, peduncle slender, and 10–15cm. The bract is foliaceous, 3-lobed, 1–2.5cm long, and yellow–brown villous. The calyx tube is tubular and 2cm long. The segments are 7mm long. The corolla is white, the segments are oblong–ovate, more or less villous, 15mm – 20mm × 10mm – 12mm; there are three stamens, filaments are 0.5mm; and the anthers are 7mm. The female flowers are solitary, on a 1–4cm-long pedicel. The gynaecium is oblong, 12mm × 5mm, yellow–brown villous, and develops three stigmas. The fruits are oblong–ovate, orange, 4–5cm, smooth, 10-ribbed, and fusiform (Figure 17.3).

17.2.2 Ethnopharmacology

In Cambodia, Laos, and Vietnam, the plant is used as a postpartum remedy. In Malaysia, a decoction of leaves is used as a drink to counteract the poisonous effects of the fruits. The juice squeezed from the leaves is used to soothe inflamed eyes.

In China, the plant is called *jin gua*. The pharmacological properties of this plant, and the *Gymnopetalum* species in general, are unexplored as of yet. Note the presence of saponins in the fruits of the *Gymnopetalum integrifolium* Kurz, including aoibaclyin and β-sitosterol-3-*O*-β-D-glucopyranoside.[1]

Figure 17.3 *Gymnopetalum cochinchinense* (Lour.) Kurz. [From: Flora of Sabah. Herbarium of the Forest Department of Sandakan. Geographical localization: District Kinabatangan. Field collectors: L. Madani et al. Seasonal swamp forest.]

17.3 *HODGSONIA MACROCARPA* (BL.) COGN.

[After Hodgson, and from *macrocarpa* = with large fruits.]

17.3.1 Botany

Hodgsonia macrocarpa (Bl.) Cogn. (*Trichosanthes macrocarpa* Bl. and *Hodgsonia capniocarpa* Ridl.) is a woody climber that grows in Bangladesh, India, Malaysia, Burma, and China to a height of 30m. The stems are glabrous. The tendrils are linear. The leaves are simple and spiral. The petiole is stout, 4–8cm long, and striate. The blade is coriaceous, 3–5-lobed, and up to 20cm × 15cm long. The flowers are yellow outside and white inside, narrowly tubular, 8cm – 10cm × 7mm – 9mm and 5-lobed; the lobes are triangular-lanceolate and 5mm long. The fruits are large, pale brown to reddish-brown, globose, 20cm × 15cm long, and contain a few woody seeds which are 7cm × 3cm (Figure 17.4).

17.3.2 Ethnopharmacology

In Malaysia, *Hodgsonia macrocarpa* (Bl.) Cogn. or *kepayang akar* (Malay) or *you zha guo* (Chinese) is used to heal nasal ulcers, smoked like a cigar with the smoke blown through the nostril. The oil of the seed is a mosquito repellent. Malays drink a decoction of leaves or inhale smoke of burned leaves to cure nose complaints. A decoction of leaves is used as a drink to bring fever down. In Borneo, the oil expressed from the fresh seeds is rubbed onto the abdomen after childbirth and is applied to the breast to deflate swelling. The seeds are known to be

Figure 17.4 *Hodgsonia macrocarpa* (Bl.) Cogn. [From: FRIM. Det.: W. J. J. O. de Wilde, February 1995. Geographical localization: Selangor, June 23, 1927.]

poisonous. The pharmacological potential of this plant and of the *Hodgsonia* species in general is unexplored.

17.4 *TRICHOSANTHES QUINQUANGULATA* A. GRAY

[From: Latin *tricho* = hairy or hair-like and Greek *anthos* = flower and *quinquangulata* = 5-angled.]

17.4.1 Botany

Trichosanthes quinquangulata A. Gray is a climber that grows in Taiwan, China, the Philippines, Indonesia, Malaysia, Burma, New Guinea, and Vietnam. The tendrils are forked. The leaves are simple and spiral. The petiole is slender and 10cm long. The blade is 5-lobed, 13cm – 22.5cm × 10cm – 20cm, membranaceous, and finely denticulate. The male flowers are arranged in lax racemes which are 15–30cm long and 8–10-flowered. The fruits are globose, 8cm long, glabrous, red, and contain several small, brownish seeds. The fruit pedicel is 3cm long (Figure 17.5).

17.4.2 Ethnopharmacology

In the Philippines, the seeds are fried and the cooked oil is used externally to calm itchiness. The seeds are reduced to powder and mixed with wine to make a drink used to assuage stomachaches. In China, the *Trichosanthes* species including *Trichosanthes quinquangulata* A. Gray or *wu jiao gua lou* have attracted a great deal of interest on account of their ability to elaborate a ribosome-inhibiting protein called trichosanthin, which has displayed encouraging signs as an anti-Human Immunodeficiency Virus (HIV) drug.[2] The *Trichosanthes* species are elaborate,

Figure 17.5 *Trichosanthes quinquangulata* A. Gray. [From: Flora of Sabah. Herbarium of the Forest Dept. of Sandakan. SAN No: 143173. Geographical localization: Upper River Miau near Mount Murut. Alt.: 1300m. April 13, 2000. Botanical identification: W. J. J. O. de Wilde, July 10, 2000.]

besides trichosanthin, trichomislin, which induces apoptosis.[3] Cytotoxic multiflorane triterpenoids including karounidol are known to occur in the *Trichosanthes* species.[4]

17.5 *TRICHOSANTHES TRICUSPIDATA* LOUR.

[From: Latin *tricho* = hairy or hair-like and Greek *anthos* = flower and *tricuspidata* = three-pointed.]

17.5.1 Botany

Trichosanthes tricuspidata Lour. is a climber that grows in Indonesia, Malaysia, Burma, Cambodia, Laos, Vietnam, and Thailand. The stems are stout, angular–striate, ribbed, and glabrous. The tendrils are forked. The leaves are simple and spiral. The petiole is 5cm long. The blade is broadly ovate–cordate, 12–13cm, papery, and undulate–denticulate at the margin. The inflorescences are axillary (Figure 17.6).

17.5.2 Ethnopharmacology

In Malaysia, the leaves of *Trichosanthes tricuspidata* Lour., *orsan jian gua lou* (Chinese), are reduced to a paste which is applied to boils. Indonesians drink the sap squeezed from the fresh leaves to stop diarrhea. The fruits are known to abound with a series of cucurbitacins, including tricuspidatin and 2-*O*-glucocucurbitacin J, which inhibit the survival of KB cells cultured *in vitro*.[5,6] Such compounds probably explain the antitumor effects of extracts of *Trichosanthes* root tubers on HepA-H

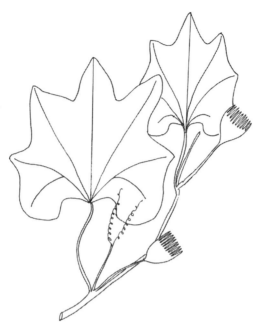

Figure 17.6 *Trichosanthes tricuspidata* Lour. [From: Flora of Sabah. Herbarium of the Forest Dept. of Sandakan. SAN No: 125603. Geographical localization: Mountain Trusmadi, District Tambuman, lowland. Aug. 26, 1982.]

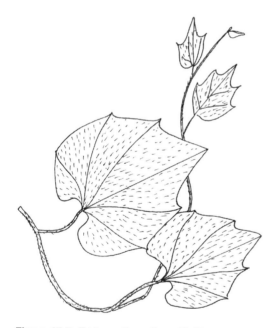

Figure 17.7 *Trichosanthes villosa* Bl. [From: Flora of Sabah. Herbarium of the Forest Herbaria of Sandakan. SAN No: 144257. Geographical localization: Mount Kinabalu Park, District Ranau. Alt.: 1500m. Field collectors: W. J. J. O. de Wilde et al., July 26, 2001.]

cells and HeLa cells reported by Dou and Li.[7,8] Cucurbitane saponins are known to occur in the leaves and stems.[9]

17.6 *TRICHOSANTHES VILLOSA* BL.

[From: Latin *tricho* = hairy or hair-like and Greek *anthos* = flower and *villosa* = hairy.]

17.6.1 Botany

Trichosanthes villosa Bl. is a climber that grows in a geographical area spanning Indonesia, Laos, Malaysia, Vietnam, Borneo, and the Philippines. The stems are stout and densely brownish-villous. The leaves are simple and spiral. The petiole is slender, up to 12.5cm long, and densely brownish-villous. The blade is broadly ovate, 11cm – 18cm × 11cm – 17cm, membranaceous, trilobate, and denticulate. The male inflorescences consist of 10–20cm-long, 15–20-flowered racemes, which are densely brownish-villous. The female flowers are solitary on a 1.5cm-long densely villous pedicel. The fruits are subglobose, brown–red, and 8–13cm long (Figure 17.7)

17.6.2 Ethnopharmacology

In Malaysia, a paste of leaves of *mi mao gua lou* (Chinese) is used externally to bring fever down and to deflate swollen legs after childbirth. The pharmacological properties of this plant are unknown.

17.7 *TRICHOSANTHES WAWRAE* COGN.

[From: Latin *tricho* = hairy or hair-like and Greek *anthos* = flower.]

17.7.1 Botany

Trichosanthes wawrae Cogn. (*Trichosanthes trifolia* auct. non [L.] Bl.) is a climber that grows to a length of 8m in Malaysia, Singapore, Indonesia, and Borneo. The stems are subglabrous and develop axillary forked tendrils. The leaves are simple and spiral. The petiole is slender, and up to 7cm long. The blade is oblong–lanceolate, membranaceous, 5cm × 9cm, with four pairs of secondary nerves. The fruits are red with a white–yellow strip at the base and 7cm × 6cm in size (Figure 17.8).

17.7.2 Ethnopharmacology

In the Southern parts of Thailand (*khi ka din*) and Malaysia a paste of leaves is applied externally to treat ague. Its pharmacological potential is unknown.

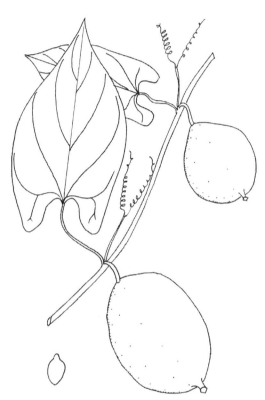

Figure 17.8 *Trichosanthes wawrae* Cogn. [From: Flora of Malaya. FRI No: 38316. Geographical localization: Kedah, Mahang, Forest Reserve, hill forest. Alt.: 700m.]

REFERENCES

1. Sekine, T., Kurihara, H., Waku, M., Ikegami, F., and Ruangrungsi, N. 2002. A new pentacyclic cucurbitane glucoside and a new triterpene from the fruits of *Gymnopetalum integrifolium*. *Chem. Pharm. Bull. (Tokyo)*, 50, 645.

2. Shaw, P. C., Lee, K. M., and Wong, K. B. 2005. Recent advances in trichosanthin, a ribosome-inactivating protein with multiple pharmacological properties. *Toxicon*. 45, 683.

3. Mi, S. L., An, C. C., Wang, Y., Chen, J. Y., Che, N. Y., Gao, Y., and Chen, Z. L. 2005. Trichomislin, a novel ribosome-inactivating protein, induces apoptosis that involves mitochondria and caspase-3. *Arch. Biochem. Biophys.*, 434, 258.

4. Akihisa, T., Tokuda, H., Ichiishi, E., Mukainaka, T., Toriumi, M., Ukiya, M., Yasukawa, K., and Nishino, H. 2001. Anti-tumor promoting effects of multiflorane-type triterpenoids and cytotoxic activity of karounidiol against human cancer cell lines. *Cancer Lett.*, 173, 9.

5. Kanchanapoom, T., Kasai, R., and Yamasaki, K., Cucurbitane, hexanorcucurbitane and octanorcucurbitane glycosides from fruits of *Trichosanthes tricuspidata*. *Phytochemistry*, 59, 215.

6. Mai le, P., Guenard, D., Franck, M., Van, T. M., and Gaspard, C. 2002. New cytotoxic cucurbitacins from the pericarps of *Trichosanthes tricuspidata* fruits. *Nat. Prod. Lett.,* 16, 15.

7. Dou, C. M. and Li, J. C. 2004. Effect of extracts of trichosanthes root tubers on HepA-H cells and HeLa cells. *World J. Gastroenterol.,* 15, 2091.

8. Dou, C. M. and Li, J. C. 2003. Preliminary study on effects of *Trichosanthes kirilowi* root on hela cells. *Zhongguo Zhong Xi Yi Jie He Za Zhi,* 23, 848.

9. Kanchanapoom, T., Kasai, R., and Yamasaki, K. 2002. Cucurbitane, hexanorcucurbitane and octan-orcucurbitane glycosides from fruits of *Trichosanthes tricuspidata. Phytochemistry,* 59, 215.

CHAPTER 18

Medicinal Plants Classified in the Family Connaraceae

18.1 GENERAL CONCEPT

The family Connaraceae consists of approximately 20 genera and 350 species of trees, shrubs, and climbers, tropical in distribution, and notably known to abound with tannins,[1,4] benzoquinones (rapanone), and nonprotein amino acids. When searching for Connaraceae in the field, one is advised to look for a woody climber or treelet with compound alternate leaves without stipules, terminal or axillary racemes, or panicles of 5-merous regular flowers, and especially the fruits which are pods, often scarlet, and enclosing a single glossy black seed embedded partially in a yellow or orange aril lode. Some compare the seed in its aril lode as "a dog-eye globe." The seeds of the Connaraceae are often poisonous because of L-methionine sulphoximine (glabrine, cnestine), which is an unusual amino acid that causes convulsion and death (Figure 18.1). For example, the crushed seeds of *Rourea glabra* mixed with corn mash have been used in tropical America for criminal purposes. About 10 species of plants classified within the Connaraceae family are used for medicinal purposes in the Asia–Pacific. It will be interesting to learn whether a more intensive study on the Connaraceae family discloses any molecules of therapeutic interest.

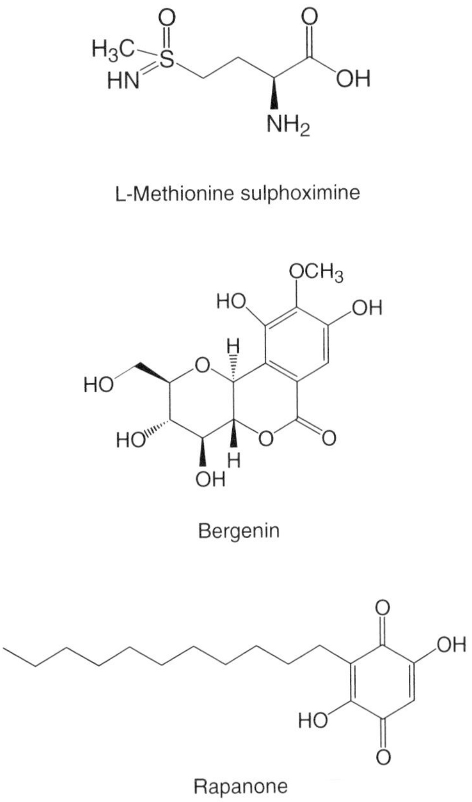

L-Methionine sulphoximine

Bergenin

Rapanone

Figure 18.1 Examples of bioactive natural products from the family Connaraceae.

18.2 *CONNARUS FERRUGINEUS* JACK

[From: Latin *Connarus* = name for a spring tree and *ferrugineus* = rusty red.]

Figure 18.2 *Connarus ferrugineus* Jack. [From: Flora of Malaya. No: 27387. Geographical localization: Legoy Forest Reserve, logging track. Alt.: 200ft. Feb. 8, 1983. Field collector and botanical identification: F. S. P. Ng.]

18.2.1 Botany

Connarus ferrugineus Jack is a woody climber which grows in open fields and secondary forests throughout Malaysia and Indonesia. The leaves are imparipinnate and exstipulate. The inflorescences are panicles of little 5-merous flowers. The sepals, petals, and stamen are punctuated by glands appearing as black dots in dried specimens. Ten stamens are united at the base. The fruits are pod-like, rusty tomentose, and up to 5cm long. The seed is solitary, glossy, black, and embedded at the base in a yellow aril lode. The folioles are obovate, bullate, densely red–brown, and hairy (Figure 18.2).

18.2.2 Ethnopharmacology

The seeds are used to kill wild dogs in Southeast Asia. The precise pharmacotoxicological mechanism involved here is unknown, but one could perhaps think of L-methionine sulphoximine, which is widespread in the Connaraceae.[1–3] Its mode of action is based on the fact that it is shaped like L-glutamic acid (an excitatory amino acid) and acts as a false substrate for glutamine synthetase, which normally converts glutamic acid into glutamine. Glutamic acid becomes excessive and causes a continuous depolarization of neurons, central nervous system disturbances, and convulsion.[4–6]

The phenolic contents of the *Connarus* species would be worth investigating for antiinflammatory potential.[7,8]

REFERENCES

1. Jeannoda, V. L., Rakoto-Ranoromalala, D. A., Valisasolalao, J., Creppy, E. E., and Dirheimer, G. 1985. Natural occurrence of methionine sulfoximine in the Connaraceae family. *J. Ethnopharmacol.*, 14, 11.
2. Jeannoda, V. L., Creppy, E. E., and Dirheimer, G. 1984. Isolation and partial characterization of glabrin, a neurotoxin from *Cnestis glabra* (Connaraceae) root barks. *Biochimie*, 66, 557.
3. Jeannoda, V. L., Creppy, E. E., Beck, G., and Dirheimer, G. 1983. Demonstration and partial purification of a convulsant from *Cnestis glabra* (Connaraceae): effect on cells in culture. *C. R. Seances Acad. Sci.*, III, 296, 335.

4. Murakoshi, I., Sekine, T., Maeshima, K., Ikegami, F., Yoshinaga, K., Fujii, Y., and Okonogi, S. 1993. Absolute configuration of L-methionine sulfoximine as a toxic principle in *Cnestis palala* (Lour.) Merr. *Chem. Pharm. Bull. (Tokyo)*, 388.

5. Ratnakumari, L., Murthy, and Ch, R. K. 1990. Effect of methionine sulfoximine on pyruvate dehydrogenase, citric acid cycle enzymes, and aminotransferases in the subcellular fractions isolated from rat cerebral cortex. *Neurosci. Lett.*, 108, 328.

6. Blizard, D. A. and Balkoski, B. 1982. Tryptophan availability, central serotoninergic function, and methionine sulfoximine-induced convulsions. *Neuropharmacology*, 21, 27.

7. Vickery, M. and Vickery, B. 1980. Coumarins and related compounds in members of the Connaraceae. *Toxicol. Lett.*, 5, 115.

8. Kuwabara, H., Mouri, K., Otsuka, H., Kasai, R., and Yamasaki, K. 2003. Tricin from a Malagasy connaraceous plant with potent antihistaminic activity. *J. Nat. Prod.*, 66, 1273.

CHAPTER 19

Medicinal Plants Classified in the Family Anisophylleaceae

19.1 GENERAL CONCEPT

The family Anisophylleaceae (Ridley, 1922) comprises the genera *Anisophyllea*, *Poga*, *Combretocarpus*, and *Polygonanthus* with approximately 40 species of trees or shrubs known to abound with tannins, including ellagic acid (Figure 19.1).

The Anisophylleaceae Family are recognized in the field by the architecture of their branches, which often consist of two ranks of closely packed alternate small leaves which are asymmetrical and membranaceous. To date, the Anisophylleaceae Family is not being pharmacologically investigated. In the Asia–Pacific, *Anisophyllea disticha* (Jack) Baill and *Halogaris disticha* Jack are medicinal. Note that the Anisophylleaceae is often incorporated in the family Rhizophoraceae.

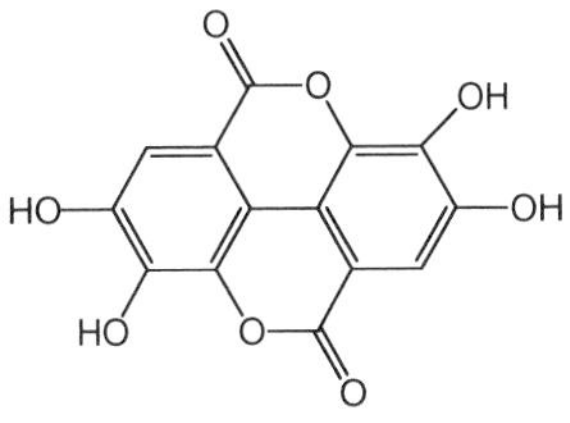

Ellagic acid

Figure 19.1

19.2 *ANISOPHYLLEA DISTICHA* HOOK. F.

[From: Greek *anisos* = unequal and *phullon* = leaf, and from Latin *disticha* = in two ranks.]

19.2.1 Botany

Anisophyllea disticha Hook. f. is a treelet that grows to a height of 7m in the lowlands to the forests in Malaysia, Sumatra, the Lingga Islands, and Borneo. The stems are hairy, fissured, brown, and zigzag-shaped with 4–5mm-long internodes. The leaves are simple, alternate, and exstipulate. The blade is papery, almost translucent, asymmetrical, and shows three nerves. The midrib is sunken above. The flowers are pinkish-white. The fruits are ovoid, glossy, and crimson, and are 1.9cm × 9mm (Figure 19.2).

19.2.2 Ethnopharmacology

Malays call the plant *raja berangkat* or *kayu ribu-ribu*, and eat the leaves to stop diarrhea and dysentery. In Sumatra, the roots are boiled with other herbs to make a drink used to relieve weariness. The pharmacological potential of this herb is unexplored. Note that the plant most likely contains

Figure 19.2 *Anisophyllea disticha* Hook f.

ellagic acid. A dichloromethane-methanol extract of *Anisophyllea apetala* provided 3′-methyl-3,4-*O*,*O*-methylidene-4′-*O*-β-D-glucopyranosyl ellagic acid, which showed some DNA damaging effect *in vitro*, and potently inhibited the survival of yeast.[1] It will be interesting to learn whether further pharmacological study discloses any cytotoxic or antiviral molecules.

REFERENCE

1. Xu, Y. M., Deng, I. Z., Ma, J., Chen, S. N., Marshall, R., Jones, S. H., Johnson, R. K., and Hecht, S. M. 2003. DNA damaging activity of ellagic acid derivatives. *Bioorg. Med. Chem.*, 11, 1593.

CHAPTER 20

Medicinal Plants Classified in the Family Rosaceae

20.1 GENERAL CONCEPT

The Rosaceae are a huge family of approximately 100 genera and 3000 species of ubiquitous medium-sized trees, shrubs, or herbs, that are well known for accumulating tannins. In this family, the leaves are alternate, simple or compound, with serrated edges, and stipulate. The flowers are characteristic in the sense that petals are often orbicular, spoon-shaped, somewhat ephemeral, and fragrant, and the androecium consists of numerous stamens attached to tiny filaments attached to a well-developed hypanthium (Figure 20.1). The fruits of several Rosaceae are edible, for example, *Malus domestica* L. (apples), *Pyrus communis* L. (pears), *Prunus armeniaca* L. (apricots), and *Rubus idaeus* L. (red raspberries).

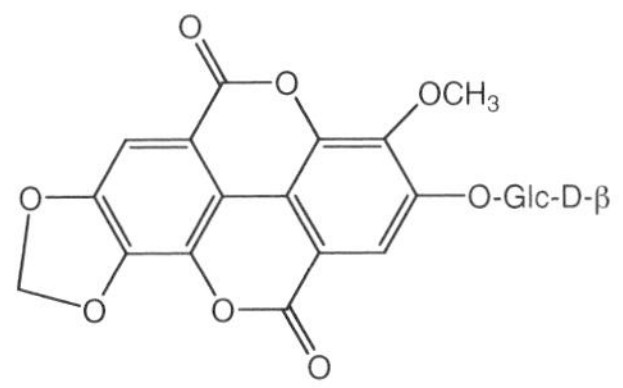

3'-Methyl-3,4-*O,O*-methylidene-4'-*O*-β-D-glucopyranosyl ellagic acid

Figure 20.1 Botanical hallmarks of Rosaceae.

The *Rosa* species, or roses, are common examples of the Rosaceae Family, and are probably the most popular ornamental plants. The apple itself has been used from long ago for food, and the pectin of the apple is known to lower serum cholesterol (Gonzales et al.).[1] *Crataegus* or aubepine (*French Pharmacopoeia*, 1965) or the dried fruits of *Crataegus oxyacantha* (*Crataegus monogyna* and *Crataegus oxyacanthoides*) have been used as a tincture (1 in 2.5 by maceration with alcohol 70%) in the treatment of heart diseases. Rose Oil (*British Pharmaceutical Codex*, 1949) is the volatile oil obtained by distillation from the fresh flowers of the Damask Rose, *Rosea damascene*. This oil is aromatic and largely employed in perfumery and in pharmaceutical technology to prepare lozenges, toothpastes, and ointments.

The petals of the red or Provins Rose, *Rosa gallica* (*British Pharmaceutical Codex*, 1949), have been employed usually as an infusion for its mild, astringent properties, and as a coloring agent. Rose fruits (*British Pharmaceutical Codex*, 1954), or the fresh ripe fruits of the various *Rosa* species, have been used in the form of syrup as a dietary supplement since it accumulates ten times more vitamin C than orange juice. Bitter Almond (*British Pharmaceutical Codex*, 1934) or the dried ripe seeds of *Prunus amygdalus* var. *amara* (*Amygdalus communis* var. *amara*) and Sweet Almond (*British Pharmaceutical Codex*, 1934), or the dried seeds of *Prunus amygdalus*

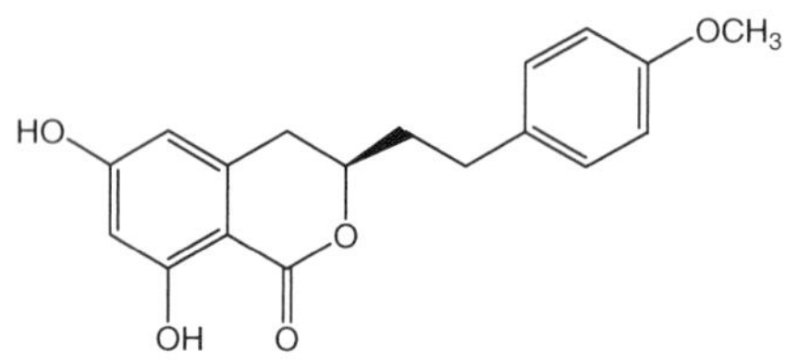

Agrimonolide

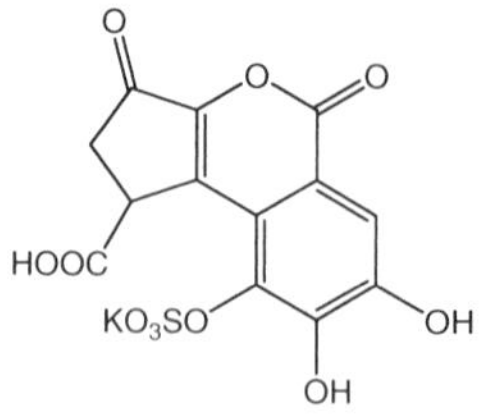

Brevifolin carboxylic acid

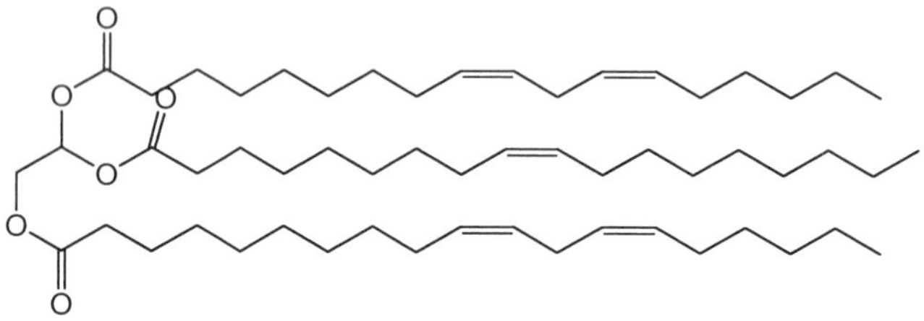

1.3-Dilinoleoyl 2 olein

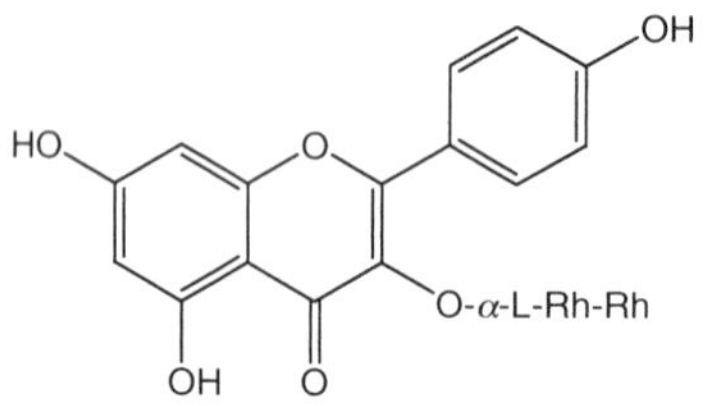

Kaempferol-3-O-α-L-rhamnopyranosyl(1-2)rhamnopyranoside

Figure 20.2 Examples of bioactive natural products from the family Rosaceae.

var. *dulcis* (*Amygdalus communis* var. *dulcis*) have been used to some extent in Western medicine. The fresh leaves of *Prunus laurocerasus* (Cherry-laurel, *British Pharmaceutical Codex*, 1949) have been used as a flavoring agent and as a sedative for nausea and vomiting. The dried rhizomes of the common tormentil (*Potentilla erecta*) have been used both internally and externally as an astringent. (*Swiss Pharmacopoeia*, 1934). With regard to the pharmacological potential of Rosaceae, the evidence currently available suggests that this family is a storehouse of cytotoxic and antiviral agents awaiting discovery. Strawberry, raspberry, and blueberry extract, for instance, prevent the survival of promyelocytic HL60 cell line cultured *in vitro*.[2] Brevifolin carboxylic acid from *Duchesnea chrysantha* (Zoll. & Moritzi) Miq. (Mock Strawberry) showed a strong cytotoxic activity against PC14 and MKN45 human cancer cell cultured *in vitro*.[3] A methanol extract of whole *Agrimonia pilosa* Ledeb. (Chinese Agrimony) inhibited the enzymatic activity of Human Immunodeficiency Virus (HIV)-1 reverse transcriptase with an IC_{50} value of 8.9µg/mL.[4] Agrimonolide from the roots showed hepatoprotective effects on both tacrine-induced cytotoxicity in human liver-derived Hep G2 cells, and tert-butyl hydroperoxide-induced cytotoxicity in rat primary hepatocytes with EC_{50} values of 88.2µM and 37.7µM, respectively.[5]

Anthocyanin mixtures from the *Amelanchier* species inhibited the enzymatic activity of cyclooxygenase *in vitro*.[6] Note that anthocyanins scavenge free radicals, hence they are a vasculoprotector and have antiaging potential (Figure 20.2). In the Asia–Pacific, about 90 species of Rosaceae are medicinal, mainly on account of their astringency.

20.2 *ERIOBOTRYA JAPONICA* (THUNB.) LINDL.

[From: Greek *erion* = wool and *botrys* = cluster, a bunch of grapes, alluding to the clustered and woolly panicles, and from Latin *japonica* = from Japan.]

20.2.1 Botany

Eriobotrya japonica (Thunb.) Lindl. (*Mespilus japonicus* Thunb., *Mespilus japonica* Thunb., and *Crataegus bibas* Lour.) is an ornamental tree native to South China and Japan, first cultivated in Europe in the 18th century. The plant grows to a height of 7m. The stems are woolly at the apex and 5mm in diameter. The leaves are simple, spiral, and stipulate. The stipule is bifid and 7mm long. The petiole is stout and short. The blade is glossy, serrate, 12.5cm × 5cm – 30cm × 10cm, dark green above, spathulate, glabrous above and woolly below, showing 15 pairs of secondary nerves. The inflorescences are terminal, 7cm long, and the panicle is golden yellow. The fruits are 2.5cm long, velvety, and golden yellow, globose, or obovate, and 1–1.5cm in diameter (Figure 20.3).

20.2.2 Ethnopharmacology

The vernacular names for *Eriobotrya japonica* (Thunb.) Lindl. are *loquat* (Japanese), plum; *nispero japonés* (Spanish); *nespola giapponese* (Italian); *néflier du Japon* (French); *ameixa do Japao* (Portuguese); *japanische mispel* (German); and *pi ba* (Chinese). In China, the leaves are used to treat bronchitis, cough, fever, and nausea. The juice squeezed from the bark is used as a drink to fight nausea and to stop vomiting. The fruits are edible and thirst quenching.

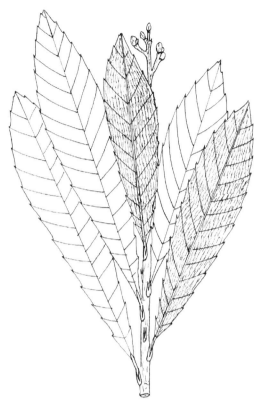

Figure 20.3 *Eriobotrya japonica* (Thunb.) Lindl. [From: Ex. Herb. Bot. Singaporense. Geographical localization: Botanic Gardens, Singapore. Sept. 1, 1929. Botanical identification: C. X. D. R. Fuitado.]

The antitussive property of the plant is probably owed to saponins, which are known to abound in the plant. A remarkable advance in the pharmacological assessment of *Eriobotrya japonica* has been provided by Ito et al.[7,8] They isolated from the leaves roseoside and procyanidin B-2, both of which inhibit the activation of early antigen Epstein–Barr virus in Raji cells by 12-*O*-tetradecanoylphorbol-13-acetate (TPA). Roseoside significantly delayed carcinogenesis *in vivo* in a two-stage carcinogenesis assay on mouse skin. They also reported an interesting series of phenolic oligomers, which stopped the growth of human squamous cell carcinoma and human salivary gland tumor cell lines cultured *in vitro*.

20.3 *PRUNUS ARBOREA* (BL.) KALKMAN

[From: Latin *prunus* = an ancient Latin name for the plum and *arborea* = tree, alluding to a tree-like habit of growth.]

20.3.1 Botany

Prunus arborea (Bl.) Kalkman (*Polydontia arborea* Bl., *Pygeum griffithii* Hk. f. *sensu* Koehne, *Prunus ovalifolium* King, *Prunus patens* Ridl., *Prunus persimile* Kurz, *Prunus rubiginsum* Ridl.,

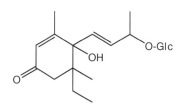

Procyanidin B -2

Roseoside

Figure 20.4 Examples of bioactive natural products characterized from the family Rosaceae.

Prunus stipulaceum King, and *Prunus parviflorum* T. et B.) is a tree that grows to 30m high and 2m in girth in the lowland rain forests of a geographical area which spans Continental Asia to Indonesia, to an altitude of up to 1300m. The stems and buds are velvety. The leaves are simple and stipulate. The blade is elliptical to oblong or ovate to lanceolate, 11cm × 5.5cm – 7cm × 3cm, membranaceous, brittle, and velvety on the midrib below. It has 8–10 pairs of secondary nerves. The secondary nerves and midrib are sunken above the blade. The petiole is velvety. The inflorescences are axillary spikes up to 5cm long. The flowers are cream. The fruits are strongly bilobed, broader than long, 1.1cm × 1.7cm (Figure 20.5).

Figure 20.5 *Prunus arborea* (Bl.) Kalkman. [From: Comm. Ex. Herb. Hort. Bot. Sing. Geographical localization: Mount Ophir via Tangkok. Alt.: 3000–4000ft.]

20.3.2 Ethnopharmacology

In Malaysia and Indonesia, the plant is known as *pepijat*. A decoction of leaves is used as a drink to precipitate childbirth during labor. The only pharmacological report available thus far on this plant is a clinical study of a phytosterol extract obtained by Cuervo Blanco et al.[9] They showed that, in the case of prostatitis, the extract is effective for more than 90% of the patients treated, and was presented as "making it a highly reliable product for use by the specialist or general practitioner." Cuervo Blanco et al.[9] made the interesting observation that the seeds of *Prunus dulcis* contain an arabinan-rich pectic polysaccharide, which stimulated T-lymphocyte activity both *in vitro* and *in vivo*. This polysaccharide stimulates the multiplication of spleen mononuclear cells and the immunostimulating potentials of *Prunus dulcis* should be further investigated.

20.4 *RUBUS MOLUCCANUM* L.

[From: Latin *rubus* = blackberry from *ruber* = red, and *moluccanum* = from the Moluccas.]

20.4.1 Botany

Rubus moluccanum L. (*Rubus glomeratus* Bl.) is a scrambling shrub that grows to a height of 3m in a geographical area which covers the Himalayas, south to India and Sri Lanka, Malaysia to Australia, China, Papua New Guinea, Solomon Islands, New Caledonia and Fiji, Hawaii and Mauritius, and Madagascar to the west. The stems are woody, clothed with short rusty or white woolly hairs, and armed with numerous prickles. The leaves are simple, spiral, and stipulate. The petiole is slender and spiny. The blade is 3- or 5-lobed, 3cm – 20cm × 2cm – 18cm, glabrous or sparsely hairy above, and densely white or rusty-colored and hairy below. The flowers are white or reddish-pink and arranged in racemes in the upper axils. The berries are about 1–1.3cm in diameter, red, and tasteless (Figure 20.6).

20.4.2 Ethnopharmacology

The vernacular name for *Rubus molucca-nus* L. includes Molucca Raspberry, Broad Leafed Bramble, Molucca Bramble; Wild Raspberry, Molucca Raspberry; *piquant lou-lou* (French Mauritius); and *wa votovotoa*

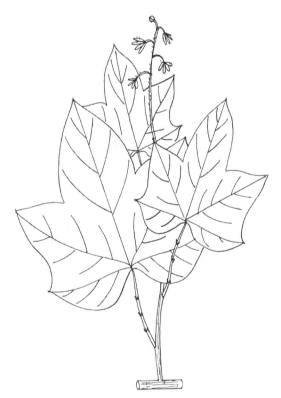

Figure 20.6 *Rubus moluccanum* L. [From: Flora of Malay Peninsula, Forest Department. Field collector: J. W. Smith, March 9, 1947. No: 56950.]

(Fiji). Papua New Guineans apply heated leaves to the abdomen to mitigate abdominal pain. Counterirritancy is most likely responsible for such a use. The plant's pharmacology is nevertheless unexplored.

REFERENCES

1. Gonzalez, M., Rivas, C., Caride, B., Lamas, M. A., and Taboada, M. C. 1998. Effects of orange and apple pectin on cholesterol concentration in serum, liver and faeces. *J. Physiol. Biochem.,* 54(2), 99–104.
2. Skupien, K., Oszmianski, J., Kostrzewa-Nowak, D., and Tarasiuk, J. 2005. *In vitro* antileukemic activity of extracts from berry plant leaves against sensitive and multidrug resistant HL60 cells. *Cancer Lett.,* in press.
3. Lee, I. R. and Yang, M. Y. 1994. Phenolic compounds from *Duchesnea chrysantha* and their cytotoxic activities in human cancer cell. *Arch. Pharm. Res.,* 17, 476.
4. Min, B. S., Kim, Y. H., Tomiyama, M., Nakamura, N., Miyashiro, H., Otake, T., and Hattori, M., 2001. Inhibitory effects of Korean plants on HIV-1 activities. *Phytother. Res.,* 15, 481.
5. Park, E. J., Oh, H., Kang, T. H., Sohn, D. H., and Kim, Y. C. 2004. An isocoumarin with hepatopro-tective activity in Hep G2 and primary hepatocytes from *Agrimonia pilosa. Arch. Pharm. Res.,* 27, 944.
6. Adhikari, D. P., Francis, J. A., Schutzki, R. E., Chandra, A., and Nair, M. G. 2005. Quantification and characterization of cyclo-oxygenase and lipid peroxidation inhibitory anthocyanins in fruits of *Amelanchier. Phytochem. Anal.,* 16, 175.
7. Ito, H., Kobayashi, E., Li, S. H., Hatano, T., Sugita, D., Kubo, N., Shimura, S., Itoh, Y., Tokuda, H., Nishino, H., and Yoshida, T. 2002. Antitumor activity of compounds isolated from leaves of *Eriobotrya japonica. J. Agric. Food. Chem.,* 50, 2400.
8. Ito, H., Kobayashi, E., Takamatsu, Y., Li, S. H., Hatano, T., Sakagami, H., Kusama, K., Satoh, K., Sugita, D., Shimura, S., Itoh, Y., and Yoshida, T. 2000. Polyphenols from *Eriobotrya japonica* and their cytotoxicity against human oral tumor cell lines. *Chem. Pharm. Bull. (Tokyo),* 48, 687.
9. Cuervo Blanco, E., Francia Bengoechea, A., and Fraile Gomez, B. 1978. Clinical study of a phytosterol extract of *Prunus arborea* and 3 amino acids: glycine, alanine and glutamic acid. *Arch. Esp. Urol.,* 31, 97.

CHAPTER 21

Medicinal Plants Classified in the Family Thymeleaceae

21.1 GENERAL CONCEPT

The family Thymeleaceae (A. L. de Jussieu, 1789 nom. conserv., the Mezereum Family) comprises 50 genera and 500 species of trees and shrubs, of which 50 species are of medicinal value in the Asia–Pacific. A common example of Thymeleaceae is *Daphne mezereum* L., which is cultivated for ornamental purposes although poisonous. Thymeleaceae are known to elaborate a series of complex and unusual diterpenoid esters of the tigliane, daphnane type such as mezerein, which impart to the fruits of *Daphne mezereum* L. its toxic effects: irritation, ulceration of mucosa, violent purgation, vomiting, headaches, convulsions, and death (Figure 21.1).

The evidence currently available suggests that mezerein, and congeners such as gnidilatimonoein, are strongly carcinogenic. Mezerein activates the enzymatic activity of serine/threonine protein kinase, which is the major receptor for a number of tumor-promoting agents including the phorbol esters of Euphorbiaceae (see Chapter 26). It has been proposed that tumor promoting phorbol esters such as 12-*O*-tetradecanoylphorbol-13-acetate (TPA) involve the stimulation of protein kinase.[1] Investigating natural products for kinase activity is a worthy task, since kinases are involved in manifold diseases. Thymeleaceae are also known to elaborate lignans of which (−)-aptosimon and (−)-diasesamin-di-γ-lactone, hinokinin, and 1α,7α,10αH-guaia-4,11-dien-3-one abrogated the survival of P-388 and HT-29 tumor cell lines cultured *in vitro*.[2]

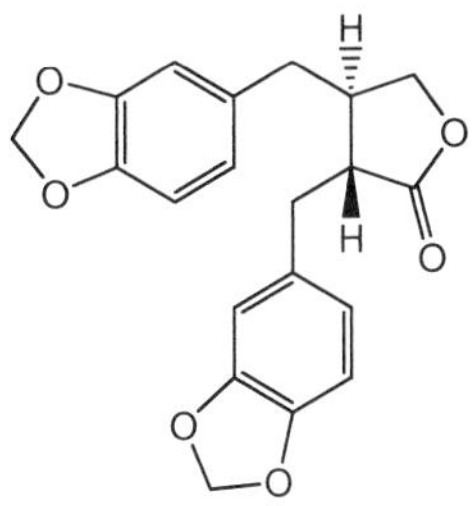

Figure 21.1 Examples of bioactive natural products from the family Thymeleaceae.

Hinokinin

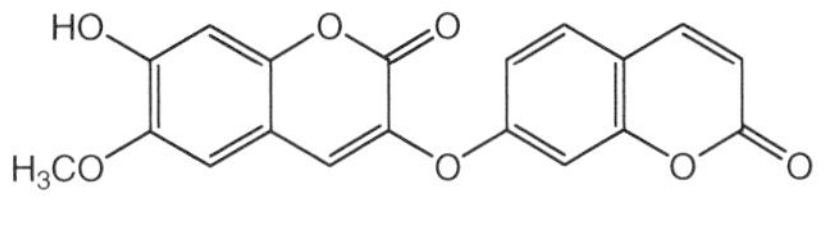

Daphnoretin

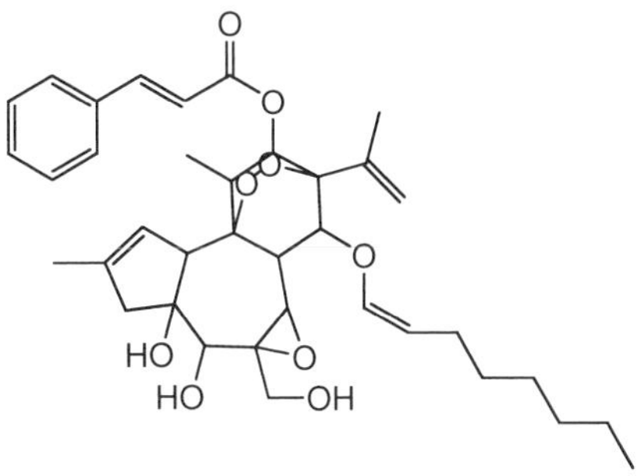

Gnidilatimonoein

h regard to the phenolic substances, Li et al.,[3] for instance, isolated from *Edgeworthia gardneri* a series of biscoumarin derivatives including 7-hydroxy-3,7′-dicoumaryl ether (edgeworin), 7-hydroxy-6-methoxy-3,7′-dicoumaryl ether (daphnoretin), and 6,7-dihydroxy-3,7′-dicoumaryl ether (edgeworthin), which inhibited the enzymatic activity of DNA polymerase β lyase with IC_{50} values of 7.3μg/mL, 43.0μg/mL, and 32.1μg/mL, respectively.

The roots of *Wikstroemia indica* elaborate an interesting series of biflavonoids, including sikokianin B and sikokianin C, which destroy the chloroquine-resistant strain of *Plasmodium falciparum* cultured *in vitro*, with IC_{50} values of 0.54μg/mL and 0.56μg/mL, respectively.[4] From the same plant, Wang et al.[5] identified a guaiane-type sesquiterpene, indicanone, which inhibits nitric oxide production by a mouse monocyte macrophage cell line, (RAW) 264.7, stimulated by lipopolysaccharide (LPS) and recombinant mouse interferon-gamma with an IC_{50} value of 9.3μM. In summary, most of the evidence which has recently emerged lends support to the idea that Thymeleaceae would be worth investigating for its pharmacology, and especially for cytotoxic and antiviral activity. An interesting development from this family and from the following plants would be the search for antiviral diterpenes.

21.2 *GONYSTYLUS CONFUSUS* AIRY SHAW

[From: Greek *gonia* = angle and *stulos* = style, and from Latin *confusus* = confused.]

Figure 21.2 *Gonystylus confusus* Airy Shaw. [From: Federated Malay States. Geographical Localization: Kuala Lumpur. June 3, 1916. Field collectors: M. S. Hamid and I. H. Burkill. Botanical identification: 1948.]

21.2.1   Botany

Gonystylus confusus Airy Shaw is a tree that reaches a height of 30m and a girth of 2.2m. The plant grows in the low undulating rain forests of Thailand, Malaysia, and Sumatra up to 600m in altitude. The plant can be found in peat swamp forests and lowland dipterocarp forests. The bark shows elongated, adherent scales. The leaves are simple and alternate exstipulate. The petiole is woody and cracked. The blade is oblong–elliptical, and shows about 10 pairs of secondary nerves, 8cm × 2cm × 15cm × 6cm, tapering at the apex, wedge-shaped at the base, and is a drying, dull purplish-brown. The margin is slightly recurved. The blade is thinly leathery. The midrib is flat or prominently raised above the blade. Inflorescences have terminal racemes up to 10cm long. The fruits are ovoid, 4–10cm across, rough, dull brown, pointed at the apex, and 3-shouldered (Figure 21.2).

21.2.2   Ethnopharmacology

A decoction of roots is used as a drink by the Malays (*ramin pinang muda*) to recover from the exhaustion of childbirth. To date, evidence of pharmacological activity from the *Gonystylus* species is virtually nonexistent. Note, however, that Fuller et al.[6] isolated a series of cytotoxic cucurbitacins in *Gonystylus keithii*. Cucurbitacins are well known to abound in the Malvales order and one might wonder if the

Gonystylus species are really at home within the Thymeleaceae. The question arises as to whether or not *Gonystylus confusus* Airy Shaw contains cucurbitacins and what are their possible cytotoxic properties?

21.3 *GONYSTYLUS MACROPHYLLUS* (MIQ.) AIRY SHAW

[From: Greek *gonia* = angle and *stulos* = style, and from Latin *macrophyllus* = large leaves.]

21.3.1 Botany

Gonystylus macrophyllus (Miq.) Airy Shaw (*Gonystylus miquelianus* Tysm. & Binn.) is a tree native to Java, Sumatra, and the Nicobar Islands. The leaves are simple and alternate exstipulate. The petiole is woody and cracked. The blade is glabrous, oblong–elliptic, and shows approximately 10 pairs of secondary nerves, arching at the margin, 15cm × 4cm × 17cm × 6 cm, tapering at the apex, wedge-shaped at the base, and dries to a dull, purplish-brown. The margin is slightly recurved. The blade is leathery with numerous prominent ter-tiary nerves on the lower surface, which are more or less at right angles to the secondary nerves. The inflorescences are terminal racemes up to 10cm long. The fruits are ovoid, 2.5cm × 5cm, rough, dull brown, pointed at the apex, and 3-shouldered (Figure 21.3).

21.3.2 Ethnopharmacology

Indonesians burn the essential oil of the wood as incense which is inhaled to relieve asthma. Pharmacological properties are unexplored.

Figure 21.3 *Gonystylus macrophyllus* (Miq.) Airy Shaw. [From: University of Illinois at Chicago. North Sulawesi, 220Km west of Manado, 50Km inland from Pangi, on the Ilang River. Primary lowland forest. Alt.: 400m, 0°41′ N, 12°40′ E. Field collectors: J. Burley et al., 3632, March 4, 1990. Botanical identification: C. Tawan, Feb. 24, 2004. Collected under sponsorship of the U.S. National Cancer Institute.]

REFERENCES

1. Geiges, D., Meyer, T., Marte, B., Vanek, M., Weissgerber, G., Stabel, S., Pfeilschifter, J., Fabbro, D., and Huwiler, A. 1997. Activation of protein kinase C subtypes alpha, gamma, delta, epsilon, zeta, and eta by tumor-promoting and nontumor-promoting agents. *Biochem. Pharmacol.*, 53, 865.
2. Lin, R. W., Tsai, I. L., Duh, C. Y., Lee, K. H., and Chen, I. S. 2004. New lignans and cytotoxic constituents from *Wikstroemia lanceolata*. *Planta Med.*, 70, 34.
3. Li, S. S., Gao, Z., Feng, X., and Hecht, S. M. 2004. Biscoumarin derivatives from *Edgeworthia gardneri* that inhibit the lyase activity of DNA polymerase beta. *J. Nat. Prod.*, 67, 1608.
4. Nunome, S., Ishiyama, A., Kobayashi, M., Otoguro, K., Kiyohara, H., Yamada, H., and Omura, S. 2004. *In vitro* antimalarial activity of biflavonoids from *Wikstroemia indica*. *Planta Med.*, 70, 76.

5. Wang, L. Y., Unehara, T., and Kitanaka, S. 2005. Anti-inflammatory activity of new guaiane type sesquiterpene from *Wikstroemia indica. Chem. Pharm. Bull. (Tokyo)*, 1, 137–139.

6. Fuller, R. W., Cardellina, J. H., Cragg, G. M., and Boyd, M. R. 1994. Cucurbitacins: differential cytotoxicity, dereplication and first isolation from *Gonystylus keithii. J. Nat. Prod.*, 57, 1442.

CHAPTER 22

Medicinal Plants Classified in the Family Melastomataceae

22.1 GENERAL CONCEPT

About 30 species of plants classified as Melastomataceae are used for medicinal purposes in the Asia–Pacific. Note that Melastomataceae are tanniferous and therefore astringent, hence their relatively frequent use to stop diarrhea and bleeding, to heal and resolve infected or wounded skin, and for postpartum invigoration. The family Melastomataceae itself (A. L. de Jussieu, 1789 nom. conserv., the Melastoma Family) is a vast taxon which includes 200 genera and approximately 4000 species of herbs, shrubs, and trees that are widespread in tropical regions. Searching for Melastomataceae in the field is guided by three botanical hallmarks: the leaves are marked with 3–9 longitudinal nerves parallel to the midrib, the flowers show numerous stamens with 2-locular, basifixed anthers that open by a single pore, and the connectives which are often appendaged, and by the fruits which are cup-shaped (Figure 22.1).

Figure 22.1 Botanical hallmarks of Melastomataceae. **(See color insert following page 168.)**

The evidence so far presented is consistent with the view that the Melastomataceae family is a vast source of pharmacologically active tannins and flavonoids awaiting experimentation.[1] With regard to the tannins, which are often skipped in biological screenings, Melastomataceae elaborate an unusual series of hydrolyzable tannin oligomers such as nobotannin B, which has exhibited anti-Human Immunodeficiency Virus (HIV) property *in vitro*[2,3] (Figure 22.2).

Both castalagin and procyanidin B-2 are ubiquitous in the family and are known to lower blood pressure in spontaneously hypertensive rats dose-dependently through decrease of sympathetic tone.[4] With regard to the flavonoids, quercitrin, isoquercitrin, rutin, and quercetin abound in the family and have displayed a broad spectrum of pharmacological properties, including the scavenging

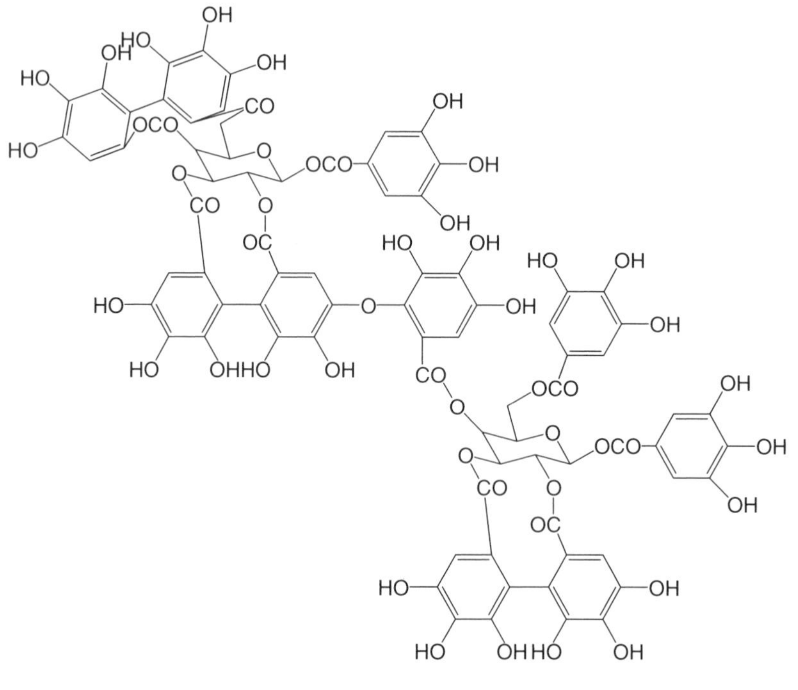

Nobotannin B

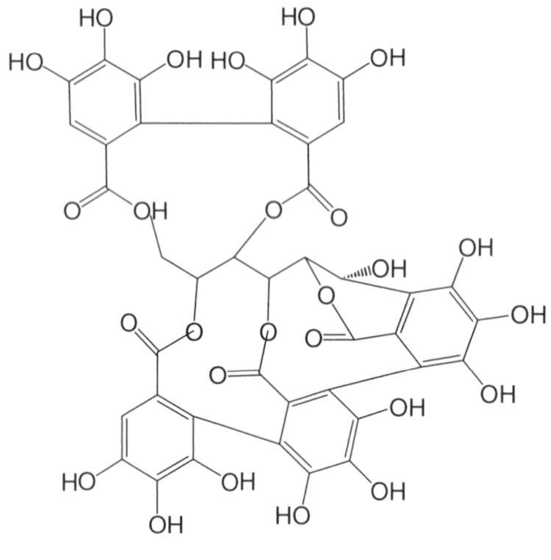

Castalagin

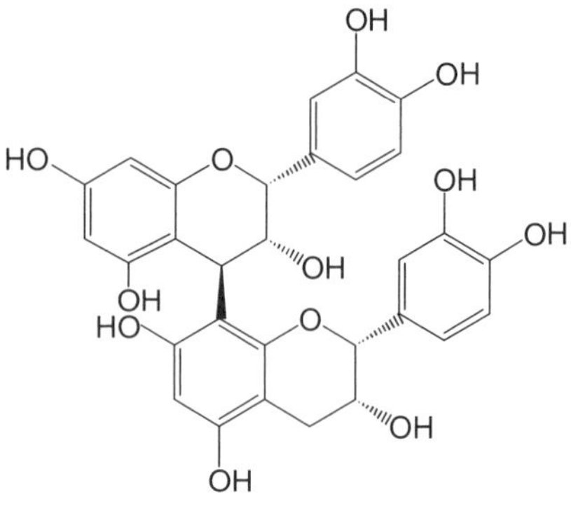

Procyanidin B-2

Figure 22.2 Examples of bioactive natural products from the Melastomataceae.

of free radicals and inhibition of mono amine oxidase (MAO)-B with IC_{50} values of 19.06μM, 11.64μM, 3.89μM, and 10.89μM, respectively.[5] The medicinal flora of the Asia–Pacific encompasses approximately 30 species of Melastomataceae, of which *Blastus cogniauxii* Stapf., *Diplectria divaricata* (Willd.) O. Ktze., *Dissochaeta annulata* Hook. f., *Dissochaeta bracteata* (Jack) Bl., *Dissochaeta punctulata* Hook. f. ex Triana, *Medinilla hasselti* Bl., *Medinilla radicans* (Bl.) Bl., *Melastoma polyanthum* Bl., *Melastoma sanguineum* Sims., *Memecylon dichotomum* C. B. Clarke,

Neodissochaeta gracilis (Jack) Bakh. f., *Osbeckia chinensis* L., and *Pternandra coerulescens* Jack are presented in this chapter.

22.2 *BLASTUS COGNIAUXII* STAPF.

[From: Latin *blastus* = brings forth and after Alfred Celestin Cogniaux, Belgian botanist (1841–1916).]

22.2.1 Botany

Blastus cogniauxii Stapf. is a treelet that grows up to 3m in the rain forests of Southeast Asia. The stems are glabrous, terete, 3mm in diameter, and have 6–6.5cm long internodes. The leaves are simple, opposite, and exstipulate. The petiole is slender. The blade is elliptic, 15.2cm × 3cm – 17cm × 4.4cm – 1cm × 3.2cm – 16cm × 3.4cm, tailed at the apex, and showing two pairs of secondary nerves. The blade shows 40 pairs of tertiary nerves below. The inflorescences are 1.2cm long. The flowers are minute with four stamens. The calyx is green and the petals are white (Figure 22.3).

22.2.2 Ethnopharmacology

The Malays use the roots to make a postpartum remedy. The pharmacological properties of this plant, and of the genus *Blastus* in general, are unexplored.

Figure 22.3 *Blastus cogniauxii* Stapf. [From: Sarawak Forest Department. Field collector: B. Lee. No: S52432. Geographical localization: Hill mixed dipterocarp forest on a slope at 900m, Tebunan Hill, Ulu Trusan, Lawas, 5th Division.]

22.3 *DIPLECTRIA DIVARICATA* (WILLD.) O. KTZE.

[From: Latin *plectrum* = a stick with which the strings of a stringed instrument were struck and from Latin *divarico* = to spread apart.]

22.3.1 Botany

Diplectria divaricata (Willd.) O. Ktze. (*Dissochaeta divaricata* [Willd.] G. Don., *Anplectrum barbatum* Wall. ex C.B. Clarke, *Anplectrum cyanocarpum* [Blume] Triana, *Anplectrum divaricatum* [Willd.] Triana, *Anplectrum glaucum* [Jack] Triana, *Anplectrum patens* Geddes, *Anplectrum stellulatum* Geddes, *Backeria barbata* [Wall. ex C.B. Clarke] Raizada, *Diplectria barbata* [Wall. ex C.B. Clarke] Franken & Roos, *Diplectria cyanocarpa* [Bl.] Kuntze, *Dissochaeta cyanocarpa* [Bl.] Bl., *Dissochaeta glauca* [Jack] Bl., *Melastoma cyanocarpon* Bl., *Melastoma divaricatum* Willd., and *Melastoma glaucum*) is a woody climber that grows in Thailand, Malaysia, and Indonesia. The stems are finely cracked and the internodes are 2.7–3cm long. The leaves are simple, opposite, and

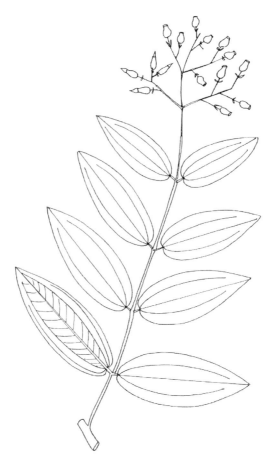

Figure 22.4 *Diplectria divaricata* (Willd.) O. Ktze. [From: Flora of Malaya. Kepong Field No: 14154. Geographical localization: East of G Bongsu Forest Reserve. In low undulating country in disturbed forest on sandy soil (shale). Alt.: 1000ft.]

exstipulate. The petiole is 4mm long. The blade is elliptic, 8.5cm × 3.2cm – 7.5cm × 3cm – 6cm × 2cm, and shows 12 pairs of secondary nerves. The flowers are arranged in large erect panicles, which are 8cm long. The calyx tube is purple and the corolla is pink. The androecium consists of four stamens alternating with the petals, and the anthers are yellow. The flower pedicels are 4mm long and show two purple bracts. The flower buds are fusiform. The gynaecium includes a white stigma. The fruits are urn-shaped, with a square depression at the apex, and the seeds are transparent (Figure 22.4).

22.3.2 Ethnopharmacology

In Malaysia, the leaves are used to make a postpartum remedy. The pharmacological properties remain unexplored, but one can reasonably envisage tannins to mediate styptic and antiinflammatory effects.

22.4 *DISSOCHAETA ANNULATA* HOOK. F.

[From: Latin *annulata* = marked by or surrounded by rings, as is the stem of this species.]

22.4.1 Botany

Dissochaeta annulata Hook. f. (*Dissochaeta annulata* var. *griffithii* [Nayar] Maxwell, *Macrolenes griffithii* [Nayar]) is a woody twining climber that grows in Thailand, Malaysia, Borneo, and Moluccas. The young stems and leaves are densely covered with rusty hairs. The mature stems are whitish and the internodes are 6–6.3cm. The nodes are swollen and show distinct interpetiolar ridges. The petiole is 5mm–3cm long and densely rusty tomentose. The blade is 6cm × 15cm – 13cm × 4.3cm, and 10cm × 4cm. The secondary nerves are prominent below and show 15 pairs of tertiary nerves. The apex of the blade develops into a 2cm-long tail. The inflorescences are many-flowered paniculate cymes in the upper leaf axils, 18–25cm long with 20–35 flowers. The fruits are urceolate berries, which are woody, about 1cm across and crowned by the four persistent calyx lobes, and densely covered with hairs (Figure 22.5).

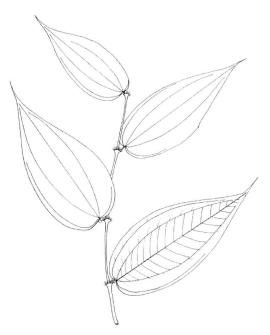

Figure 22.5 *Dissochaeta annulata* Hook. f. [From: Negeri Sembilan, Jelebu District. Alt.: 200m. Foothill logged rain forest. Field collectors: E. Gardette et al., Oct. 1, 1996. EG 2279.]

22.4.2 Ethnopharmacology

The Malays use the plant to make a postpartum protective remedy against infection. The pharmacological properties are unexplored. Tannins are very probably responsible for the medicinal use.

22.5 *DISSOCHAETA BRACTEATA* (JACK) BL.

[From: Latin *bracteate* = bearing bracts.]

22.5.1 Botany

Dissochaeta bracteata (Jack) Bl. is a woody climber that grows in the rain forests of Borneo, the Philippines, Malaysia, Java, and Sumatra. The stems show conspicuous nodes, with 5.3cm-, 4cm-, and 3.5cm-long internodes. The blade is broadly lanceolate, 11cm × 5cm – 13cm × 5.4cm – 10cm × 4.9cm, is cordate at the blade, tailed at the apex, and shows two pairs of secondary nerves and about 17 pairs of tertiary nerves that are prominent below. The flowers are lilac with a white line. The stamens are yellow. The inflorescences consist of 4cm-long axillary panicles. The fruits are purplish-blue, 1.3cm × 7mm on 1.2cm-long pedicels (Figure 22.6).

22.5.2 Ethnopharmacology

The plant is used to make a postpartum protective remedy against infection in Malaysia. The pharmacological properties are unexplored. Like the *Diplectria divaricata* (Willd.) O. Ktze., tannins are probably involved here.

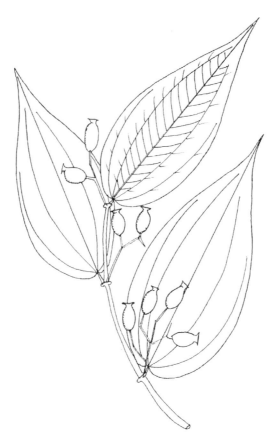

Figure 22.6 *Dissochaeta bracteata* (Jack) Bl. [From: Flora of Pahang. Comm. Ex. Herb. Hort. Bot. Sing. In jungle path to Serundom Mountain Forest, Kuantan. Alt.: 2500–3400ft. Oct. 26, 1975. Botanical identification: J. F. Maxwell, March 1983. Field collector: M. Shah.]

22.6 *DISSOCHAETA PUNCTULATA* HOOK. F. EX TRIANA

[From: Latin *punctum* = a prick, little hole, or puncture.]

22.6.1 Botany

Dissochaeta punctulata Hook. f. ex Triana is a woody climber that grows to a height of 12m in the rain forests of Malaysia. The stem is rusty and scaly. The leaves are simple, opposite, and exstipulate. The blade is conspicuously dotted and grayish below, 7.2cm × 4.1cm – 6.5cm × 4cm – 7cm × 4.2cm, and broadly elliptical. The blade shows two pairs of secondary nerves and about 18 pairs of tertiary nerves that are prominent below. The inflorescences are terminal with 7cm-long panicles. Parts of the flowers are white or purple. The fruits are rusty brown, cup-shaped, and 8mm × 5mm (Figure 22.7).

22.6.2 Ethnopharmacology

The plant is used to make a postpartum protective remedy against infection. The pharmacological properties of this plant are still unexplored.

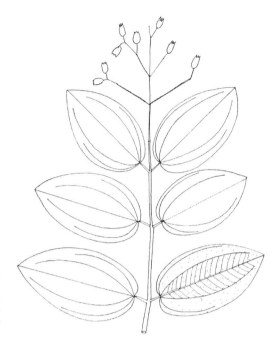

Figure 22.7 *Dissochaeta punctulata* Hook. f. ex Triana. [From: Flora of Malaya. Comm. Ex. Herbarium, Botanic Gardens, Singapore. No: MS 3650. March 13, 1975. Field collector: M. Shah. Geographical localization: Water catch area Mount Ophir.]

22.7 *MEDINILLA HASSELTI* BL.

[After van Hasselt, a Dutch botanist in Indonesia, 19th century.]

22.7.1 Botany

Medinilla hasselti Bl. (*Medinilla rubicunda* [Jack] Bl. var *hasselti* [Bl.] Bakh. f.) is a woody climber that grows in the rain forests of Malaysia. The stems are fissured longitudinally. The leaves are simple, opposite, and show a pair of lateral lobes at the base. The petiole is woody, rough, 9mm × 2 mm, and scurfy. The blade is elliptical, thick, 8.5cm × 2.9cm – 9.2cm × 3cm – 8.5cm × 2.4cm – 11.8cm × 3.6cm – 13.6cm × 3.6cm, and develops a 1.2cm-long tail at the apex. The inflorescences are 1cm long. The androecium comprises about 10 anthers, which are endowed with a connective spur and a lobe at the base of each locule. The fruits are 6mm long (Figure 22.8).

22.7.2 Ethnopharmacology

The leaves are used externally to mitigate headaches in Malaysia. The pharmacological properties of this plant are unexplored.

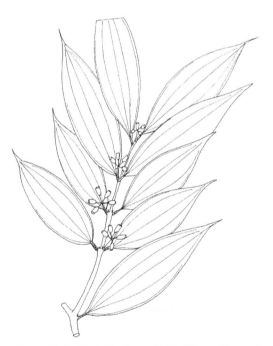

Figure 22.8 *Medinilla hasselti* Bl. [From: Flora of Malaya. Geographical localization: Endau Rompak Expedition, 1985. South Plateau, Endau Johore, in a mossy forest among large sandstone boulders. Bushy plant. Field collector: R. Kiew, April 17, 1986. No: RR2171.]

22.8 *MEDINILLA RADICANS* (BL.) BL.

[From: Latin *radicans* = with rooting stems.]

Figure 22.9 *Medinilla radicans* (Bl.) Bl. [From: Flora of Malaya. Kep. Field No: 98231. Geographical localization: Pahang. Field collector: Chelliah, Jan. 21, 1966. Botanical identification: J. F. Maxwell, March 1, 1983. Compt 39, Gadin Forest Reserve. Primary forest ridge top. Alt.: 2700ft.]

22.8.1 Botany

Medinilla radicans (Bl.) Bl. (synonym: *Melastoma radicans* Blume) is a woody climber that grows on trunks to a height of 18m in the rain forests of Thailand, the Malay Peninsula, Sumatra, and Java. The bark is striated. The stems are pustulate, light gray at the apex, lenticelled, cracked, swollen, and rooting. The leaves are simple, exstipulate, and in whorls of four or five. The petiole is 2–3.2cm × 2mm long. The blade is obovate–elliptical, 14.2cm × 5.7cm – 11.7cm × 5.3cm – 11.7cm × 5.3cm – 11cm × 4.8cm, coriaceous, acute or acuminate at the apex and decurrent, at the base decurrent. The secondary nerves are raised on both the surface and the blade, which shows 11 pairs of tertiary nerves visible only from above. The inflorescences are 1.5cm-long, umbelliform short cymes with 6–8 little flowers with 4.8–10mm-long, oblong, thickened white petals, and eight stamens with connective appendages and a pink calyx. The fruits are 7mm × 5mm, thistle-shaped pink to purple–red (Figure 22.9).

22.8.2 Ethnopharmacology

The Indonesians eat the leaves with a bit of salt to remove blood from feces, most likely because of tannins, which are styptic. The pharmacological properties are as of yet unknown.

22.9 *MELASTOMA POLYANTHUM* BL.

[From: Greek *melastoma* = black mouth, and from *poly* = many and *anthemon* = flower.]

22.9.1 Botany

Melastoma polyanthum Bl. (*Melastoma affine* D. Don., *Melastoma malabathricum* L. ssp. *malabathricum* L., *Melastoma malabathricum* var. *polyanthum* (Bl.) Benth., *Melastoma malabathricum* var. *grandiflorum* Craib, and *Melastoma imbricatum* var. *longipes* Craib) is a shrub that grows to a height of 2.5m in a geographical zone which covers Southeast Asia to Papua New Guinea, New Hebrides, and Australia. Its stem is brown, hairy, and quadrangular, with 4.6cm-long internodes. The leaves are simple, exstipulate, and opposite. The petiole is hairy and 8mm long. The blade is elliptical, dark green above, pale green below, and hairy, 12cm × 3.9cm – 12cm ×

3.5cm – 7cm × 2.3cm – 10.4cm × 2.5cm. The inflorescences are few, flowered, and terminal or in the upper leaf axils. The flowers are 5-merous. The calyx is pinkish, red, and hairy. The petals are bright purple, the style is pink, and the stamen is purplish. The fruits are 1cm × 7mm and hairy (Figure 22.10).

22.9.2 Ethnopharmacology

The Indonesians drink a decoction of the leaves, or swallow the sap squeezed from the leaves, to stop diarrhea. The plant is also used to counteract putrefaction of the genitals, and to heal thrush, burns, ulcers, smallpox sores, wounds, and piles. In the Solomon Islands, the sap is used as a drink to treat infection of the genitals. The pharmacological properties of this plant are unexplored as of yet.

22.10 *MELASTOMA SANGUINEUM* SIMS.

[From: Greek *melastoma* = black mouth and Latin *sanguineum* = bloody, referring to the fruits.]

22.10.1 Botany

Melastoma sanguineum Sims. (*Melastoma sanguineum* Sims., *Melastoma decemfidum* Roxb., *Melastoma dendrisetosum* C. Chen, and *Melastoma sanguineum* var. *latisepalum* C. Chen) is a shrub that grows to a height of 3m in Southeast Asia, China, and Hawaii. The stems are 3mm in diameter, quadrangular, and hirsute at the nodes. The internodes are 2cm long. The leaves are simple, opposite, and exstipulate. The petiole is 1.7cm × 1.5 mm and pilose. The blade is 13cm – 10.5cm – 8.5cm × 3cm – 1.2cm – 1.4cm × 4.3cm. The secondary nerves are prominent below the blade, hairy, and sunken above. No tertiary nerves are visible from above. The flowers are 5-merous, mauve, solitary axillary, or arranged in few-flowered cymes. The petals are 2–5cm long and membranous. The androecium consists of 10 stamens with connectives. The fruits are 2cm long and red, covered with spreading bristles, and a 1.5–2.5cm-long dehiscent capsule (Figure 22.11).

Figure 22.10 *Melastoma polyanthum* Bl. [From: Harvard University Herbaria, Herbarium Bogoriense. Plants of Indonesia. Botanical identification: G. Paoli, April 1994. Geographical localization: Borneo West Kalimantan, Bukit Baka National Park, East Camp along bank of Ella River. 0°37′ S, 112°15′ E. Alt.: 290m. In mixed dipterocarp forest.]

Figure 22.11 *Melastoma sanguineum* Sims. [From: Institute of Systematic Botany, University of Mainz, Germany. Flora of Johor. Comm. Ex. Herbarium, Botanic Gardens Singapore. MS No: 3648. Botanical identification: K. Meyer, 1996. Field collector: M. Shah. Geographical localization: Water catchment area at a damp, Ophir Mount. Alt.: 1000–1500ft.]

22.10.2   Ethnopharmacology

In Cambodia, Laos, and Vietnam, a decoction of the aerial parts is used to treat diarrhea, dysentery, and genital infection. In Cambodia, the roots are used to invigorate, and are boiled to make a drink that is used for vertigo and weakness, and they are also used as an ingredient in a treatment for jaundice. The pharmacological potential of this plant is unexplored as of yet.

22.11 *MEMECYLON DICHOTOMUM* C.B. CLARKE

[From: Greek *Memecylon* = name for *Arbutus unedo* L., the European strawberry tree, because of the resemblance of the fruit, and from Latin *dichotomum* = divided or branched in pairs, in reference to the inflorescences.]

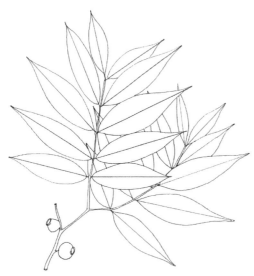

Figure 22.12 *Memecylon dichotomum* C.B. Clarke. [From: Flora of Malaya. FRI No: 0925. Geographical localization: New road of Genting Highlands. Alt.: 3800ft. Botanical identification: J. F. Maxwell, 1980.]

22.11.1   Botany

Memecylon dichotomum C.B. Clarke (*Memecylon ridleyi* Cogn., *Memecylon elegans* Kurz var. dichotoma Cl., *Memecylon eugeniiflora* Ridl., and *Memecylon dichotomum* [Cl.] King var. *eugeniiflorum* [Ridl.] Ridl.) is a tree that grows to a height of 12m with a girth of 30cm. The crown is spreading and the bole is straight or irregular with many nodes. The bark is finely fissured, gray, thin, and scaly. The stems are slender, 1mm in diameter with 2cm-long internodes, and is sharply quadrangular at the nodes. The leaves are simple, opposite, and exstipulate. The blade is 6.7cm × 2.2cm – 4.4cm × 1.9cm – 7.5cm × 3.1cm, and shows a midrib sunken above and five pairs of secondary nerves which are inconspicuous. The flowers are white–pink and arranged in divided or branched pairs. The fruits are 1cm in diameter on 2mm-long pedicles. They are yellow, flushed red, 6mm in diameter, and sweet to eat (Figure 22.12).

22.11.2   Ethnopharmacology

The roots are used as a postpartum remedy in Malaysia where the plant is known as *nipit kulit*. Note that the methanol extracts of *Memecylon malabaricum* leaves inhibited the growth of both Gram-positive and Gram-negative bacteria, and also fungi.[6] Oral administration of an alcoholic extract of the leaves of *Memecylon umbellatum* lowered the serum glucose levels of normal and alloxan-induced diabetic mice.[7] Are tannins involved here? Probably.

22.12 *NEODISSOCHAETA GRACILIS* (JACK) BAKH. F.

[From: Latin *gracilis* = slender.]

22.12.1   Botany

Neodissochaeta gracilis (Jack) Bakh. f. (*Neodissochaeta gracilis* [Jack] Bl., *Dissochaeta gracilis* [Jack] Bl., and *Melastoma gracile* Jack) is a woody climber that grows up to 3–4m long in the rain forests of Thailand, Malaysia, Sumatra, and Borneo. The bark is pale gray. The stems are 4mm in diameter, sparsely to densely covered with red–brown stellate hairs, and show prominent interpetiolar ridges. The internodes are 6cm, 5.2cm, 4.9cm, and 5cm long. The leaves are simple, opposite, and exstipulate. The petiole is hairy and 1.4cm long. The blade is membranaceous, glabrous above, glabrous or with scattered stellate hairs below, acuminate at the apex, 20cm × 4.2cm – 18.3cm × 3cm – 8cm × 4.1cm – 7.2cm × 1.8cm, and shows 18 pairs of tertiary nerves. The inflorescences are 8cm long with many flowered paniculate cymes. The flowers are small, 4-merous, and whitish in the buds. The calyx is brownish. The fruits are 4mm-long berries with a vestigial calyx (Figure 22.13).

22.12.2   Ethnopharmacology

Malays drink a decoction of leaves to counteract the poisonous effects of *Antiaris toxicaria*. The pharmacological properties of this plant are unexplored.

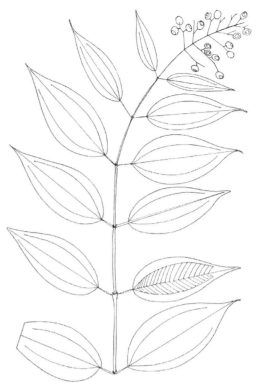

Figure 22.13 *Neodissochaeta gracilis* (Jack) Bakh. f. [From: Sarawak Forest Department. Field collectors: P. C. Yii et al. No: S48867. Geographical localization: Ebau River, Dataran Tinggi Merurong, Jelalong River, 4th Division. In mixed dipterocap forest along ridge at 350m, above sea level.]

22.13 *OSBECKIA CHINENSIS* L.

[After Ozbeck and from Latin *chinensis* = from China.]

22.13.1   Botany

Osbeckia chinensis L. (*Osbeckia japonica* [Naud.], *Osbeckia angustifolia* D. Don., *Osbeckia chinensis* var. *angustifolia* [D. Don.] C. Y. Wu & C. Chen, *Osbeckia parva* Geddes, and *Osbeckia watanae* Craib) is a perennial herb that grows to a height of 70cm in Southeast Asia, China, Japan, New Guinea, and Australia from sea level up to 1600m altitude in grassy areas and deciduous forests. The stems are minutely hairy, squared, and reddish, and the internodes are 20cm. The leaves are opposite up to 3.5cm long, simple, and exstipulate. The blade is narrowly oblong or lanceolate, 1.6cm × 4cm – 3.1cm × 2cm, and hairy with a midrib which is sunken with several prominent nerves running the length of the blade. The flowers are 4-merous, arranged in terminal heads. The corolla is 1.2–1.7cm long and purple. The androecium consists of eight yellow stamens. The corolla is ephemeral and drops on collection. The fruits are campanulate or urceolate capsules, which are 3–5mm long (Figure 22.14).

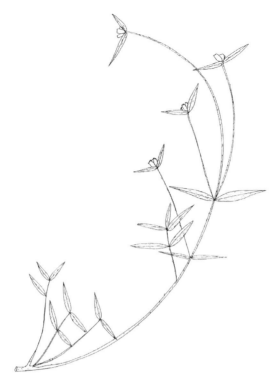

Figure 22.14 *Osbeckia chinensis* L. [From: Forestry and Timber Bureau. Atherton, Queensland, Australia. Field collector: B. Hyland. No: 7015. Oct. 26, 1973. Geographical localization: 10 miles North of Archer River on Kennedy Road. 13°25′ S, 142°50′ E.]

22.13.2   Ethnopharmacology

The plant is known as Chinese *Osbeckia*. In Taiwan, a decoction of the aerial part is used as a drink to treat dysentery. The Filipinos swallow the juice of the roots to alleviate cough and to remove blood from saliva. In Papua New Guinea, the plant affords a remedy for toothache. The plant is known to elaborate a series of hydrolyzable tannins, including casuarinin, casuariin, punicacortein A, and degalloyl-punicacortein A, which showed some levels of antioxidant activity.[8] It would be interesting to assess the plant for any hepatoprotective and immunomodulating properties given that an aqueous extract of the leaves of *Osbeckia aspera* has displayed hepatoprotective effects *in vitro* and *in vivo*. It has also shown inhibitory effects on the complement system and on *in vitro* phagocytosis by polymorphonuclear cells. Nicholl et al.[9] investigated the effect of *Osbeckia aspera* on lymphocyte proliferation using mitogens and antigens and showed that the inhibitory principles in the aqueous extract might act on antigen-presenting cell function. Are tannins or flavonoids involved here?

22.14   *PTERNANDRA COERULESCENS* JACK

[From: Greek *pterna* = heel and *aner* (*andros*) = man, referring to the heel-like extension of the anther connective, and from *coerulescens* = bluish, referring to the petals.]

22.14.1   Botany

Pternandra coerulescens Jack (*Pternandra coerulescens* Jack var. *jackiana* Cl., *Pternandra jackiana* [Cl.] Ridl., *Pternandra capitellana* Jack, *Pternandra coerulescens* Jack var. *capitellata*

[Jack] King, *Pternandra coerulescens* Jack var. *paniculata* [Miq.] King and *Pernandra paniculata* Benth. ex Cl.) is a tree that grows to a height of 20m with a girth of 2.70m in China, Malaysia, Thailand, Borneo, Sumatra, Celebes, Moluccas, Papua New Guinea, and Australia. The bark is finely fissured, thin, and gray to brownish. The inner bark is white and the wood is yellow. The stems are terete. The leaves are simple, opposite, and exstipulate. The petiole is 7mm × 3mm, glabrous, and woody. The blade is 10.5cm × 5.4cm – 11cm × 6cm – 11.2cm × 6.4cm – 9cm × 5.5cm –14cm × 8.2cm; strychnos-like or extremely variable in texture, size, and shape; glabrous; and membranaceous. The secondary nerves are sunken above. The inflorescences are axillary cymes. The corolla comprises four bluish petals. The fruits are 4mm long and cup-shaped, with whitish-green to bluish patterns (Figure 22.15).

Figure 22.15 *Pternandra coerulescens* Jack. [From: Rijskherbarium Leiden, October 1977. Flora of Malaya. FRI No: 13586. Field collector: B. Everett, Oct. 30, 1969.]

22.14.2   Ethnopharmacology

The roots are used by the Malays to make a protective remedy. The pharmacological properties have not yet been explored. Note that tannins are probably responsible for the medicinal use. Tannins abound in the family Rhizophoraceae, which are described in the next chapter.

REFERENCES

1. Yoshida, T., Ito. H., and Hipolito, I. J. 2005. Pentameric ellagitannin oligomers in melastomataceous plants—chemotaxonomic significance. *Phytochemistry*, in press.
2. Yoshida, T., Amakura, Y., Yokura, N., Ito, H., Isaza, J. H., Ramirez, S., Pelaez, D. P., and Renner, S. S. 1999. Oligomeric hydrolysable tannins from *Tibouchina multiflora*. *Phytochemistry*, 52, 1661.
3. Yoshida, T., Arioka, H., Fujita, T., Chen, X. M., and Okuda, T. 1994. Monomeric and dimeric hydrolysable tannins from two melastomataceous species. *Phytochemistry*, 37, 863.
4. Cheng, J. T., Hsu, F. L., and Chen, H. F. 1993. Antihypertensive principles from the leaves of *Melastoma candidum*. *Planta Med.*, 59, 405.
5. Lee, M. H., Lin, R. D., Shen, L. Y., Yang, L. L., Yen, K. Y., and Hou, W. C. 2001. Monoamine oxidase B and free radical scavenging activities of natural flavonoids in *Melastoma candidum* D. Don. *J. Agric. Food. Chem.*, 49, 5551.
6. Hullatti, K. K. and Rai, V. R. 2004. Antimicrobial activity of *Memecylon malabaricum* leaves. *Fitoterapia*, 75, 409.
7. Amalraj, T. and Ignacimuthu, S. 1998. Evaluation of the hypoglycaemic effect of *Memecylon umbellatum* in normal and alloxan diabetic mice. *J. Ethnopharmacol.*, 62, 247.
8. Jeng-De, Su., Toshihiko, O., Kawakishi, S., and Namiki, M. 1988. Tannin antioxidants from *Osbeckia chinensis*. *Phytochemistry*, 27, 1315.
9. Dawn, S., Nicholl, Daniels, H. M., Thabrew, M. I., Grayer, R. J., Simmonds, M. S. J., and Hughes, R. D. 2001. *In vitro* studies on the immunomodulatory effects of extracts of *Osbeckia aspera*. *J. Ethnopharmacol.*, 78, 39.

CHAPTER 23

Medicinal Plants Classified in the Family Rhizophoraceae

23.1 GENERAL CONCEPT

The family Rhizophoraceae (R. Brown in Flinders, 1814 nom. conserv., the Red Mangrove Family) consists of approximately 14 genera and 100 species of tropical trees that are often of mangrove habit (tribe Rhizophoraceae). Members of this family are well known to be tanniferous and to elaborate some series of pyrrolidine, pyrrolizidine, and tropane alkaloids. Rhizophoraceae are easily identified in the field with their stilt roots and viviparous fruits in mangroves which present a long body known as hypocotyls (Figure 23.1). To date, the Rhizophoraceae has received little attention from pharmacologists. There have been a few studies on the pharmacological properties of Rhizophoraceae, and this little body of evidence lends support to the interesting fact that Rhizophoraceae have anti-Human Immunodeficiency Virus (HIV) potential. A significant advance in this regard has been provided by the work of Premanathan et al.[1,2] They screened mangrove plants *in vitro* against Human Immunodeficiency Virus and observed that most of the active plants are from the family Rhizophoraceae. Further study led to the identification of a polysaccharide

Figure 23.1 Fruit of Rhizophoraceae. **(See color insert following page 168.)**

from the leaf of *Rhizophora apiculata* Bl., which inhibited the replication of HIV-1 cultured *in vitro*, blocked the expression of HIV-1 antigen in MT-4 cells, abolished the production of HIV-1 p24 antigen in peripheral blood mononuclear cells, and blocked the binding of HIV-1 virions to MT-4 cells. Another interesting fact about this family is the presence of kaurane and ent-beyerane diterpenes which have exhibited some levels of cytotoxicity.[3] The traditional systems of medicine of the Pacific Rim use about 20 species of Rhizophoraceae of which *Bruguiera sexangula* (Lour.)

Poir, *Carallia brachiata* (Lour.) Merr., *Carallia suffruticosa* King, *Ceriops tagal* (Pers.) C.B. Rob., *Gynochtodes axillaris* Bl., *Rhizophora apiculata* Bl., and *Rhizophora mucronata* Lamk. are presented in this chapter.

23.2 *BRUGUIERA SEXANGULA* (LOUR.) POIR.

[After J. G. Bruguieres, 1750–1798, who worked for the *Encyclopedia Methodica Lamarck,* and from *sexangula* = six-angled.]

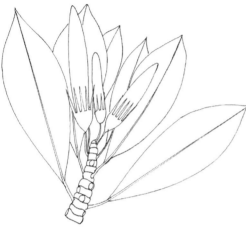

Figure 23.2 *Bruguiera sexangula* (Lour.) Poir. [From: Leiden, Sept. 19, 1966. Flora of Malay Peninsula. Botanical identification: D. Hou. Field collector: B. H. Barnard, Jan. 3, 1923.]

23.2.1 Botany

Bruguiera sexangula (Lour.) Poir. (*Bruguiera eripetala* W. & A. ex Arn.) is a tree that grows up to 33m in height with a girth of 180cm in the mangroves, from Sri Lanka to Papua New Guinea. The bole is buttressed and shows stilt roots. The bark is pale with 2.5cm-diameter lenticels. The stem is rough with scars of leaves, 5mm in diameter. The leaves are simple, decussate, and stipulate, the stipules are lanceolate and 2–4cm long. The petiole is 1.5cm long. The blade is elliptical to elliptical–oblong, 12cm × 8cm, 10.5cm × 3.7cm, black-dotted below. The flowers are yellow, 1–1.2cm, and the calyx is 10–12cm long. The fruits are angular hypocotyls, 7cm × 15cm, and 6–8cm long with blunt ends. The calyx is persistent, 1.4cm × 1.7cm, with 12 triangular lobes which are 1.8cm × 2mm and red. (Figure 23.2).

23.2.2 Ethnopharmacology

The plant is known in Malaysia as *mata buaya* or *tumu puteh* and the fruits are used externally to treat shingles, whereas the roots and leaves are used to treat burns. The pharmacological potential of this plant is unexplored. Are antiviral oligosaccharides involved here? Note that the *Bruguiera* species are known to elaborate a series of diterpenes such as 15(R)-ent-pimar-8(14)-en-1,15,16-triol (Figure 23.3).[4] What are the pharmacological properties of such diterpenes? Cytotoxic?

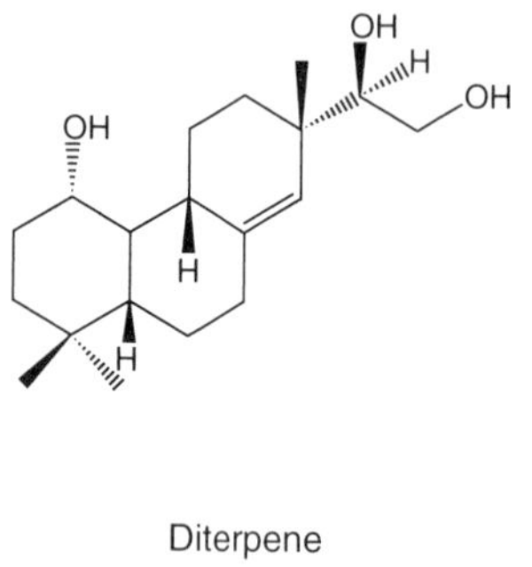

Diterpene

Figure 23.3 Diterpene of *Bruguiera* — 15(R)-ent-pimar-8(14)-en-1,15,16-triol.

23.3 *CARALLIA BRACHIATA* (LOUR.) MERR.

[From: Tamil *karalli* = Indian plant name and from Latin *brachiata* = joined.]

23.3.1 Botany

Carallia brachiata (Lour.) Merr. (*Carallia integerrima* DC., *Carallia lucida* Roxb., *Carallia scortechinii* King, and *Carallia spinulosea* [Ridl.]) is a tree that grows to 33m in height with a

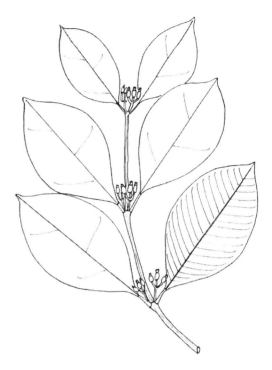

Figure 23.4 *Carallia brachiata* (Lour.) Merr. [From:
Rijskherbarium Leiden, March 22, 1943.
No: 63365. Botanical identification: D.
Hou, Sept. 19, 1966.]

girth of 210cm, and is found in forest swamps and lowland forests from Madagascar to Australia,
including China. The bole shows buttresses. The bark is reddish-brown with gray patches. The
inner bark is yellow–brown with yellow sapwood. The stems are glabrous, smooth, lenticelled,
slightly fissured longitudinally, and show conspicuous nodes with horizontal rings. The internodes
are 3–6cm long. The blade is spathulate–elliptical, 8.2cm × 4.4cm, with 17 pairs of secondary
nerves. The midrib is sunken above, and the margin of the blade is minutely serrate. The flower
pedicels are 4mm long. The inflorescences are 1cm. The fruits are ovoid, open at the apex, and
4mm long (Figure 23.4).

23.3.2 Ethnopharmacology

In Cambodia, Laos, and Vietnam, the plant is used to treat scabies. In Malaysia, the leaves are
used to make a tea that is used in treating septicemia, and the bark is used to treat itch. The plant
is known to elaborate a series of megastigmanes such as 3-hydroxy-5,6-epoxy-β-ionol-3-*O*-β-apio-
furanosyl-(1→6)-β-glucopyranoside (Figure 23.5), flavonoids, hygroline, and tannins. An interesting
development with this plant would be to investigate its potential as a source of antibacterial agents.

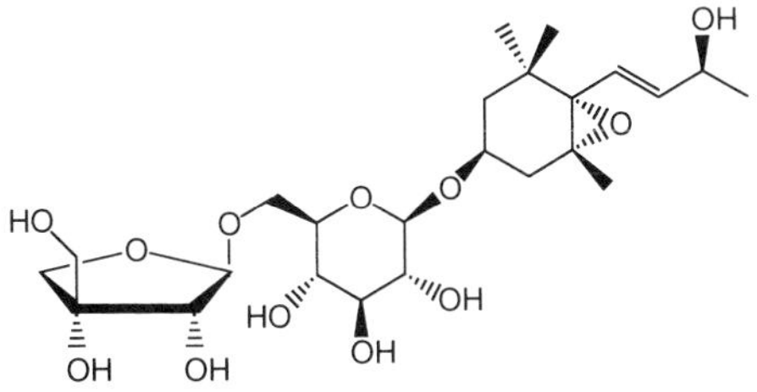

3-hydroxy-5,6-epoxy-β-ionol-3-*O*-β-apiofuranosyl-(1→6)-β-glucopyranoside

Figure 23.5

Figure 23.6 *Carallia suffruticosa* King. [From: Singapore. Field No: 40136. Distributed from The Botanic Gardens Singapore. Geographical localization: Klang Gates, Selangor. Botanical identification: J. Sinclair, Nov. 12, 1953. Field collector: J. Sinclair.]

23.4 *CARALLIA SUFFRUTICOSA* KING

[From: Tamil *karalli* = Indian plant name and from Latin *suffruticosa* = shrubby.]

23.4.1 Botany

This plant is a tree that grows up to 9m in Cambodia, Laos, Vietnam, and Malaysia. The bark is greenish-brown with prominent lenticels. The stems are slender and 2mm in diameter. The leaves are decussate, simple, and stipulate; the stipule is lanceolate. The petiole is 4–6mm long and grooved above. The blade is papery and lanceolate, 15.4cm × 5.7cm – 13cm × 4.7cm – 9cm × 3.3cm, toothed, and has 10 pairs of secondary nerves, a few of which are tertiary. The midrib is sunken above. The flowers are yellow, in cymes, and small. The fruits are red. The fruit is axillary, 7mm × 4mm, and the stipule is 1.1cm × 3 mm (Figure 23.6).

23.4.2 Ethnopharmacology

The Malays call the plant *tulang daeng*. They mix its leaves with water and apply the paste to boils. It also reduces fever. A decoction is used as a drink to expel worms from the intestines, and to recover from the exhaustion of childbirth. The pharmacological potential of this plant is unexplored. The plant is rare and might disappear soon.

23.5 *CERIOPS TAGAL* (PERS.) C.B. ROB.

[From: Greek *ceriops* = horn bearing, referring to the extended hypocotyls, and from Filipino *tagal* = plant name.]

23.5.1 Botany

Ceriops tagal (Pers.) C.B. Rob. (*Ceriops candolleana* Arn., *Ceriops timoriensis,* and *Ceriops boiviniana*) is a tree that grows up to 35m high and 20cm in diameter. It is found in the mangroves of a zone that spans East Africa to Micronesia. The leaves are simple, decussate, stipulate, and gathered at the apex of stems. The petiole is 1.5–2.5cm long. The blade is obovate to spathulate, 2.8cm × 1.2cm – 7.5cm × 4.2cm – 6.2cm × 3.2cm. The inflorescences consist of clusters of 2–10 flowers. The hypocotyls are club-shaped, angled, and 35cm × 5mm (Figure 23.7).

23.5.2 Ethnopharmacology

Yellow Mangrove, or Spur Mangrove, is used as an astringent and for tanning in tropical Asia. In the Philippines, the plant is called *tangal* or *tagal* and used is to treat diabetes. In Malaysia,

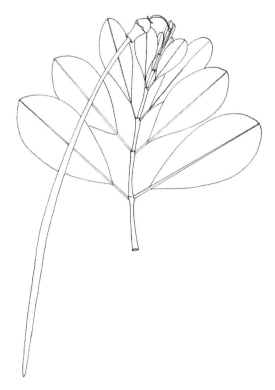

Figure 23.7 *Ceriops tagal.* (Pers.) C.B. Rob. [From: Flora of Malaya. KLU Herbarium No: 42. Nov. 28, 1987. Geographical localization: Sementa, Kelang Selangor.]

an infusion of the barks (*tengar*) is used to assuage abdominal pains after childbirth. The plant is known to contain a series of diterpenes of which tagalsin A-H did not affect the survival of HL-60, Bel-7402, and HeLa cells cultured *in vitro* (Figure 23.8).[6]

23.6 *GYNOTROCHES AXILLARIS* BL.

[From: Greek *gune* = woman, *trochos* = wheel, from the shape of the stigma, from and Latin *axillaris* = axillary, positioned in the leaf axils, referring to the flowers.]

23.6.1 Botany

Gynotroches axillaris Bl. is a tree that grows to a height of 36m with a girth of 1.8m in the lowland and swamp rain forests of Burma, Thailand, south to Australia and the Pacific Islands at an altitude of up to 1400m. The bole is buttressed. The bark is grayish, smooth, lenticelled, cracking, and finely fissured. The inner bark is reddish and fibrous. The sapwood is yellowish. The wood is moderately hard. The stems are hollowed, 3mm in diameter, swollen at the nodes, with 4–5cm-long internodes. The leaves are simple, decussate, and stipulate. The petiole is 1.4cm long. The stipules are lanceolate, 1.5cm long, imbricate with one margin free. The blade is lanceolate, 16cm × 6.8cm – 12.5cm × 5cm, and shows 6–8 pairs of secondary nerves. The tertiary nerves and reticulations are raised below. The margin is wavy to faintly toothed when young. The apex is pointed or blunt, and the base is acute to rounded. The inflorescences are axillary clusters of little greenish-white flowers. The calyx is deeply 4–5 lobed. The corolla consists of 4–5 petals, obovate or elliptic, divided into filamentous appendages at the apex. The flower shows a nectary disc which

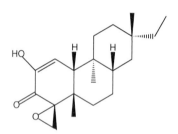

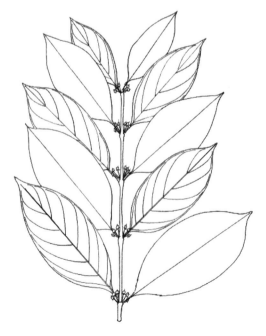

Figure 23.8 Dolabrane-type diterpene derivatives of *Ceriops tagal* (Pers.) C.B. Rob.

is 8–10 lobed. The gynaecium is 4–5 celled. The stigma is discoid and 4–8 lobed. The flower pedicel is 5mm long. The fruits are globose berries which are 5mm × 7mm – 3mm × 5mm, red, ripening black, and glossy with persistent calyx lobes (Figure 23.9).

23.6.2 Ethnopharmacology

In Malaysia, the plant is called *sebor chet* (Jahut) and *mata keli* (Malay). Its leaves are used externally to reduce fever and to assuage headache. It would be interesting to evaluate the pharmacological properties of this common medicinal plant.

Figure 23.9 *Gynotroches axillaris* Bl. [From: Flora of Malaya. April 1, 2005. Geographical localization: Selangor, Gombak, FRIM, Kapur road in plantation forest. 3°14′ N, 10°38′ E. Alt.: 50m.]

23.7 *RHIZOPHORA APICULATA* BL.

[From: Greek *rhizo* = root, *phoros* = bearing, and from Latin *apiculata* = ending somewhat abruptly in a short or sharp point or apex, referring to the blade.]

23.7.1 Botany

Rhizophora apiculata Bl. (*Rhizophora cadelaria* DC. and *Rhizophora conjugata* [non L.] Arn.) is a tree that grows to a height of 30m with a girth of 1.5m throughout the mangroves and deep soft-mud estuaries of the Asia–Pacific. The bole produces stilt roots. The bark is gray with shallow horizontal cracks. The stems are somewhat swollen, marked by conspicuous leaf scars, 1cm in diameter, and annular. The stipules are 4–8cm long. The petiole is 1.5–3cm long and reddish. The blade is elliptical–oblong, 7cm × 18cm × 3cm × 8cm, and reddish at the margin. The base is cuneate and the apex is apiculate. The flowers are green to yellowish, and arranged in pairs on 5mm × 1.5cm-long pedicels with the pedicel arising from the leafless parts of the stems. The hypocotyls are club-shaped, cylindrical with a blunt tip, smooth, and up to 40cm long and 1.5cm in diameter, brown with a few large lenticels (Figure 23.10).

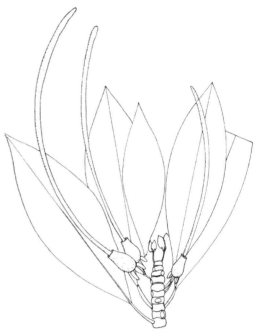

Figure 23.10 *Rhizophora apiculata* Bl. [From: Royal Botanic Gardens, Kew, July 19, 1977. Flora of Malaya. Field FRI No: 16178. Geographical localization: Cap Rachedo Forest Reserve, Negeri Sembilan.]

23.7.2 Ethnopharmacology

The vernacular name for this species is Tall-Stilted Mangrove. In Indonesia and Malaysia, the plant is called *bakau minyak* and is used to treat dysentery, probably on account of its astringency. The plant contains polysaccharides with anti-HIV activity *in vitro*.[1,2]

23.8 *RHIZOPHORA MUCRONATA* LAMK.

[From: Greek *rhizo* = root, *phoros* = bearing, and from Latin *mucronata* = sharply pointed, with regard to the blade.]

23.8.1 Botany

Rhizophora mucronata Lamk. is a tree that grows to a height of 30m with a girth of 2.1m throughout the mangroves of the Asia–Pacific and East Africa. The bark is blackish with grid cracks or horizontal fissures. The stems are 7mm in diameter, regularly annulate, and conspicuously marked with leaf scars. The leaves are decussate, simple, and stipulate. The stipules are 5.5–8.5cm long. The blade is elliptic and 16cm × 8.7cm, 15cm × 8cm, 11.5cm × 5.3cm. The inflorescences are 4.6cm + 1.5cm + 1cm. The flowers exist in pairs on 2.5–0–5cm-long pedicels. The petals are velvety outside. The flower buds are 1.5cm × 4mm. The hypocotyls are 60cm × 1.5cm, cylindrical, and warty (Figure 23.11).

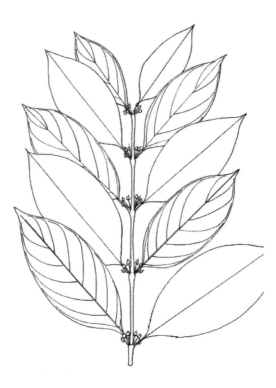

Figure 23.11 *Rhizophora mucronata* Lamk. [From: Forest Department Malaya. Geographical localization: Klang Offshore Island. FRI No: 119971. Field collector: F. S. P. Ng, Feb. 17, 1966. Botanical identification: F. S. P. Ng, April 24, 1966.]

23.8.2 Ethnopharmacology

In Burma, the bark is used to remove blood from urine. In Japan and China, a decoction of the bark affords a treatment for diarrhea. In Cambodia, Laos, and Vietnam, the plant is used to check bleeding. The Malays drink a decoction of the leaves after childbirth to counteract infection. The plant would be worth investigation for anti-HIV activity *in vitro*.[1] It is known to produce a series of secolabdane and beyerane diterpenes as well as sesquiterpenes and triterpenes, although the pharmacological properties are still unexplored.[7–9]

REFERENCES

1. Premanathan, M., Nakashima, H., Kathiresan, K., Rajendran, N., and Yamamoto, N. 1996. *In vitro* anti-human immunodeficiency virus activity of mangrove plants. *Indian J. Med. Res.*, 103, 278.
2. Premanathan, M., Arakaki, R., Izumi, H., Kathiresan, K., Nakano, M., Yamamoto, N., and Nakashima, H. 1999. Antiviral properties of a mangrove plant, *Rhizophora apiculata* Blume, against human immunodeficiency virus. *Antiviral Res.*, 44, 113.
3. Han, L., Huang, X., Sattler, I., Dahse, H. M., Fu, H., Lin, W., and Grabley, S. 2004. New diterpenoids from the marine mangrove *Bruguiera gymnorrhiza. J. Nat. Prod.*, 67, 1620.
4. Subrahmanyam, C., Rao, B. V., Ward, R. S., Hursthouse, M. B., and Hibbs, D. E. 1999. Diterpenes from the marine mangrove *Bruguiera gymnorhiza. Phytochemistry*, 51, 83.
5. Ling, S. K., Takashima, T., Tanaka, T., Fujioka, T., Mihashi, K., and Kouno, I. 2004. A new diglycosyl megastigmane from *Carallia brachiata. Fitoterapia*, 75, 785.

6. Zhang, Y., Deng, Z., Gao, T., Proksch, P., and Lin, W. 2005. Tagalsins A–H, dolabrane-type diterpenes from the mangrove plant, *Ceriops tagal. Phytochemistry*, 66, 1465.
7. Anjaneyulu, A. S. and Rao, V. L. 2001. Rhizophorin A, a novel secolabdane diterpenoid from the Indian mangrove plant *Rhizophora mucronata. Nat. Prod. Lett.*, 15, 13.
8. Anjaneyulu, A. S., Anjaneyulu, V., and Rao, V. L. 2002. New beyerane and isopimarane diterpenoids from *Rhizophora mucronata. J. Asian Nat. Prod. Res.*, 4, 53.
9. Laphookhieo, S., Karalai, C., and Ponglimanont, C. 2004. New sesquiterpenoid and triterpenoids from the fruits of *Rhizophora mucronata. Chem. Pharm. Bull. (Tokyo)*, 52, 883.

CHAPTER 24

Medicinal Plants Classified in the Family Olacaceae

24.1 GENERAL CONCEPT

The family Olacaceae (Mirbel ex A. P. de Candolle, 1824 nom. conserv., the Olax Family) consists of approximately 30 genera and 250 species of plants widespread in tropical and subtropical regions. Of these plants, *Olax scandens* Roxb., *Anacolosa griffithii* Mast., *Ochanostachys amentacea* Mast., *Scorodocarpus borneensis* Becc., *Strombosia philippinensis* (Baill.) Rolfe, and *Ximenia americana* L. are of medicinal value in the Asia–Pacific. Olacaceae are classically known to produce tannins, cyanogenetic glycosides, polyacetylenic fatty acids, flavonoids, and an unusual series of polysulfides (Figure 24.1).

Polysulfides are particularly abundant in the *Scorodocarpus*, *Olax*, *Ochanostachys*, and *Ximenia* species. These include 2,4,5,7-tetrathiaoctane 4,4-dioxide, which imparts the plant's pungency and garlic odor. These polysulfides are somewhat similar to that of the *Allium* species (garlic, onions) and are antimicrobial and cytotoxic. Examples of commercial medicinal products from Olacaceae are the roots of *Ptychopetalum olacoides* Benth. (PO), known as *muira*

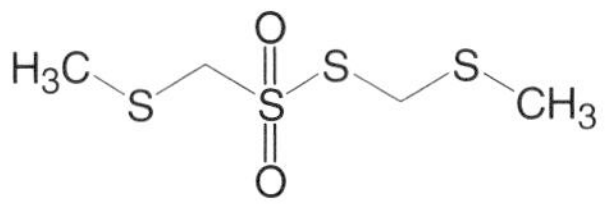

2,4,5,7-Tetrathiaoctane 4,4-dioxide

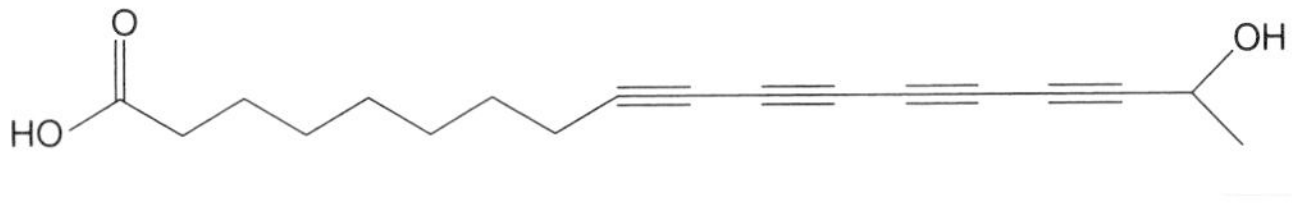

Minquartynoic acid

Figure 24.1 Examples of bioactive natural products from the family Olacaceae.

puama in the Brazilian Amazon where it is used as a tonic. The plant is included among phytopharmaceutical products which are claimed to enhance physical and mental performance although displaying anticholinesterase activity.[1,2] The evidence so far presented suggests that the Olacaceae represent a reserve of chemicals the antiviral potential of which would be worth investigating.

24.2 *OCHANOSTACHYS AMENTACEA* MAST.

[From: Greek *okanos* = shield handle and *stachys* = spike, and from Latin *amentacea* = in the form of a *catkin*.]

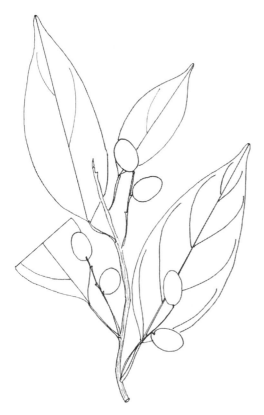

Figure 24.2 *Ochanostachys amentacea* Mast. [From: Federated Malay States. Geographical localization: Public Garden, Kuala Lumpur. Alt.: 100ft. July 2, 1918. Field collector: A. B. Ahmad. No: 2466.]

24.2.1 Botany

Ochanostachys amentacea Mast. is a tree that grows to a height of 30m with a girth of 1.8m in the rain forests of Malaysia. Its wood is very hard and durable, so it is useful as timber. The bole is fluted and buttressed. The bark is pale brown with a purplish tinge, and has round to oblong, thin, adherent scales. The inner bark is yellow to orange with minute drops of milky latex. The stems are fissured longitudinally. The leaves are simple, spiral, and exstipulate. The petiole is 2–2.5cm long. The blade is 11cm × 3.5cm – 7cm × 3.9cm. The base is broadly wedged, and the apex is pointed. The blade shows five pairs of secondary nerves sunken above with tertiary nerves. The inflorescences are 4cm-long axillary spikes. The fruits are 2.3cm × 1.5cm, ovoid, and glossy (Figure 24.2).

24.2.2 Ethnopharmacology

In Malaysia, the plant is known as *petaling*. A decoction of the bark is used to bathe the body after childbirth and to reduce fever. A remarkable advance in the pharmacology of *Ochanostachys amentacea* Mast. has been provided by the work of Rashid et al.[3] They isolated polyacetylenic acid, 17-hydroxy-9,11,13,15-octadecatetraynoic acid, or minquartynoic acid, which protected human lymphoblastoid cells cultured *in vitro* against killing by Human Immunodeficiency Virus (HIV)-1 with an IC_{50} value of 2–5μg/mL. 17-hydroxy-9,11,13,15-octadecatetraynoic acid (minquartynoic acid) is cytotoxic against a panel of human tumor cell lines.[4]

24.3 *XIMENIA AMERICANA* L.

[From: Latin *americana* = from America.]

24.3.1 Botany

Ximenia americana L. is a treelet that grows up to 4m tall in sandy areas behind beaches and along seashores, and up to an altitude of 100m in India, Indonesia, Malaysia, Myanmar, the Philippines, Sri Lanka, Thailand, Africa, America, Australia, and the Pacific Islands. The stems are fissured, spiny, and showy. The leaves are simple, spiral, and exstipulate. The petiole is 3–5mm long. The blade is 4cm × 2.8–2.7cm × 1.5cm, spathulate, and thick, with six pairs of secondary nerves. The inflorescences are cymes or racemes which are axillary, 1.5–2.5cm long, and 3–6-flowered. The corolla comprises four or five, white or greenish petals, which are 5–7mm oblong. The petals are green outside, and yellow inside with white whiskers. The androecium comprises 8–10 stamens. The fruits are green, orange, ovoid drupes which are 2–3cm in diameter. The fruits are dispersed both by birds eating the succulent mesocarp and by water, as the endocarp contains air spaces and is able to float for a long period of time (Figure 24.3).

24.3.2 Ethnopharmacology

In Indonesia, the roots are reduced to a paste which is used to treat colic. The seeds are eaten to induce purgation of the bowels. *Ximenia americana* L. contains hydrocyanic acid which causes cyanide poisoning. Voss et al.[5] drew attention to the fact that the plant elaborates a series of proteins which inhibit the survival of a broad spectrum of cancer cells cultured *in vitro* from rodent colorectal cancer. These proteins could be responsible for the anti-HIV properties displayed by a stem bark extract of the plant.[6]

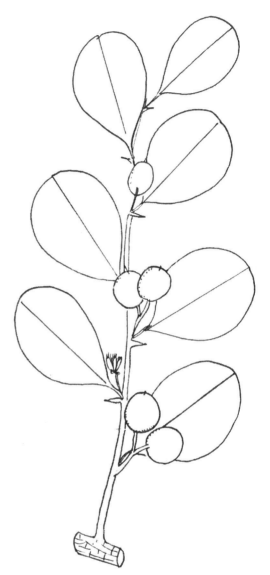

Figure 24.3 *Ximenia americana* L. [From: Comm. Ex. Herb. Hort. Bot. Sing. Geographical localization: Jerjak Island. Feb. 2, 1976. Seaside.]

24.4 *STROMBOSIA PHILIPPINENSIS* (BAILL.) ROLFE

[From: Greek *strombos* = circle, referring to the fruits, and from Latin *philippinensis* = from the Philippines.]

24.4.1 Botany

Strombosia philippinensis (Baill.) Rolfe is a tree that grows to a height of 20m and a diameter of 24cm in the rain forests of the Philippines. The bark is chocolate with light brown irregular specks. The wood is used for timber and tinder. The petiole is 1.2–1.6cm long. The leaves are simple, spiral, and exstipulate. The blade is 13.8cm × 6cm – 10cm × 7cm and elliptic. The apex is acuminate, with 4–6 pairs of secondary nerves. The corolla is caducous and the fruits are drupes (Figure 24.4).

24.4.2 Ethnopharmacology

In the Philippines the plant is named *tamoya* and used as an antidote for Lophopetalum poisoning. The pharmacological potential of *Strombosia philippinensis* (Baill.) Rolfe and the *Strombosia* species in general is unknown.

Figure 24.4 *Strombosia philippinensis* (Baill.) Rolfe. [From: Plants of The Philippines. Luzon, Sierra Madre Mountain Range (East Foothills), Isabella Province, Palanan Municipality, Narangy, San Isisdoro, Sitio Diago. 17°0.7'7" N, 12°30.9' E. In a lowland dipterocarp forest. Alt.: 10–20m.]

REFERENCES

1. Siqueira, I. R., Fochesatto, C., da Silva, A. L., Nunes, D. S., Battastini, A. M., Netto, C. A., and Elisabetsky, E. 2003. *Ptychopetalum olacoides,* a traditional Amazonian "nerve tonic," possesses anticholinesterase activity. *Pharmacol. Biochem. Behav.,* 75, 645.
2. Silva, A. L., Bardini, S., Nunes, D. S., and Elisabetsky, E. 2002. Anxiogenic properties of *Ptychopetalum olacoides* Benth. (Marapuama). *Phytother. Res.,* 16, 223.
3. Rashid, M. A., Gustafson, K. R., Cardellina, J. H., and Boyd, M. R. 2001. Absolute stereochemistry and anti-HIV activity of minquartynoic acid, a polyacetylene from *Ochanostachys amentacea. Nat. Prod. Lett.,* 15, 21.
4. Ito, A., Cui, B., Chavez, D., Chai, H. B., Shin, Y. G., Kawanishi, K., Kardono, L. B., Riswan, S., Farnsworth, N. R., Cordell, G. A., Pezzuto, J. M., and Kinghorn, A. D. 2001. Cytotoxic polyacetylenes from the twigs of *Ochanostachys amentacea. J. Nat. Prod.,* 64, 246.
5. Voss, C., Eyol, E., and Berger, M. R. 2005. Identification of potent anticancer activity in *Ximenia americana* aqueous extracts used by African traditional medicine. *Toxicol. Appl. Pharmacol.,* in press.
6. Asres, K., Bucar, F., Kartnig, T., Witvrouw, M., Pannecouque, C., and De Clercq, E. 2001. Antiviral activity against human immunodeficiency virus type 1 (HIV-1) and type 2 (HIV-2) of ethnobotanically selected Ethiopian medicinal plants. *Phytother. Res.,* 15, 62.

CHAPTER 25

Medicinal Plants Classified in the Family Icacinaceae

25.1 GENERAL CONCEPT

The family Icacinaceae (Miers, 1851 nom. conserv., the Icacina Family) consists of approximately 50 genera and 400 species of trees, shrubs, and woody climbers of tropical distribution, which are known to produce purine and monoterpenoid isoquinoline alkaloids, iridoids, saponins, and proanthocyanins. An interesting feature of the family is the presence of monoterpenoid alkaloids, which are well known for their emetic, amebicidal, antiviral, and cytotoxic properties (Figure 25.1). Camptothecin is a cytotoxic monoterpene indole alkaloid produced by *Nothapodytes foetida* (Wight) Sleumer, *Pyrenacantha klaineana* Pierre ex Exell & Mendonca, *Merrilliodendron megacarpum* (Hemsl.) Sleumer, elaborate *O*-acetylcamptothecin, camptothecin, and 9-methoxycamptothecin, all of which inhibit the survival of KB cells at very small doses.[1] Camptothecin is planar and alleviates the enzymatic activity of topoisomerase normally responsible for the isomerization of DNA during replication. It is the precursor of irinotecan (Campto®), which is used in the treatment of cancers of the lung, colon, cervix, and ovaries. It is presently the treatment of choice used in combination with fluoropyrimidines as first-line therapy for patients with advanced colorectal cancer, cervical cancer, ovarian cancer, and malignant gliomas.[2]

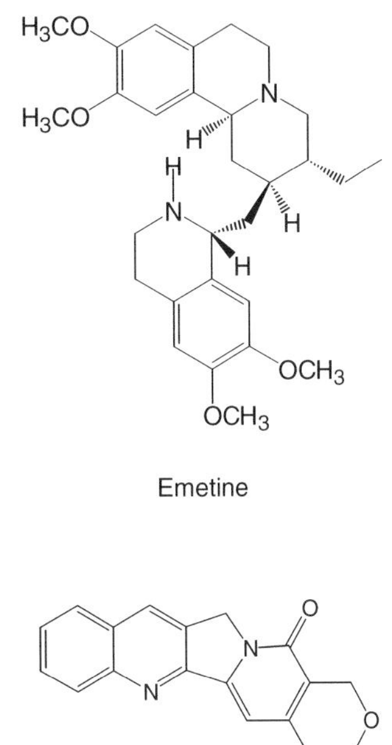

The combined sales of irinotecan and topotecan (Hycamtin®) were expected to reach $1 billion.[3] The question arises as to whether other camptothecin-like alkaloids are present in other members of the family, and an interesting development would be to focus on alkaloids of Icacinaceae for cytotoxic and antiviral activity. *Gonocaryum subrostratum* Pierre, *Gonocaryum gracile* Miq., *Gonocaryum calleryanum* (Baill.) Becc., *Gomphandra quadrifida* (Bl.) Sleumer var. *angustifolia* (King) Sleumer, *Gomphandra quadrifida* (Bl.) Sleumer var. *ovalifolia* (Ridl.) Sleumer, and *Rhyticarium* sp. are used for medicinal purposes in the Asia–Pacific.

Emetine

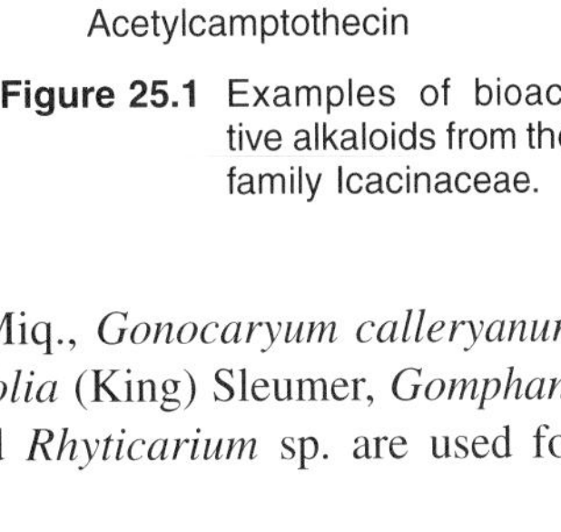

Acetylcamptothecin

Figure 25.1 Examples of bioactive alkaloids from the family Icacinaceae.

25.2 *GONOCARYUM GRACILE* MIQ.

[From: Greek *gonia* = corner and *karion* = nut, and from Latin *gracile* = slender.]

Figure 25.2 *Gonocaryum gracile* Miq. [From: Flora of Malaya. FRI No: 8679. Geographical localization: Johor, Kluang Forest. Field collector: T. C. Whitmore, May 10, 1968. In Virgin Jungle Reserve, Banks of Sungai Pahang.]

25.2.1 Botany

Gonocaryum gracile Miq. (*Gonocarium longeracemosum* King) is a treelet that grows to a height of 12m in the rain forests of Sumatra, Malaysia, and Borneo. The leaves are simple, alternate, and exstipulate. The petiole is yellowish and wrinkled transversely, it is 8mm × 1cm long. The blade is elliptical–oblong, 8cm × 18cm × 3.5cm × 6.5cm, pointed at the apex, cuneate at the base, and showing 4–6 pairs of secondary nerves. The flowers are arranged in spikes, which are axillary. The corolla is tubular. The fruits are ovoid, pointed at the apex, and obtusely trigonous, each of the three faces with two longitudinal ribs (Figure 25.2).

25.2.2 Ethnopharmacology

The Malays apply a paste of fruits to the forehead to mitigate headaches. The pharmacology is unexplored. Counterirritancy is probable here.

REFERENCES

1. Wu, T. S., Leu, Y. L., Hsu, H. C., Ou, L. F., Chen, C. C., Chen, C. F., Ou, J. C., and Wu, Y. C. 1995. Constituents and cytotoxic principles of *Nothapodytes foetida. Phytochemistry*, 39, 383.
2. Lorence, A. and Nessler, C. L. 2004. Camptothecin, over four decades of surprising findings. *Phytochemistry*, 65, 2735.
3. Oberlies, N. H. and Kroll, D. J. 2004. Camptothecin and taxol: historic achievements in natural products research. *J. Nat. Prod.*, 67, 129.

CHAPTER 26

Medicinal Plants Classified in the Family Euphorbiaceae

26.1 GENERAL CONCEPT

The family Euphorbiaceae (A. L. de Jussieu, 1789 nom. conserv., the Spurge Family) consists of 300 genera and approximately 7500 species of trees, shrubs, herbs, climbers, and even cactus-shaped plants, often exuding a milky poisonous latex, and known to produce aporphine, pyridine, indole, and tropane-type alkaloids, lignans, phloroglucinol derivatives, various sorts of terpenes, ellagitannins, proanthocyanins, cyanogen glycosides, anthraquinones, and fatty acid epoxides. When collecting Euphorbiaceae, one is advised to look for laticiferous plants with conspicuously 3-lobed capsules (Figure 26.1), but caution must be taken as members of the *Aleurites*, *Croton*, *Euphorbia*, *Excoecaria*, *Hippomane*, *Hura*, and *Jatropha* species elaborate complex diterpenoid esters of the tigliane, ingenane, or daphnane type which are drastic cathartics, which cause intense contact inflammation, and are both tumor-promoting and antitumor agents. One such compound is 12-*O*-tetradecanoylphorbol-13-acetate, which is one of the most potent known inducers of skin tumor in mice. As a pharmacological tool it is valuable because it activates the phosphorylation enzyme, protein kinase C. Note that these diterpenes display interesting anti-Human Immunodeficiency Virus (HIV) activity *in vitro* (Figure 26.2).[1]

A classical example of a pharmaceutical product used in Western medicine is the oil expressed from the seeds of *Ricinus communis* L. (Castor Oil, *British Pharmacopoeia*, 1963), which has been used for a very long time to relieve the bowels from constipation and to induce labor. Euphorbiaceae contain proteins (phytoxin), which are among the most violent existing poisons, such as curcin from *Jatropha curcas*, and ricin from *Ricinus communis* L. The fatal dose of ricin by intravenous injection of animals has been reported to be as low as 0.3μg/Kg. (*Extra Pharmacopoeia*, Martindale, 25th Edition). About 150 species of plants classified within the Euphorbiaceae family are used for medicinal purposes in the Pacific Rim, mostly to relieve the bowels from costiveness, to promote urination, to soothe inflammation, and to promote expectoration. It will be interesting to learn whether a more intensive study of Euphorbiaceae discloses any molecules of therapeutic interest. Note that hydrolyzable tannins and diterpenes are predominantly responsible for the medicinal properties of Euphorbiaceae.

Figure 26.1 Fruits of Euphorbiaceae. **(See color insert following page 168.)**

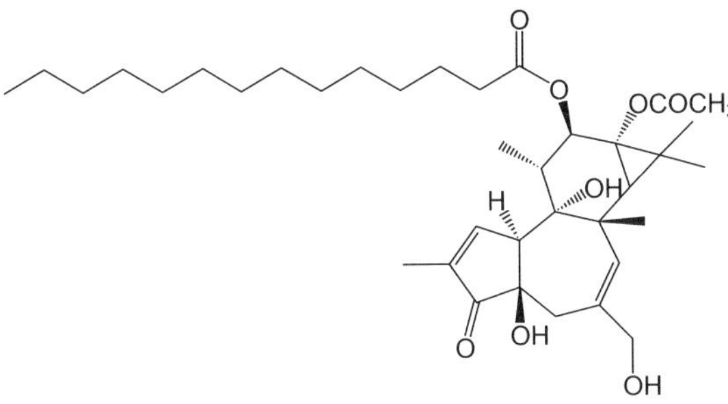

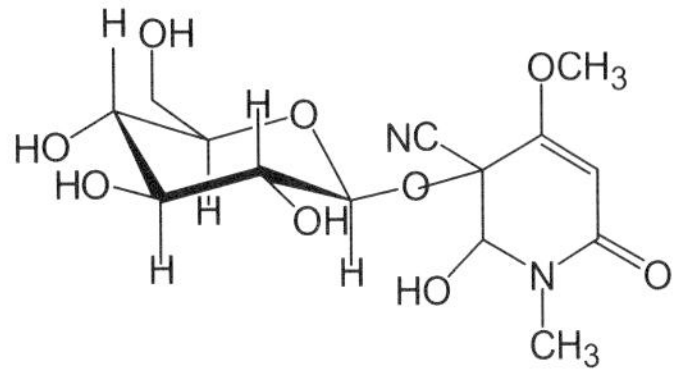

12-*O*-tetradecanoylphorbol-13-acetate   Acalypphine

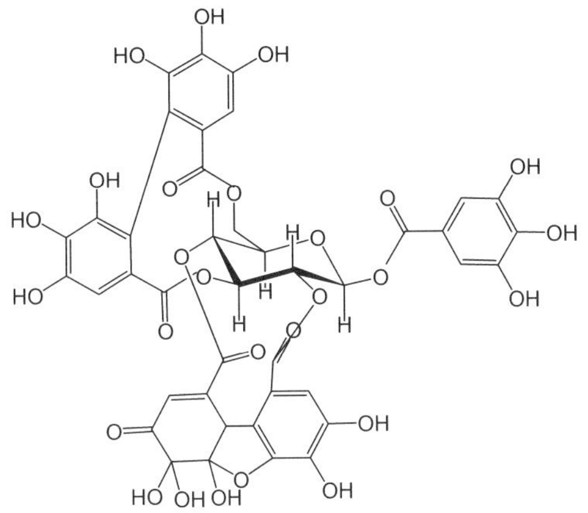

Geraniin

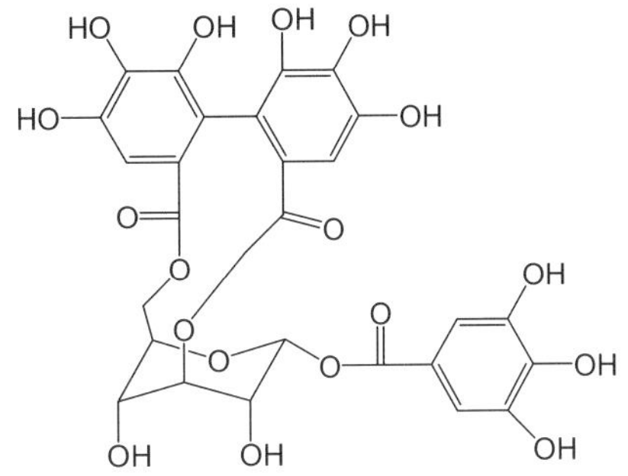

Isocorilagin

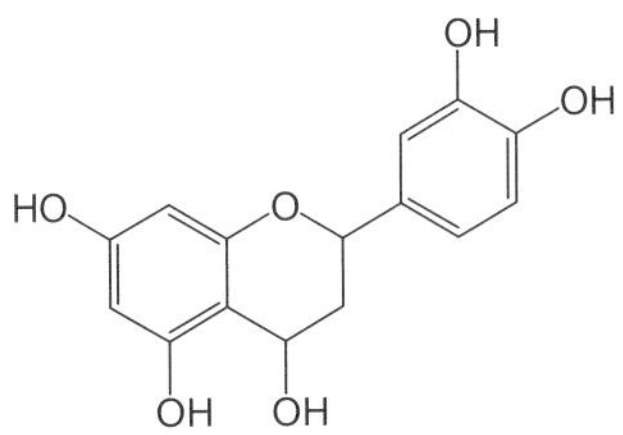

Luteoforol

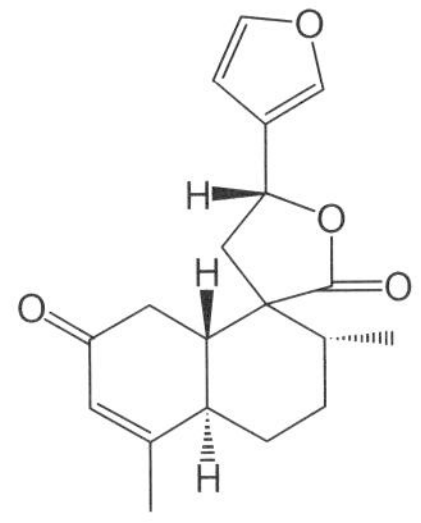

Dehydrocrotonin

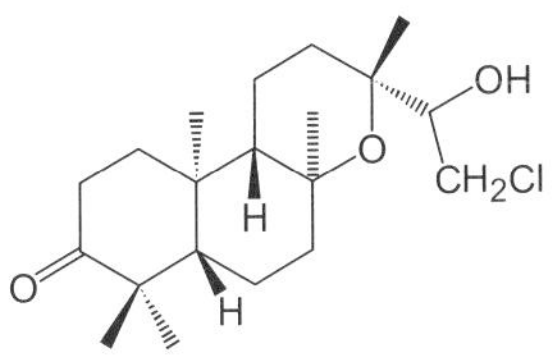

Agallochin A

Figure 26.2 Examples of bioactive natural products from the family Euphorbiaceae.

26.2 *ANTIDESMA GHAESEMBILLA* GAERTN.

[From: Greek *anti* = against and *desma* = a constriction.]

26.2.1 Botany

Antidesma ghaesembilla Gaertn. (*Antidesma frutescens* Jack, *Antidesma paniculatum* Bl., and *Antidesma pubescens* Roxb.) is a treelet that grows to a height of 6m in the secondary forests near the rivers of India, Ceylon, Southeast Asia, China, the Himalayas, Nepal, Australia, and the Pacific Islands. The leaves are simple, spiral, and stipulate. The stipules are small. The blade is ovate to oblong–ovate, 5cm × 3.5cm, and show four pairs of secondary nerves. The inflorescences consist of spikes which are 2.5–5cm long (Figure 26.3).

26.2.2 Ethnopharmacology

In Cambodia, where the plant is called *choi moi*, a decoction of the bark is used as a drink to treat diarrhea, to promote menses, to recover from childbirth, and to invigorate. The leaves are used externally to assuage headaches in children. In Malaysia, a paste of the leaves is applied externally to relieve headache, skin diseases, and abdominal swelling. The leaves are also used to make a water bath to reduce fever. In

Figure 26.3 *Antidesma ghaesembilla* Gaertn. [From: Flora of Malacca. Distributed from the Herbarium Botanic Gardens Singapore. Field collector: H. M. Burkill. Botanical identification: H. M. Burkill. Geographical localization: Five miles south of Malacca, Berendam Road. Tidal freshwater, Gelam Swamp.]

the Philippines, the leaves are used to promote the healing of wounds. The pharmacological property of this plant is unexplored. Note that the *Antidesma* species are known to elaborate a series of unusual quinoline and cyclopeptide alkaloids (Figure 26.4).[2] An example of a quinoline alkaloid is antidesmone, which inhibits the growth of *Cladosporium cucumerinum* cultured *in vitro*.[3–6]

Note that the general chemical profile of antidesmone and congeners strongly suggest some potential as cytotoxic agents via mitochondrial inhibition as well as central nervous system activity.

26.3 *EUPHORBIA THYMIFOLIA* L.

[Named for Euphorbus, Greek physician of Juba II, King of Mauretania, and from Latin *thymifolia* = thyme-like leaves.]

26.3.1 Botany

Euphorbia thymifolia L. (*Chamaesyce thymifolia* [L.] Millsp.) is an annual herb that grows on roadsides, grasslands, and vacant plots of land in the Asia–Pacific. The stems are slender, and up to 20cm long, thin, usually prostrate, and finely articulate. The roots are fibrous and long. The leaves are simple, opposite, and stipulate. The stipules are lanceolate, 1–1.5mm long, and caducous. The petiole is 1mm long. The blade is asymmetrical, membranaceous, cordate at the base, serrulate,

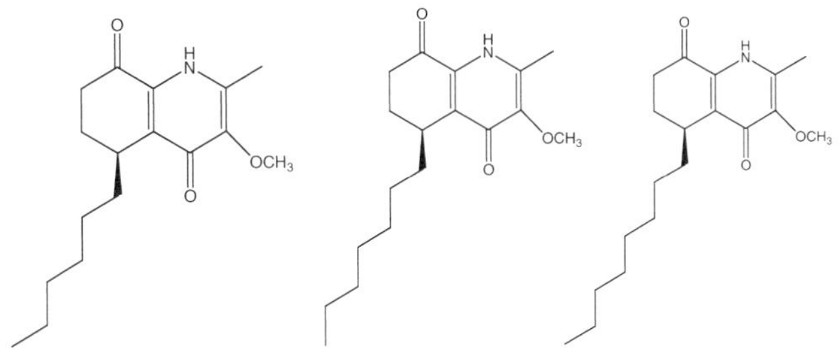

17,18-*bis*-nor-antidesmone 18-nor-antidesmone Antidesmone

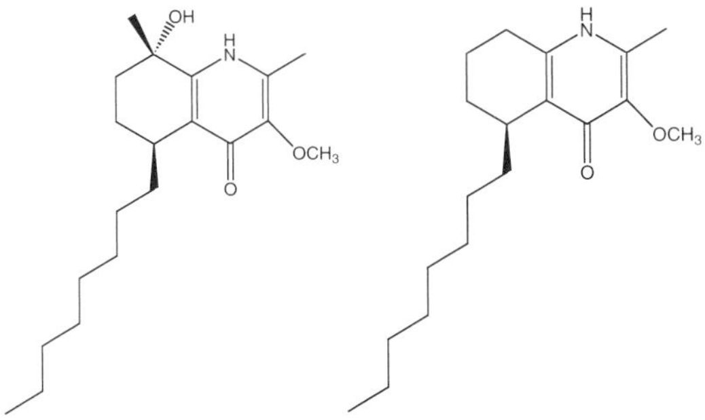

8-Dyhydroantidesmone 8-Deoxoantidesmone

Figure 26.4 Unusual quinoline alkaloids of *Antidesma* species.

and occasionally the entirety of both surfaces is pubescent. The inflorescences are axillary clusters of minute flowers. The fruits are capsular, ovoid–trigonous, 1.5mm × 1.2mm – 1.5mm, and pubescent (Figure 26.5).

26.3.2 Ethnopharmacology

The vernacular names for this herb include red caustic creeper, thyme-leafed spurge, *qian gen cao* (Chinese), and *dudhi khurd* (Unani). In Taiwan, the plant is used to promote urination and to empty the bowels. In Cambodia, Laos, and Vietnam, a decoction of the plant is used to empty the bowels and to expel worms from the intestines. The Malays use the plant externally for skin diseases and dislocated bones. A decoction of the plant is used as a drink to alleviate abdominal pain. In Indonesia, a decoction of the plant is used as a drink to stop diarrhea; the plant is also applied to wounds. In the Philippines, the plant is used to heal wounds and the latex is used to clear vision.

The healing and antidiarrheal properties of *Euphorbia thymifolia* are most likely owed to

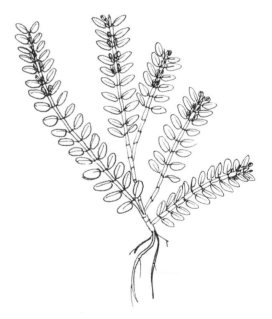

Figure 26.5 *Euphorbia thymifolia* L. [From: Herb. Hort. Bot. Sing. Field collector: Taeon, March 13, 1941. Botanical identification: J. Sinclair, April 21, 1954. Flora of North Borneo. Geographical localization: District Eloper, Suan Lambah. Alt.: 20ft.]

tannins, which are known to abound in the plant.[7] When 3-*O*-galloyl-4,6-(*S*)-hexahydroxydiphe-noyl-D-glucose, rugosin B, and 1,3,4,6-tetra-*O*-galloyl-K-β-D-glucose were extracted from the plant, they showed antioxidant activities, while 3-*O*-galloyl-4,6-(*S*)-hexahydroxydiphenoyl-D-glu-cose inhibited the replication and infectivity of Herpes Simplex Virus (HSV)-2 cultured *in vitro*.[8,9] Is 3-*O*-galloyl-4,6-(*S*)-hexahydroxydiphenoyl-D-glucose involved in the antibacterial properties reported by Khan et al.?[10]

26.4 *MACARANGA TANARIUS* MUELL.-ARG.

[From: Macaranga = a native name from Madagascar, and from Latin *tanarius* = Macaranga, parasol tree.]

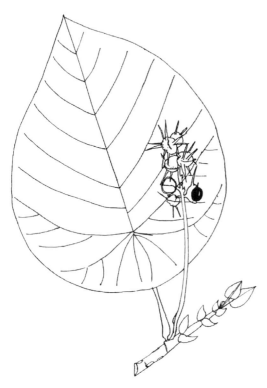

26.4.1 Botany

Macaranga tanarius Muell.-Arg. (*Rici-nus tanarius* L. and *Mappa tanarius* [L.] Bl.) is a dioecious bushy treelet that grows to a height of 9m in the secondary forests of Andamans, Nicobars, South China, Formosa, Ryukyu, Thailand, and the Pacific Islands. The stems are glaucous and strongly con-stricted. The leaves are simple, spiral, and exstipulate. The petiole and blade below are softly hairy, and velvety to the touch. The blade is round and broadly pointed, not lobed. The stipules are membranaceous, ful-vous–pubescent, oblong, 1–3cm long, decid-uous, entirely or shallowly wavy, and dentate. The petiole is 6–30cm long. The blade is thin and leathery, palmate, and shows 8–9 pairs of secondary nerves. The flowers are very small. The male ones consist of three sepals that are 1mm long, with 4–10 stamens, the anthers of which are 4-locular. The female flowers are 2–3-lobed, pubescent and com-prise a 3-locular ovary. The fruits are capsu-lar, 1–1.5cm long, glaucous green, and cov-ered with fleshy spines (Figure 26.6).

Figure 26.6 *Macaranga tanarius* Muell.-Arg. [From: Flora of Malaysia. FRI No: 0280. Geo-graphical localization: Penang Hill by upper tunnel station. Secondary forest. Alt.: 2000ft.]

26.4.2 Ethnopharmacology

In the Philippines, juice squeezed from the bark is applied to wounds to promote healing. The roots are used to induce vomiting and to remove blood from saliva. Indonesians drink a decoction of the bark to stop dysentery. In Malaysia, the leaves are used to heal wounds and the roots offer a treatment for fever. The medicinal properties of *Macaranga tanarius* Muell.-Arg. are most likely owed to tannins. *Macaranga tanarius* Muell.-Arg. is known to elaborate a series of diterpenes, including macarangonol as well as a series of prenylated flavonoids.[11–13] A remarkable advance in

Figure 2.1 Fruits of Annonaceae with club-shaped ripe carpels.

Figure 5.1 Botanical hallmarks of Piperaceae: cordate leaf.

Figure 12.1 Botanical hallmarks of Bombacaceae.

Figure 14.1 Botanical hallmarks of Capparaceae showing protruding androecium.

Figure 17.1 Botanical hallmarks of Cucurbitaceae include membraneous flowers and cucumber-like berries.

Figure 22.1 Botanical hallmarks of Melastomataceae.

Figure 23.1 Fruit of Rhizophoraceae.

Figure 26.1 Fruits of Euphorbiaceae.

Figure 27.1 Fruits of Sapindaceae.

Figure 28.1 Fruits of Anacardiaceae.

Figure 29.1 Botanical hallmarks of Simaroubaceae.

Figure 30.1 Fruits of Meliaceae.

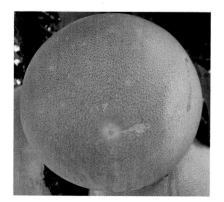

Figure 31.1 Botanical hallmarks of Rutaceae.

Figure 34.1 Botanical hallmarks of Apocynaceae.

Figure 35.1 Botanical hallmarks of Asclepiadaceae.

Figure 36.1 Botanical hallmarks of Solanaceae.

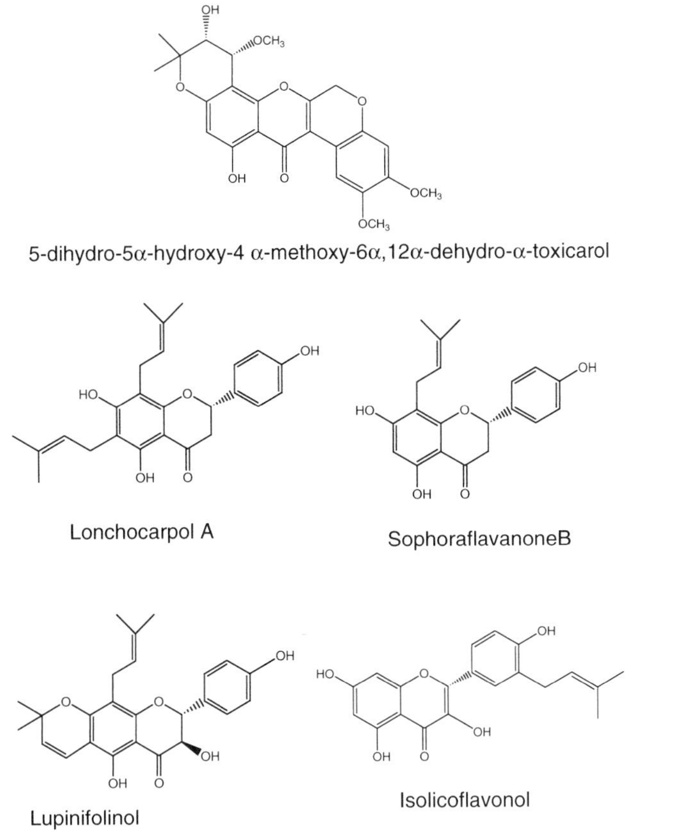

5-dihydro-5α-hydroxy-4 α-methoxy-6α,12α-dehydro-α-toxicarol

Lonchocarpol A

SophoraflavanoneB

Lupinifolinol

Isolicoflavonol

Figure 26.7 Prenylated flavonoids which inhibit the enzymatic activity of cyclooxygenase from the *Macaranga* species.

the exploration of the pharmacological potential of the *Macaranga* species has been provided by the work of Jang et al.[14,15] Using *in vitro* inhibition of cyclooxygenase-guided fractionation, they isolated a series of prenylated flavonoids which inhibit the enzymatic activity of cyclooxygenase from *Macaranga triloba* and *Macaranga conifera* (Figure 26.7). Such flavonoids include 4,5-dihydroxy-40, α-methoxy-6, and α12α-dehydro-α-toxicarol, which is possibly involved in the antiinflammatory property of *Macaranga tanarius* Muell.-Arg., but it remains to be confirmed by experimentation. Lonchocarpol A, sophoraflavanone B, lupinifolinol, and isolicoflavonol inhibited the enzymatic activity of cyclooxygenase I with IC_{50} values of 16.9μM, 72.6μM, 12.8μM, and 10.4μM, respectively.

REFERENCES

1. El-Mekkawy, S., Meselhy, R., Nakamura, N., Hattori, M., Kawahata, T., and Otake, T. 2000. Anti-HIV-1 phorbol esters from the seeds of *Croton tiglium. Phytochemistry,* 53, 457.
2. Arbain, D. and Taylor, W. C. 1993. Cyclopeptide alkaloids from *Antidesma montana. Phytochemistry,* 3, 1263.
3. Buske, A., Schmidt, J., and Hoffmann, P. 2002. Chemotaxonomy of the tribe Antidesmeae (Euphorbiaceae): antidesmone and related compounds. *Phytochemistry,* 60, 5, 489.
4. Buske, A., Busemann, S., Muhlbacher, J., Schmidt, J., Porzel, A., Bringmann, G., and Adam, G. 1999. Antidesmone, a novel isoquinoline alkaloid from *Antidesma membranaceum* (Euphorbiaceae). *Tetrahedron,* 55, 1079.

5. Bringmann, G., Schlauer, J., Rischer, H., Wohlfarth, M., Mühlbacher, J., Buske, A., Porzel, A., Schmidt, J., and Adam, G. 2000. Revised structure of antidesmone, an unusual alkaloid from tropical *Antidesma* plants (Euphorbiaceae). *Tetrahedron*, 56, 3691.

6. Garain, A. K., Chakravarti, N. N., and Chakrabartty, T. 1973. Chemical investigation of *Antidesma ghaesembilla* Gaertn. *Bull. Calcutta Sch. Trop. Med.*, 21, 26.

7. Lee, S. H., Tanaka, T., Nonaka, G. I., and Nishioka, I. 1990. Hydrolysable tannins from *Euphorbia thymifolia. Phytochemistry*, 29, 3621.

8. Lin, C. C., Cheng, H. Y., Yang, C. M., and Lin, T. C. 2002. Antioxidant and antiviral activities of *Euphorbia thymifolia* L. *J. Biomed. Sci.*, 9, 656.

9. Yang, C. M., Cheng, H. Y., Lin, T. C., Chiang, L. C., and Lin, C. C. 2005. *Euphorbia thymifolia* suppresses herpes simplex virus-2 infection by directly inactivating virus infectivity. *Clin. Exp. Pharmacol. Physiol.*, 32, 346.

10. Khan, N. H., Rahman, M., and Nur-e-Kamal, M. S. 1988. Antibacterial activity of *Euphorbia thymifolia* Linn. *Indian J. Med. Res.*, 87, 395.

11. Hui, K. K., Ng, N., Fukamiya, M., Koreeda, K., and Nakanishi, K. 1971. Isolation and structure of macarangonol, a diterpene ketol from *Macaranga tanarius. Phytochemistry*, 10, 1617.

12. Phommart, S., Sutthivaiyakit, P., Chimnoi, N., Ruchirawat, S., and Sutthivaiyakit, S. 2005. Constituents of the leaves of *Macaranga tanarius. J. Nat. Prod.*, 68, 927.

13. Hui, W. H., Li, M. M., and Ng, K. K. 1975. Terpenoids and steroids from *Macaranga tanarius, Phytochemistry*, 14, 816.

14. Jang, D. S., Cuendet, M., Pawlus, A. D., Kardono, L. B. S., Kawanishi, K., Farnsworth, N. R., Fong, H. H. S., Pezzuto, J. M., and Kinghorn, A. D. 2004. Potential cancer chemopreventive constituents of the leaves of *Macaranga triloba. Phytochemistry*, 65, 345.

15. Jang, D. S., Cuendet, M., Hawthorne, M. E., Kardono, L. B. S., Kawanishi, K., Fong, H. H. S., Mehta, R. G., Pezzuto, J. M., and Kinghorn, A. D., 2002. Prenylated flavonoids of the leaves of *Macaranga conifera* with inhibitory activity against cyclooxygenase-2. *Phytochemistry*, 61, 867.

CHAPTER 27

Medicinal Plants Classified in the Family Sapindaceae

27.1 GENERAL CONCEPT

The family Sapindaceae (A. L. de Jussieu, 1789 nom. conserv., the Soapberry Family) consists of approximately 140 genera and 1500 species of tropical trees, shrubs and woody climbers, generally tanniferous and saponiferous. When searching for Sapindaceae in the field, one should look for trees with smooth bark, not uncommonly with exudate when cut, with pinnate leaves, tiny flowers, and fleshy fruits covered with numerous fleshy appendices. The seeds are large and embedded in a fleshy sarcotesta or aril (Figure 27.1). The aril is edible, hence the cultivation of Sapindaceae as a fruit tree such as *rambutan* (*Nephelium lappaceum* L.), *longan* (*Euphoria longan* [Lour.] Steud.), and *litchi* (*Litchi sinensis* Sonn.).

A common example of ornamental Sapindaceae is *Koelreuteria paniculata* Laxm., the Golden Rain Tree of temperate regions. Of pharmaceutical interest is *Paullinia cupuna*

Figure 27.1 Fruits of Sapindaceae. **(See color insert following page 168.)**

H.B.K., the Guarana, an important crop in Amazonian Brazil, where the seeds are used in the preparation of a caffeine-rich carbonated drink. Guarana (*British Pharmaceutical Codex*, 1934) has been used for the treatment of headache and as an astringent in diarrhea, usually as a tincture (1 in 4 dose: 2–8mL). The evidence available so far on pharmacologically active principles from this large family is surprisingly small and one can reasonably envisage this family as a *terra incognita* for pharmacologists. The traditional systems of medicine of the Asia–Pacific use about 50 species of Sapindaceae, mainly to promote the healing of wounds.

27.2 *DODONAEA VISCOSA* (L.) JACQ.

[After R. Dodoens (1517–1585), a Dutch physician and botanist, and from Latin *viscosa* = viscous.]

Figure 27.2 *Dodonaea viscosa* (L.) Jacq. [From: Singapore. Field No: 37952. Distributed by The Botanic Gardens Singapore. Geographical localization: Malay Peninsula, Kedah, near Sanitorium Langkawi. Nov. 13, 1941. Field collector: J. C. Naeur. Botanical identification: M. R. Henderson in sand near sea.]

27.2.1 Botany

Dodonaea viscosa (L.) Jacq. (*Dodonaea jamaicensis* DC., *Dodonaea ehrenbergii* Schlecht, *Dodonaea eriocarpa* Sm., *Dodonaea microcarya* Small, *Dodonaea sandwicensis* Sherff, *Dodonaea elaeagnoides* Rudolph ex Ledeb. & Alderstam, *Dodonaea spathulata* Sm., *Dodonaea stenoptera* Hbd., and *Ptelea viscosa* L.) is a pantropical treelet that grows to a height of 6m on sandy shores. The bark is ridged and fissured. The stems are angular and green. The leaves are simple, glabrous, and pseudo-sessile. The blade is membranaceous and spathulate. The blade is without visible secondary nerves and measures $10.5cm \times 3cm$, $4.8cm \times 1cm$, $7cm \times 1.6cm$, and $7.4cm \times 2cm$. The fruits are cordate, flattened capsules with two lobes, each with a rounded membranous wing, notched between the lobes, green to light brown, and measuring $1.7cm \times 1.3cm – 1.2cm \times 5mm$. The fruit pedicels are slender and up to 2cm long. Each lobe is dehiscent and exposes one or two black seeds (Figure 27.2).

27.2.2 Ethnopharmacology

The vernacular names of the plant include Hopseed Bush and Hop Bush. The plant is used in Burma to make an external remedy. The leaves are heated and applied to the skin. In Palau and Taiwan, the leaves are used to reduce fever. In the Philippines, a decoction of barks is used to reduce fever, to treat eczema, and to heal ulcers. The plant exhibited some levels of activity against *Streptococcus pyogenes* and *Staphylococcus aureus*, and possessed strong activity against Coxsackievirus B3 (CVB3) and Influenza A virus.[1] Note that an aqueous extract of *Dodonaea angustifolia* L. protected rodents against the pain and fever caused by acetic acid writhing and hot plate tests, and by lipopolysaccharide (LP)-induced pyrexia tests.[2] The seeds contain saponins dodonosides A and B, which exhibited immunomodulating and molluscicidal properties.[3] Methanol extract of *Dodonaea angustifolia* Lf. extract inhibited the replication of Human Immunodeficiency Virus (HIV)-1 and HIV-2, and protected cells against the cytopathic effect of the virus.[4] Are tannins involved here?

Using bioassay-directed fractionation of the chloroform–methanol (1:1) extract of *Dodonaea viscosa* (L.) Jacq., Rojas et al.[5] isolated a series of molecules, including sakuranetin, 6-hydroxykaempferol, 3,7-dimethyl ether, hautrivaic acid, and ent-15,16-epoxy-9αH-labda-13(16)14-diene-3β,8-α-diol which inhibited spontaneous and electrically induced contractions of guinea-pig ileum. Sakuranetin and the ent-labdane inhibited ileum contractions induced by acetylcholine, histamine, and calcium.[5] The plant elaborates clerodane diterpenes (Figure 27.3).[6]

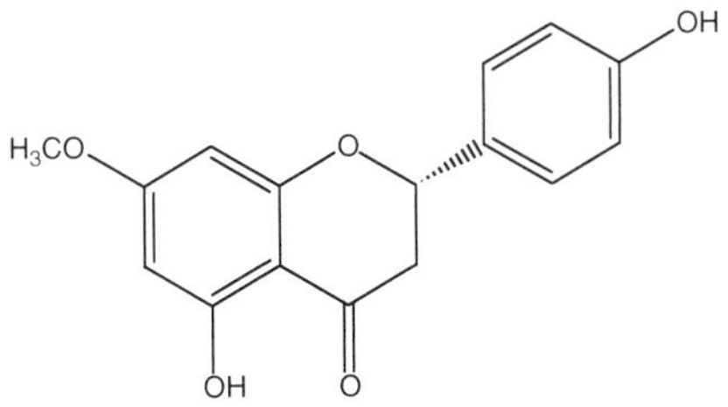

Sakuranetin

Figure 27.3 Sakuranetin, a flavonoid from *Dodonaea viscosa* (L.) Jacq.

27.3 *LEPISANTHES TETRAPHYLLA* (VAHL) RADLK.

[From: Greek *lepis* = scale, *anthos* = flower, *tetra* = four, and *phyllon* = leaf.]

27.3.1 Botany

Lepisanthes tetraphylla (Vahl) Radlk. (*Lepisanthes longifolia* Radlk., *Lepisanthes kunstleri* King, *Lepisanthes cuneata* Hiern, *Lepisanthes scortechinii* King, *Lepisanthes scortechinii* King var. *hirta* Radlk., *Lepisanthes granulata* Radlk., *Lepisanthes tetraphylla* [Vahl] Radlk. var. *cambodiana* Pierre, *Lepisanthes tetraphylla* [Vahl] Radlk. var. *indica* Pierre, *Lepisanthes poilanei* Gagnep., and *Molinaea canescens* Roxb.) is a shrub that grows to 8m tall from India to China through Southeast Asia. It has yellow sap. The stems are somewhat hairy. The leaves are pinnate and exstipulate. The rachis is 22.5cm, grooved, and has 2–5 pairs of folioles. The petiolules are swollen and grooved. The folioles are elliptical and lanceolate, and show 12 pairs of secondary nerves; they measure 21cm × 5.8cm – 26cm × 6cm – 25cm × 7cm. The flowers are white to pink, fragrant, hairy, and measure 10cm long. The

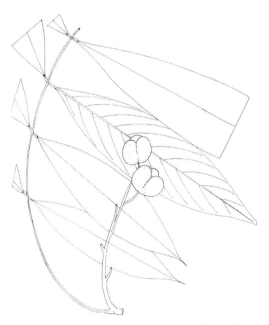

Figure 27.4 *Lepisanthes tetraphylla* (Vahl) Radlk.

sepals are hairy outside. The androecium consists of eight stamens. The gynaecium is hairy. The fruits are smooth to warty, green turning yellow, velvety, with 3 trilobed capsules which are 3.75cm long and enclose three seeds. A small persistent calyx is present (Figure 27.4).

27.3.2 Ethnopharmacology

In Malaysia, the juice squeezed from pounded leaves is used to alleviate cough, while the leaves themselves are used to make a cooling bath. The plant is used as an ingredient for dart poison. The pharmacotoxicological properties of this plant are as of yet unexplored. The hypothesis is conceivable that the plant's antitussive and poisonous properties are due to its content of saponins which are surface-acting agents and cytotoxic.[7]

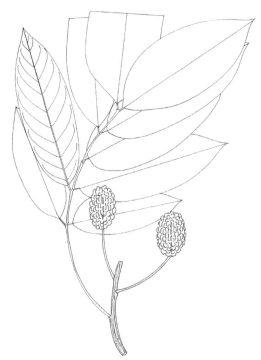

27.4 *NEPHELIUM JUGLANDIFOLIUM* BL.

[From: Greek *nephelion* = a little cloud and from Latin *juglandifolium* = leaves like the walnut, *Juglans*.]

27.4.1 Botany

Nephelium juglandifolium Bl. (*Nephelium altissimum* Teijs. & Binn. and *Nephelium tuberculatum* Radlk.) is a tree that grows in the lowland rain forests of Sumatra, Java, and Malaysia to a height of 30m with a girth of 90cm. The bole is straight with slight buttresses. The bark is smooth and dull red in color. The stems are thick to 1.2cm in diameter, with a few hairs. The leaves are pinnate, alternate, and exstipulate. The rachis is 23cm long and shows four to five pairs of folioles which are dark green above, and measure 11.3cm × 4.4cm – 16cm × 5cm – 16.5cm × 6.4cm –16cm × 6cm. The folioles are pointed at the apex, thinly coriaceous, with about 11 pairs of secondary nerves. The inflorescences are axillary and the terminal panicles are 10cm long. The flowers are minute and apetalous, with seven to eight stamens and a bilobed ovary. The calyx is partially tubular. The fruits are 4cm × 2.5cm and covered with a red pericarp and coarsely tuberculate. The sarcotesta is thin and edible (Figure 27.5).

Figure 27.5 *Nephelium juglandifolium* Bl. [From: H. S. Lenggong, along Jelebu-Seremban Road. Valley bottom of primary forest. Alt.: 300m. FRI No: 3347. Field collector: S. R. Yap, July 29, 1987.]

27.4.2 Ethnopharmacology

The seeds are known to induce narcosis if eaten. To date the pharmacotoxicological properties of *Nephelium juglandifolium* Bl. are unexplored. One might look into the antiviral potential of this plant, given that a water extract from the pericarp of *Nephelium lappaceum* L. exhibited anti-HSV-1 activity *in vitro* and *in vivo*.[11] The bark probably abounds with saponins.[9]

27.5 *POMETIA PINNATA* FORST.

[After Pierre Pomet (1558–1699), a French writer, and from Latin *pinnatus* = pinnate, the leaves.]

27.5.1 Botany

Pometia pinnata Forst. is a tree that grows to a height of 40m with a girth of 70cm in the primary rain forests of Sri Lanka, Andamans, Thailand, Cambodia, Laos, Vietnam, South China, Taiwan, and Indonesia. The crown spreads from a straight to a buttressed bole. The bark is rusty red and greenish with abundant red exudate when cut.

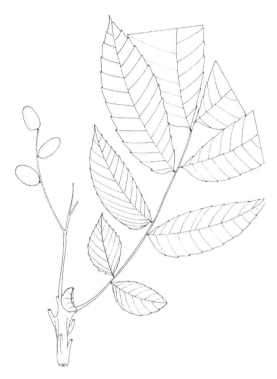

Figure 27.6 *Pometia pinnata* Forst. [From: FRI No: 32951. Botanical identification: S. L. Guan.]

The leaves are pinnate and comprise 3–10 pairs of folioles which are 5.5cm × 3.3cm – 11cm × 6cm – 15cm × 6cm – 19cm × 7 cm. The folioles are serrate, coriaceous, sessile, and show about 10 pairs of secondary nerves. The inflorescences are panicles of variable length. The fruits are 2.1cm × 1.7 cm, ovoid, purple, and glossy (Figure 27.6).

27.5.2 Ethnopharmacology

The plant is called *kasai* by the Malays and Indonesians who use the bark externally to counteract the putrefaction of wounds. The leaves and bark are used to make a bath for fever. The antiseptic property of the plant is probably owed to saponins.[10,11]

REFERENCES

1. Getie, M., Gebre-Mariam, T., Rietz, R., Höhne, C., Huschka, C., Schmidtke, M., Abate A., and Neubert, R. H. H. 2003. Evaluation of the anti-microbial and anti-inflammatory activities of the medicinal plants *Dodonaea viscosa, Rumex nervosus* and *Rumex abyssinicus*. *Fitoterapia*, 74, 139.
2. Amabeoku, G. J., Eagles, P., Scott, G., Mayeng, I., and Springfield, E. 2001. Analgesic and antipyretic effects of *Dodonaea angustifolia* and *Salvia africana-lutea*. *J. Ethnopharmacol.*, 75, 117.
3. Wagner, H., Ludwig, C., Grotjahn, L., and Khan, M. S. Y. 1987. Biologically active saponins from *Dodonaea viscosa*. *Phytochemistry*, 26, 697.
4. Asres, K., Bucar, F., Kartnig, T., Witvrouw, M., Pannecouque, C., and De Clercq, E. 2001. Antiviral activity against human immunodeficiency virus type 1 (HIV-1) and type 2 (HIV-2) of ethnobotanically selected Ethiopian medicinal plants. *Phytother. Res.*, 15, 62.
5. Rojas, A., Cruz, S., Ponce-Monter, H., and Mata, R. 1996. Smooth muscle relaxing compounds from *Dodonaea viscosa*. *Planta Med.*, 62, 154.

6. Abdel-Mogib, M., Basaif, S. A., Asiri, A. M., Sobahi, T. R., and Batterjee, S. M. 2001. New clerodane diterpenoid and flavonol-3-methyl ethers from *Dodonaea viscosa*. *Pharmazie*, 56, 830.

7. Adesanya, S. A., Martin, M. T., Hill, B., Dumontet, V., Tri, M. V., Sévenet, T., and Païs, M. 1999. Rubiginoside, a farnesyl glycoside from *Lepisanthes rubiginosa*. *Phytochemistry*, 51, 1039.

8. Nawawi, A., Nakamura, N., Hattori, M., Kurokawa, M., and Shiraki, K. 1999. Inhibitory effects of Indonesian medicinal plants on the infection of herpes simplex virus type 1. *Phytother. Res.*, 13, 37.

9. Ito, A., Chai, H. B., Kardono, L. B., Setowati, F. M., Afriastini, J. J., Riswan, S., Farnsworth, N. R., Cordell, G. A., Pezzuto, J. M., Swanson, S. M., and Kinghorn, A. D. 2004. Saponins from the bark of *Nephelium maingayi*. *J. Nat. Prod.*, 67, 201.

10. Voutquenne, L., Guinot, P., Thoison, O., and Lavaud, C. 2003. Oleanolic glycosides from *Pometia ridleyi*. *Phytochemistry*, 64, 781.

11. Jayasinghe, L., Shimada, H., Hara, N., and Fujimoto, Y. 1995. Hederagenin glycosides from *Pometia eximia*. *Phytochemistry*, 40, 891.

CHAPTER 28

Medicinal Plants Classified in the Family Anacardiaceae

28.1 GENERAL CONCEPT

The family Anacardiaceae (Lindley, 1830 nom. conserv., the Sumac Family) consists of approximately 60 genera and 600 species of tropical trees, known to produce tannins and several sorts of phenolic compounds. When searching for Anacardiaceae in tropical rain forests, one is advised to look for resinous trees with simple leaves and without stipules, often thick spathulate, and showing straight secondary nerves, panicles of tiny flowers, and principally drupaceous often kidney-shaped berries. Some of which are highly poisonous, others (*Mangifera indica* L.) are edible, others (*Gluta* species) have enlarged persistant sepals, while others (*Anacardium* species) have an enlarged and succulent pedicel (Figure 28.1).

Extra care must be taken with some members of this family, including notably *Anacardium melanorrhoea* (*rhengas* tree) and *Gluta rhengas* L. The sap contains an unusual series of long-chain phenolic substances such as urushiol, cardol, and anacardic acid which, being lipophilic, penetrate the skin quickly and cause a great deal of discomfort to the plant collector, including edema, pruritus, burning, stinging, erythematous macules, papules, vesicles, exudation, crusting, and death with anaphylactic shock (Figure 28.2). Other well-known examples of toxic Anacardiaceae are *Toxicodendron vernis* (Poison Sumac) and *Toxicodendron radicans* (Poison Ivy), are currently responsible for life-threatening allergic reactions.[1]

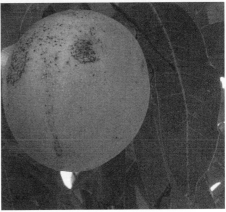

Figure 28.1 Fruits of Anacardiaceae. **(See color insert following page 168.)**

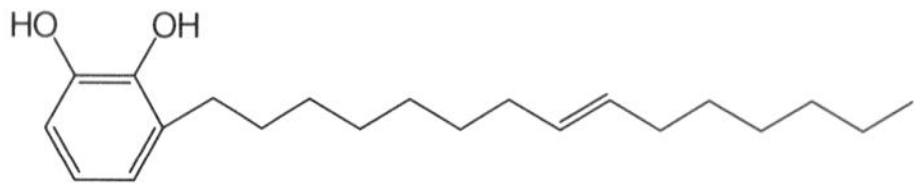

Urushiol

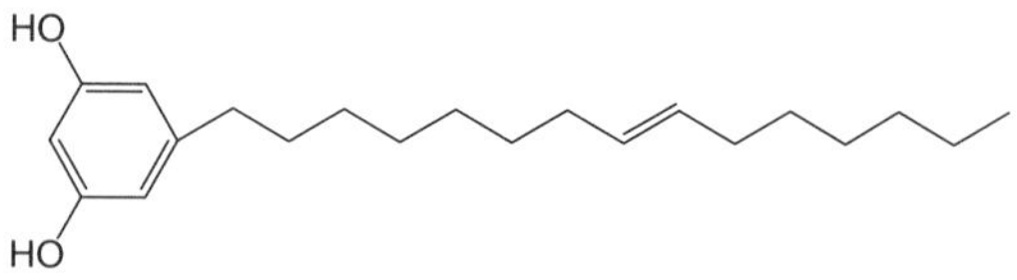

Cardol

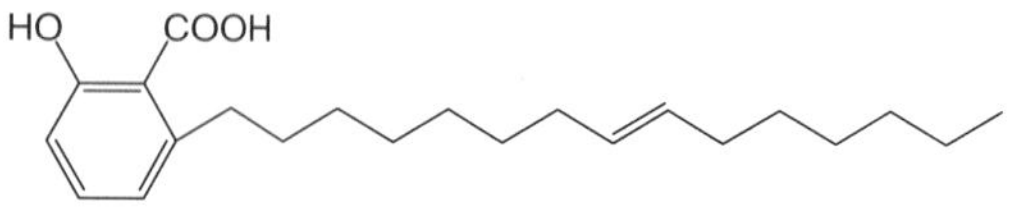

Anacardic acid

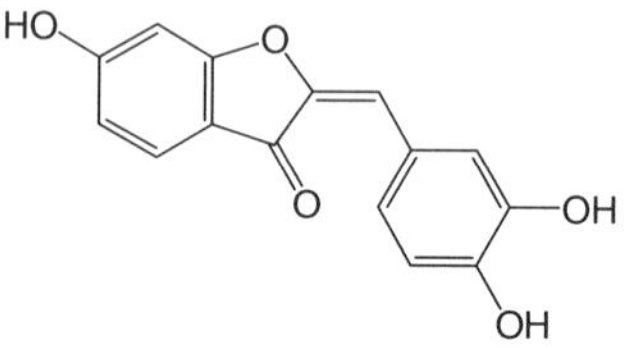

Sulfuretin

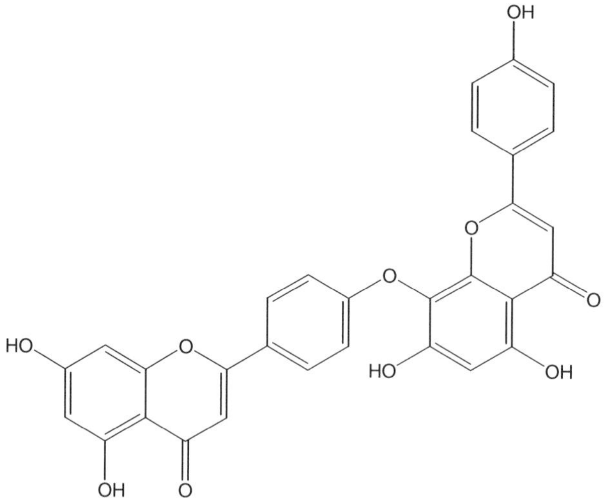

Lanaroflavone 6

Figure 28.2 Examples of bioactive phenolic compounds from the family Anacardiaceae.

Examples of commercial products of anacardiaceous origin are *Pistacia lentiscus* var. *chia* that produces mastic, *Rhus coriaria* (dyeing and tanning sumac), *Rhus succedanea* (Japanese Wax Tree), and the edible *Pistacia vera* (Pistachio Nut). The dried berries of *Rhus glabra* (Pennsylvanian Sumac) were formerly used as a decoction or a liquid extract mixed with glycerin, water, and potassium chlorate (Rhus, *British Pharmaceutical Codex*, 1934) to wash the mouth out.

The pharmacological evidence so far presented clearly indicates that urushiol and congeners inhibit the enzymatic activity of several sorts of enzymes including phospholipase A2, cyclooxygenase, 5-lipooxygenase, and prostaglandine synthetase, which mediate inflammation.[2] Other principles of interest in this family are flavonoids such as tetrahydroamentoflavone and lanaroflavone.[3] Tetrahydroamentoflavone, from *Semecarpus anacardium*, inhibits the enzymatic activity of cyclooxygenase, with an IC_{50} value of 29.5µM (COX-1).[4] Lanaroflavone, from *Campnosperma panamaense*, inhibited *Plasmodium falciparum* K1 chloroquine-resistant strain and *Leishmania donovani* cultured *in vitro* with IC_{50} values of 0.2g/mL and 3.9g/mL, respectively.[5] Corthout et al.[6] made the interesting observation that 2-*O*-caffeoyl-(+)-allohydroxycitric acid and chlorogenic acid butyl ester from *Spondias monbin* showed antiviral activities against Coxsackie and Herpes Simplex Viruses, respectively. In the Pacific Rim, about 20 species of Anacardiaceae are of medicinal value of which *Dracontomelon dao* (Blanco) Merr. & Rolfe, *Gluta rhengas* L., *Melanochyla auriculata* Hook. f., and *Pentaspadon officinalis* Holmes are presented in this chapter..

28.2 *DRACONTOMELON DAO* (BLANCO) MERR. & ROLFE

[From: Greek *drakan* = dragon, melon = a tree fruit, and from Filipino *dao* = *Dracontomelon dao* (Blanco) Merr. & Rolfe.]

28.2.1 Botany

Dracontomelon dao (Blanco) Merr. & Rolfe (*Dracontomelon mangiferum* Bl. and *Spondias dulcis*) is a resinous tree that grows in the rain forests of Southeast Asia especially on riverbanks and in swampy areas. The plant reaches a height of 36m with a girth of 2.4m. The crown is rounded and dense. The bole is straight and buttressed. The bark is grayish-brown, and the inner bark is pink. The stems are covered with a few rusty hairs at the apex. The leaves are spiral, imparipinnate, and exstipulate. The rachis is 30–45cm long and shows 5–8 pairs of folioles which are 4cm – 22.5cm × 2.5cm × 7.5 cm. The apex is pointed, the base is round and shows 10–15 pairs of secondary nerves with hairy domatia at the axil. The petiolules are 3m long. The inflorescences are up to 60cm hanging in lax panicles. The flowers are tiny, 5-lobed, white, and fragrant. The androecium comprises 10 stamens opposite the sepals. The gynaecium consists of five carpels which are partially united. The fruits are globose, 2.5–3.8cm in diameter, with green drupes turning yellow with oval markings on the upper side of the fruit (Figure 28.3).

28.2.2 Ethnopharmacology

In China, the plant is known as *j'n mien tz*. The fruits stewed in honey are relished. Chinese compare the seed to a man's face and children use them as toys. The kernels are mixed in tea to give them a fragrant and mucilaginous sweet taste. The fruits are used to cool, to calm itchiness, to cure internal ulceration, and as an antidote for poisoning. It is believed that holding a seed in the right hand on odd days and in the left hand on even days will precipitate childbirth. The fruits are also used to soothe sore throat and inflammation of the skin. Indonesians boil the bark in water to make a drink which will expel the membrane enveloping the fetus in the womb. Khan and Omoloso[7] studied the antibacterial activity of the plant and showed that the dichloromethane fraction of the leaves inhibits the growth of a broad spectrum of bacteria cultured *in vitro*.

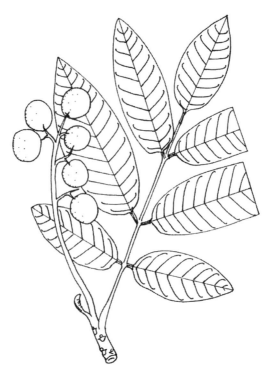

Figure 28.3 *Dracontomelon dao* (Blanco) Merr. & Rolfe. [From: Flora of Malaya. Kepong Field No: 9453. Geographical localization: Larut Hill, Perak. Alt.:1800ft. Field collector: W. S. Hing, May 31, 1968. Botanical identification: K. M. Kochummen, October 1962.]

28.3 *GLUTA RHENGAS* L.

[From: Latin *gluten* = glue and from Malay *rengas* = *Gluta rhengas* L.]

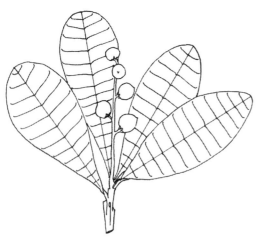

Figure 28.4 *Gluta rhengas* L. [From: Botanic Gardens Singapore. SM 26.]

28.3.1 Botany

Gluta rhengas L. is a tree that grows to 30m with a light gray bole which is often multiple stemmed. The plant grows in the lowland swampy rain forests and freshwater tidal reaches of Malaysia. The bark exudes a white sap. The leaves are simple, spiral, and exstipulate. The petiole is to 1.25cm long and winged. The blade is spathulate, leathery, 8cm × 15cm × 4cm – 8cm, and shows 17–30 pairs of conspicuous secondary nerves raised on both surfaces. The flowers are white and minute, in axillary panicles. The flower pedicels are reddish. The fruits are globose nuts that are 3.5–5cm in diameter, brown scurfy with irregular crests and protuberances, and five small spreading wings about 8mm × 2mm (Figure 28.4).

28.3.2 Ethnopharmacology

The plant is called *renghas* in Malaysia, where the sap is greatly feared by locals as it induces dangerous allergic reactions. It probably explains why only a few pharmacologists have studied the plant.[8]

28.4 *MELANOCHYLA AURICULATA* HOOK. F.

[From: Greek *melas* = black and *chulas* = sap, and from Latin *auriculata* = ear-like, referring to the petiole.]

28.4.1 Botany

Melanochyla auriculata Hook. f. is a tree that rarely reaches higher than 24m and a 1.2m girth. The bole is short and buttressed. The bark is grayish-brown, smooth, and exudes a black sap. The inner bark is pinkish. The stems are stout and 1cm in diameter. The leaves are simple, spiral, and exstipulate. The petiole is 1cm long. The blade is spathulate, leathery, and large, 22–62cm × 6–15cm, and glossy on both sides. The base is cordate or auriculate. The blade shows 25–35 pairs of secondary nerves. The flowers are minute, white, and arranged in panicles which are 25–60cm long. The fruits are light brown, fawn, smooth, ovoid, and 2cm – 3.5cm × 2cm – 2.5cm. A black sap exudes from the fruits after cutting (Figure 28.5).

28.4.2 Ethnopharmacology

Like *Gluta rhengas* L., this is called *renghas* in Malaysia where its sap is equally feared. Its pharmacotoxicology is unexplored. Similar to *Gluta rhengas*, which was mentioned earlier, the poisonous properties are most likely owed to anacardic acids and congeners. A remarkable advance in the understanding of urushiol toxicity has been provided by the work of Xia et al.[10] They showed that urushiols in the skin undergo a lipoxigenase-induced polymerization. Note that anacardic acids and congeners mediate in membrane potential and pH gradient across liposomal membranes, and inhibit the growth of methicillin-resistant strains of *Staphylococcus aureus*. Anacardic acid from the bark of *Ozoroa insignis* has inhibited Hep-G2 (human

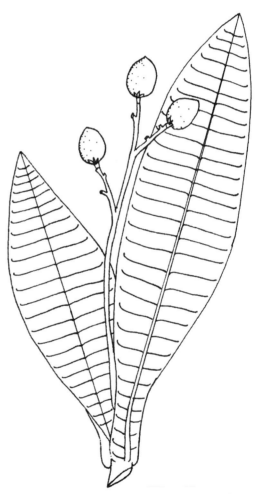

Figure 28.5 *Melanochyla auriculata* Hook. f. [From: Geographical localization: Chering River, Pahang. Oct. 10, 1928. No: 15703. Botanical identification: D. Hou, June 1979.]

hepatocellular carcinoma), MDA-MB-231 (human mammary adenocarcinoma), and 5637 (human primary bladder carcinoma).[10,11]

28.5 *PENTASPADON OFFICINALIS* HOLMES

[From: Greek *pente* = five and spadon = eunuch, from the five sterile stamens, and from Latin *officinalis* = sold as an herb.]

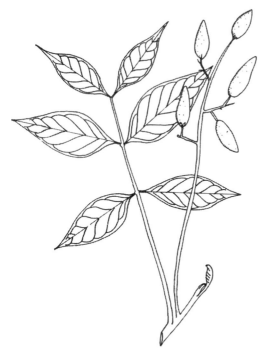

Figure 28.6 *Pentaspadon officinalis* Holmes. [From: Federated Malay States. No: 26. Geographical localization: Lipur Village, Malaysia. Field collector: M. S. Hamid, Nov. 6, 1916.]

28.5.1 Botany

Pentaspadon officinalis Holmes (*Pentaspadon motleyi* Hook f.) is a tree that grows to a height of 36m with a girth of 2.1m in the lowland rain forests of Indonesia, Malesia, and the Solomon Islands. The bole is straight and buttressed. The bark is grayish-white or grayish-brown, scaly, and lenticelled. The inner bark is pink with droplets of white sap turning brown. The leaves are spiral, crowded at the apex of stems, and exstipulate. The rachis is 10–30cm long and holds 7–9 pairs of folioles which are 7.5cm × 3.5cm – 5cm × 13cm – 2cm × 6cm. The base is cuneate and the apex is pointed. The folioles show 6–10 pairs of arching secondary nerves with hairy domatia in the axils of secondary nerves. The inflorescences are axillary lax panicles. The flowers are minute. When in flower the tree is conspicuous with full bloom and without leaves. The fruits are fusiform, 2.5cm – 5cm × 1cm – 2.75cm (Figure 28.6).

28.5.2 Ethnopharmacology

In Malaysia, the resin obtained from the stems is a counterirritant used to calm itchiness of the skin. Are urushiols involved here?

REFERENCES

1. Beaman, J. H. 1986. Allergenic Asian Anacardiaceae. *Clin. Dermatol.*, 4, 191.
2. Grazzini, R., Hesk, D., Heininger, E., Hildenbrandt, G., Reddy, C. C., Cox-Foster, D., Medford, J., Craig R., and Mumma, R. O. 1991. Inhibition of lipoxygenase and prostaglandin endoperoxide synthase by anacardic acids. *Biochem. Biophys. Res. Commun.*, 176, 775.
3. Son, Y. O., Lee, K. Y., Lee, J. C., Jang, H. S., Kim, J. G., Jeon, Y. M., Jang, Y. S., and Beaman, J. H. 2005. Selective antiproliferative and apoptotic effects of flavonoids purified from *Rhus verniciflua* Stokes on normal versus transformed hepatic cell lines. *Toxicol. Lett.*, 155, 115.
4. Selvam, C. and Jachak, S. M. 2004. A cyclooxygenase (COX) inhibitory biflavonoid from the seeds of *Semecarpus anacardium*. *J. Ethnopharmacol.*, 95, 209.

5. Weniger, B., Vonthron-Sénécheau, C., Arango, G. J., Kaiser, M., Brun, R., and Anton, R. 2004. A bioactive biflavonoid from *Campnosperma panamense*. *Fitoterapia*, 75, 764.

6. Corthout, J. P. L., Claeys, M., Vanden Berghe, D., and Vlietinck, A. 1992. Antiviral caffeoyl esters from *Spondias mombin*. *Phytochemistry*, 31, 1979.

7. Khan, M. R. and Omoloso, A. D. 2002. Antibacterial and antifungal activities of *Dracontomelon dao*. *Fitoterapia*, 73, 327.

8. Lin, C. R. and Whittow, G. C. 1960. Pharmacological activity of an aqueous extract of the leaves of the Malayan *rengas* tree *Gluta renghas*. *Br. Pharm. Chemother.*, 15, 440.

9. Muroi, H., Nihei, K., Tsujimoto, K., and Kubo, I. 2004. Synergistic effects of anacardic acids and methicillin against methicillin resistant *Staphylococcus aureus*. *Bioorg. Med. Chem.*, 12, 583.

10. Xia, Z., Miyakoshi, T., and Yoshida, T. 2004. Lipoxygenase-catalyzed polymerization of phenolic lipids suggests a new mechanism for allergic contact dermatitis induced by urushiol and its analogs. *Biochem. Biophys. Res. Commun.*, 315, 704.

11. Toyomizu, M., Okamoto, K., Akiba, Y., Nakatsu, T., and Konishi, T. 2002. Anacardic acid-mediated changes in membrane potential and pH gradient across liposomal membranes. *Biochem. Biophys. Acta (BBA) — Biomembranes*, 1558, 54.

CHAPTER 29

Medicinal Plants Classified in the Family Simaroubaceae

29.1 GENERAL CONCEPT

The family Simaroubaceae consists of approximately 25 genera and 150 species of tropical trees and shrubs known to elaborate a series of oxygenated triterpenoids, such as quassinoids and limonoids, which make the bark, wood, and seeds very bitter. The main field characteristics to note when looking for Simaroubaceae are slender trees without latex or sap, a massive crown of long pinnate leaves on top of a slender bole, and racemes of little drupes or berries. The wood is light yellow to white (Figure 29.1). Classical examples of Simaroubaceae are the ornamental *Ailanthus altissima* (Mill.) Swingle (Tree of Heaven), and the bitter tonic *Quassia amara* L. (*Surinam quassia*). The dried stem wood of *Picrasma excelsa* (*Aeschrion excelsa* and *Picraena excelsa*) was used as an infusion of 1 in 20 in cold water to promote digestion, to stimulate appetite, expel intestinal worms, and to treat pediculosis (*Quassia, British Pharmaceutical Codex*, 1973). A decoction or infusion (1 in 20) of the dried root bark of *Simaruba amara* (*Simaruba officinalis*) has been used to stimulate appetite and to stop diarrhea (*Simaruba, British Pharmaceutical Codex*, 1934).

The evidence for the existence of quassinoids of chemotherapeutic value in the Simaroubaceae is strong and it seems likely that molecules of clinical value will be derived from this family in the near future.[1] Quassinoids are par-

Figure 29.1 Botanical hallmarks of Simaroubaceae. **(See color insert following page 168.)**

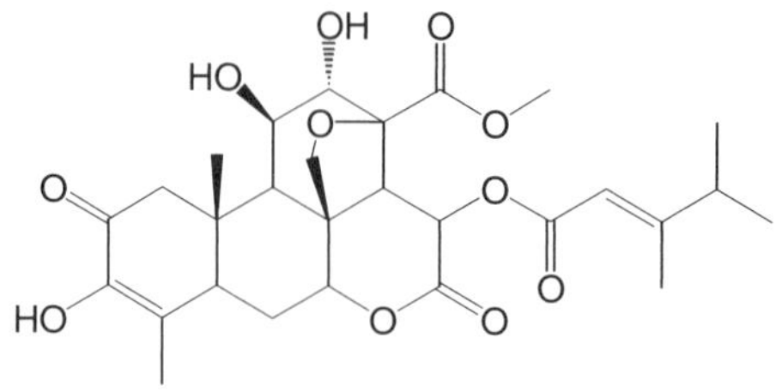

Bruceantin

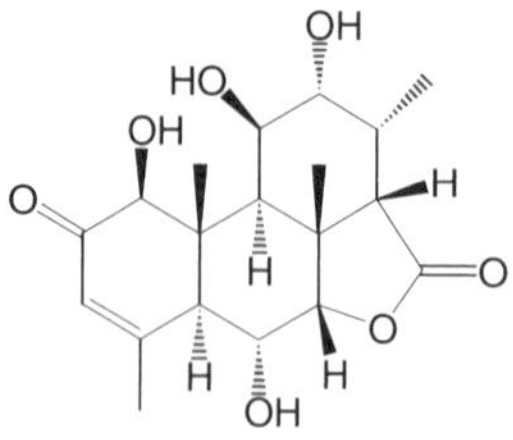

Longilactone 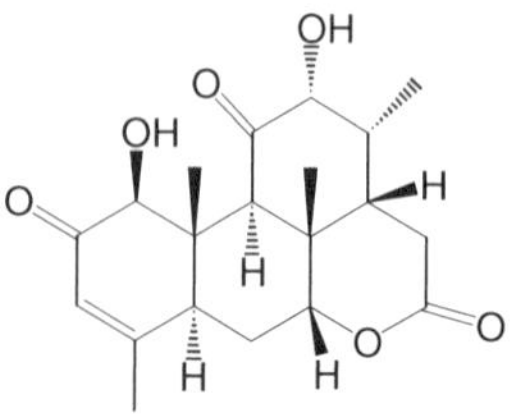   11-Dehydroklaineanone

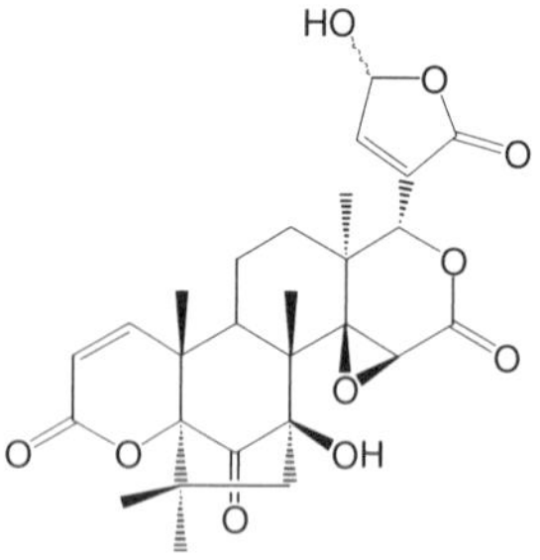

Perforatinolone 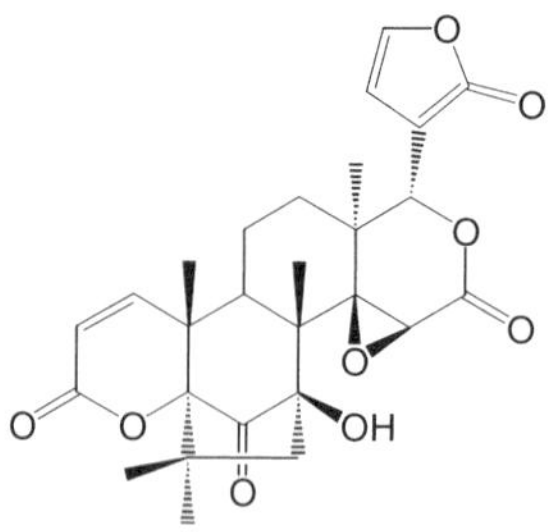   Perforatin 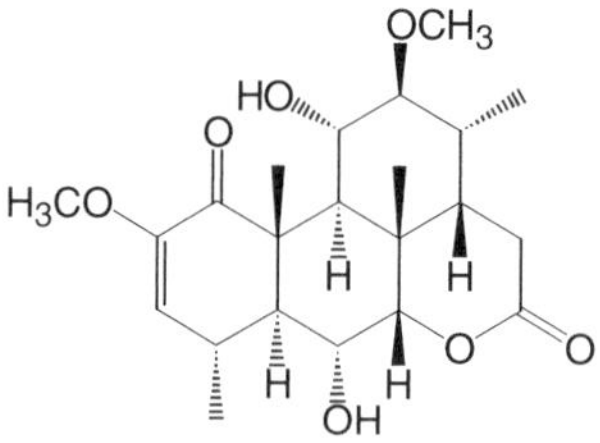   Javanicin Z

Figure 29.2 Quassinoids and limonoids of Simaroubaceae.

ticularly abundant in *Brucea, Ailanthus, Quassia, Simarouba, Castela,* and *Simaba* (Figure 29.2). Examples of cytotoxic quassinoids are bruceantin, longilactone, and 11-dehydroklaineanone. Bruceantin from *Brucea javanica* (L.) Merr., which is medicinal in the Pacific Rim, has attracted a great deal of interest on account of its ability to prevent the survival of a broad spectrum of cancer cells.[2,3] Longilactone from *Eurycoma longifolia* is cytotoxic and prevents the survival of the *Schistosoma* species at a dose of 200mg/mL.[4,5] 11-Dehydroklaineanone from the same plant has inhibited the growth of the *Plasmodium* species cultured *in vitro* with an IC_{50} value as low as 2µg/mL.[5]

In the Pacific Rim, *Ailanthus altissima* (Mill.) Swingle (*Ailanthus glandulosa* Desf., *Ailanthus giraldii* Dode), *Brucea javanica* (L.) Merr. (*Brucea amarissima* [Lour.] Desv. ex Gomes, *Brucea sumatrana* Roxb., *Gonus amarissimus* Lour.), *Eurycoma longifolia* Jack, *Eurycoma apiculata* Benn., *Harrisonia perforata* (Blco.) Merr. (*Harrisonia paucijuga* [Benn.] Oliv.), *Picrasma javanica* Bl., *Picrasma quassinoides* (D. Don.) Benn. (*Picrasma alanthoides* [Bge.] Planch.), *Quassia indica* (Gaertn.) Nootebom (*Samadera indica* Gaertn.), and *Soulamea amara* Lamk. are of medicinal value. Note that these bitter plants are often used to treat amoebiasis and malaria, to counteract putrefaction, and to reduce fever.

29.2 *EURYCOMA APICULATA* BENN.

[From: Greek *eurus* = broad and kome = hairs of the head, and from Latin *apiculata* = ending somewhat abruptly in a short or sharp point or apex.]

29.2.1 Botany

Eurycoma apiculata Benn. is a shrub that grows up to 1.5m from a stout tape root in the rain forests of Southeast Asia. The bark is a slightly roughened fawn with gray color. The leaves are crowned on top of the bole, growing up to 55cm long, and are handsomely imparipinnate. The rachis is slender and sustains about 19 pairs of folioles which are sessile, measuring 11.25cm × 3.75cm, thin, rounded at the base, and subacute at the apex with reticulate venations. The folioles show 16 pairs of secondary nerves which are discrete. The flowers are pink and minute. The four petals are glabrous inside, and several times as long as wide. The stigma is sessile on a short ovary. There are up to five fruits, that are orange, short-stalked, ellipsoid, ovoid, and 1cm × 5mm – 1.7cm × 1.2cm (Figure 29.3).

29.2.2 Ethnopharmacology

The Malays call the plants *therung, tongkat baginda, penawar serama,* or *bedara pahit.* The plant is very similar in shape to *Eurycoma longifolia* Jack, but has a much higher repute. A few pieces of the root bark are boiled in water to make a drink, which is taken as an aphrodisiac, and as a tonic, to mitigate pain in the bones, and to reduce fever. A decoction of the leaves is used to calm itchiness of the skin. The bark is applied externally to heal wounds and ulcers, and to mitigate headaches. In Indonesia, a decoction of the roots is used as a drink to reduce fever, diarrhea, and to deflate swelling.

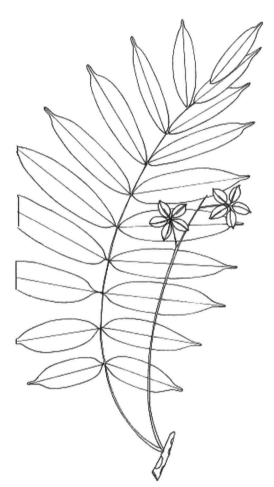

Figure 29.3 *Eurycoma apiculata* Benn. [From: Flora of Peninsular Malaysia. Forest Department. Geographical localization: Lalang River, Forest Reserve of Kajang. March 10, 1930. No: 22743. Field collector: C. F. Symington. Botanical identification: H. P. Nootebom, November 1960.]

The aphrodisiac properties of the plant have attracted a great deal of interest and to date there is a large body of work trying to substantiate this effect in the rodent.[6–8] However, there is only a little evidence which clearly assesses the toxicity of the plant. Can we reasonably expect a plant abounding with quassinoids to be devoid of toxic effects? Probably not. The very little toxicological evidence presented thus far is consistent with the view that the plant is poisonous and should not be used internally.[9] Does it promote prostate tumors, one wonders?

29.3 *QUASSIA INDICA* (GAERTN.) NOOTEBOM

[After the name of an African slave in Surinam who reported the medicinal properties of the wood to Dahlberg, a friend of Linnaeus, and from Latin *indica* = from India.]

29.3.1 Botany

Figure 29.4 *Quassia indica* (Gaertn.) Nootebom. [From: Ex Herbario Kewensis. FRI No: 35423. Geographical localization: In peat swamp forest, Baging River. Oct. 6, 1989. Botanical identification: H. P. Nootebom, 1977.]

Quassia indica (Gaertn.) Nootebom (*Samadera indica* Gaertn.) is a tree that grows up to 20m high. It is abundant in the tidal swamps of Madagascar, Ceylon, Burma, Andamans, Cambodia, Laos, and Vietnam, to the Bismarcks and the Solomons. It has pale yellow, brownish, fairly rough bark, and white to pale yellow wood. The leaves are laurel green, glossy above, and pale lime green below. Its petiole measures 2.2cm, and is grooved, cracked transversally, and woody. The blade is 17.5cm × 5.9cm – 20cm × 6.3cm, oblong–lanceolate, and shows about 10 pairs of secondary nerves. Its inflorescence is 12cm, umbelliform, with about 10 little flowers. The flower pedicel is 1.5cm long. The fruits are grouped in whorls of 1–4 carpels, and are green, blotched red, ovoid, and flattened. The seeds are white and inconspicuous. The fruit measures 6.3cm × 3.2cm (Figure 29.4).

29.3.2 Ethnopharmacology

The bark of *Quassia indica* (Gaertn.) Nootebom is used to reduce fever in Burma, Indonesia, and the Philippines. In the Philippines, chips of wood are put in coconut oil which is used as a drink as a purgative, to reduce fever. The plant is also used as a tonic and insecticide and it is applied externally as a liniment for rheumatism and bruises. The plant elaborates an interesting series of quassinoids, including samaderines, indaquassin, 2-*O*-glucosylsamaderine C, and simarinolide, which prevent the survival of the chloroquine-resistant K1 strain of *Plasmodium falciparum*, and exhibit cytotoxic properties, particularly the inhibition of endothelial cell–neutrophil leukocyte adhesion as well as antiinflammatory activity,[10,11] especially against ants. In the Solomon Islands the seeds are used to make a drink, which is taken to reduce fever. In Burma and Indonesia, the seeds are used externally to treat rheumatism. The leaves are used to kill vermin in Indonesia and in the Solomon Islands. In Borneo, the wood is used for making knife handles, and the seeds are used as an emetic and for treating fever.

REFERENCES

1. Guo, Z., Vangapandu, S., Sindelar, R. W., Walker, L. A., and Sindelar, R. D. 2005. Biologically active quassinoids and their chemistry: potential leads for drug design. *Curr. Med. Chem.*, 12, 173.
2. Anderson, M. M., O'Neill, M. J., Phillipson, J. D., and Warhurst, D. C. 1991. *In vitro* cytotoxicity of a series of quassinoids from *Brucea javanica* fruits against KB cells. *Planta Med.*, 57, 62.

3. Cuendet, M. and Pezzuto, J. M. 2004. Antitumor activity of bruceantin: an old drug with new promise. *J. Nat. Prod.*, 67, 269.
4. Chan, K. L., Choo, C. Y, 2002. The toxicity of some quassinoids from *Eurycoma longifolia, Planta Med.*, 68, 662.
5. Jiwajinda, S., Santisopasri, V., Murakami, A., Sugiyama, H., Gasquet, M., Riad, E., Balansard, G., and Ohigashi, H. 2003. *In vitro* anti-tumor promoting and anti-parasitic activities of the quassinoids from *Eurycoma longifolia*, a medicinal plant in Southeast Asia. *J. Ethnopharmacol.*, 85, 173.
6. Ang, H. H., Ngai, T. H., and Tan, T. H. 2003. Effects of *Eurycoma longifolia* Jack on sexual qualities in middle aged male rats. *Phytomedicine*, 10, 590.
7. Ang, H. H. and Lee, K. L. 2002. Effect of *Eurycoma longifolia* Jack on libido in middle-aged male rats. *J. Basic Clin. Physiol. Pharmacol.*, 13, 249.
8. Ang, H. H., Lee, K. L., and Kiyoshi, M. 2004. Sexual arousal in sexually sluggish old male rats after oral administration of *Eurycoma longifolia* Jack. *J. Basic. Clin. Physiol. Pharmacol.*, 15, 303.
9. Chan, K. L. and Choo, C. Y. 2002. The toxicity of some quassinoids from *Eurycoma longifolia*. *Planta Med.*, 68, 662.
10. Kitagawa, I., Mahmud, T., Yokota, K., Nakagawa, S., Mayumi, T., Kobayashi, M., and Shibuya, H. 1996. Indonesian medicinal plants. XVII. Characterization of quassinoids from the stems of *Quassia indica*. *Chem. Pharm. Bull. (Tokyo)*, 44, 2009.
11. Koike, K. and Ohmoto, T. 1994. Quassinoids from *Quassia indica*. *Phytochemistry*, 35, 459.

CHAPTER 30

Medicinal Plants Classified in the Family Meliaceae

30.1 GENERAL CONCEPT

The family Meliaceae (A. L. de Jussieu, 1789 nom. conserv., the Mahogany Family) comprises 51 genera and 550 species of tropical trees, which can be recognized in the field by their leaves, which are like the Simaroubaceae, elongated and compound, and by their flowers, which include a pseudo-tubular androecium of 6–10 stamens, and principally by their fruits, which are capsular with winged seeds, showy berries, or drupes containing several seeds arranged in a whorl, which look like a little pumpkin (Figure 30.1). The wood of several trees in this family is valuable as timber, for instance, with *Swietenia mahogani* (L.) Jacq. (mahogany) and *Cedrela odorata* L.

With regard to the pharmacological properties of Meliaceae, there is a massive body of evidence to support the concept that limonoids which abound in this taxon might provide in the near future, if enough work is devoted to it, antineoplastic agents of clinical value (Figure 30.2).

Figure 30.1 Fruits of Meliaceae. **(See color insert following page 168.)**

Perhaps no other single species of Meliaceae has aroused more interest in the field of chemistry and pharmacology research into limonoids than *Azadirachta indica* A. Juss. (*Melia azedarach* L.) (*Azadirachta, Indian Pharmaceutical Codex*, 1953), or *Margosa neem*, the stem bark, root bark, and leaves of which have been used for a very long time.[1,2] In addition, the following are all of medicinal value in the Asia–Pacific and would repay investigation: *Aglaia odorata* Lour., *Aphanamixis rohituka* (Roxb.) Pierre, *Aphanamixis grandifolia* Bl. (*Amoora aphanamixis* Roem. & Schult.), *Azadirachta indica* A. Juss. (*Melia azedarach* L.), *Sandoricum koejape* (Burm. f.) Merr. (*Sandoricum indicum* Cav., *Sandoricum nervosum* Bl.), *Toona sinensis* (Juss.) Roem., *Toona sureni* (Bl.) Merr., *Trichilia connaroides* (Wight & Arn.) var. *microcarpa* (Pierre) Bentvelzen, *Chukrasia*

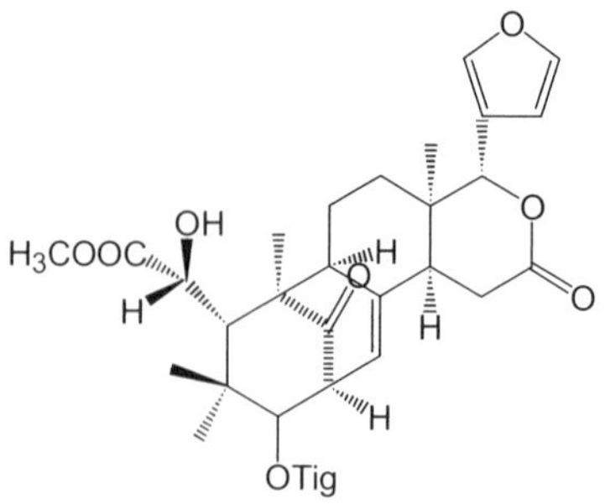

Swietenin

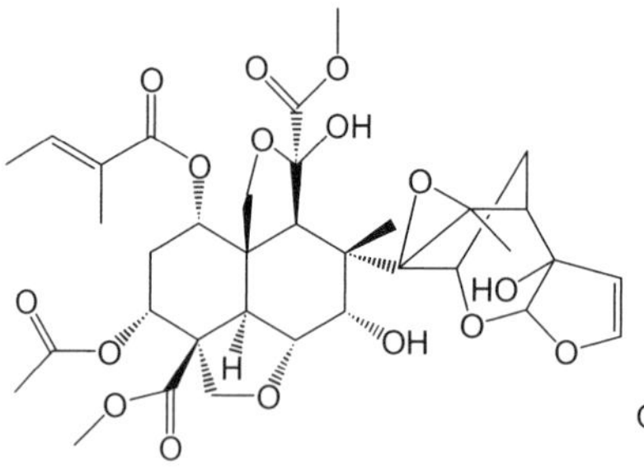

Azadirachtin

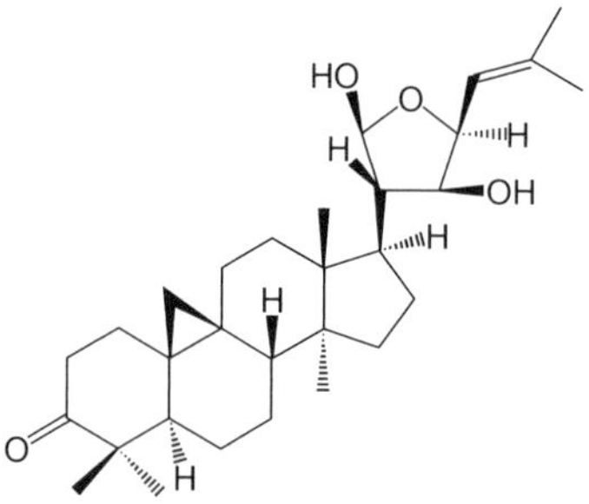

Argenteanone A

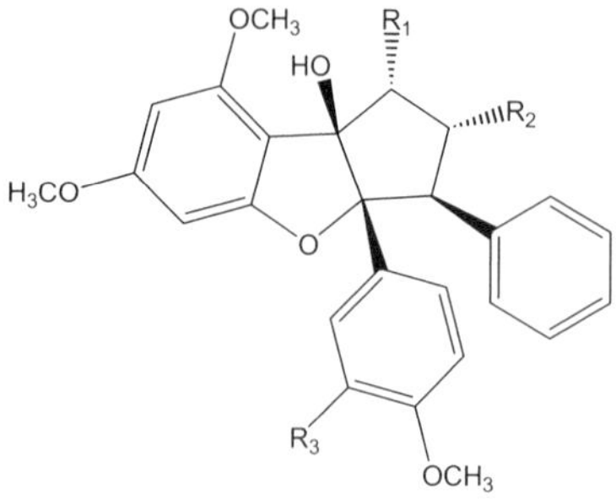

Rocaglamide

Figure 30.2 Examples of bioactive natural products from the family Meliaceae.

tabularis A. Juss., *Walsurea elata* Pierre, *Chisocheton penduliflorus* Planch. ex Hiern, and *Swietenia mahogani* (L.) Jacq.

30.2 *AGLAIA ODORATA* LOUR.

[From: Greek *Aglaia* = one of the Graces who presided over the Olympic Games, referring to the beauty of the *Aglaia* species, and from Latin *odorata* = fragrant.]

30.2.1 Botany

Aglaia odorata Lour. is a tree native to Southeast Asia and grown as an ornamental tree throughout the Pacific Rim. The plant grows to a height of 6m. The bole is grayish and crooked. The bark is light and ash-colored. The stems are lenticelled and 2.5mm in diameter. The leaves comprise two pairs of folioles and a terminal one. The rachis is finely winged and measures 4.3cm. The folioles are 3.2cm × 2.4–5.7cm × 3–4.7cm × 2.2cm and show 4–6 pairs of secondary nerves. The inflorescence is a 12.5cm-long light yellow lax panicle of tiny flowers (Figure 30.3).

30.2.2 Ethnopharmacology

Known as Mock Lemon, Chinese Perfume Plant, and Chinese Rice Flower, *Aglaia odorata* Lour. is an antipyretic remedy in the Philippines, Vietnam, and Malaysia. In the Philippines, a decoction of roots is used as a drink to reduce fever. In Cambodia, Laos, and Vietnam, the leaves and roots are used to invigorate, to reduce fever, and to calm itchiness. In Malaysia, the flowers are infused in water to make a drink taken to reduce fever. In Indonesia, a decoction of leaves is used as a drink to reduce menses and to treat venereal diseases. The antipyretic properties have not yet been substantiated exper-

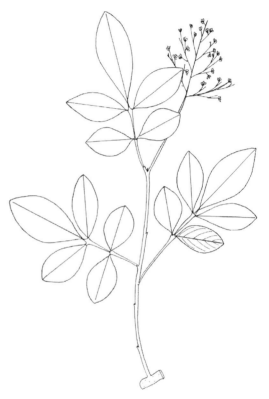

Figure 30.3 *Aglaia odorata* Lour. [From: Flora of Johor. Comm. Ex. Herb. Hort. Sing. Geographical localization: Kota Tinggi. July 1929. Field collector: T. Teruya. No: 779. Botanical identification: C. M. Pannel, May 6, 2005.]

imentally. In China, the flowers are used to scent tea, and the tender leaves are eaten as vegetables. The plant is known there as *san yeh lan* or *lan hwa mi*. In the last decade the genus *Aglaia* has attracted considerable attention as a source of cytotoxic cycloartane limonoids and cyclopentatetrahydrobenzofuran lignans flavaglines including aglafolin, rocaglamide, desmethylrocaglamide, didesmethyl-rocaglamide, and aglaiastatin.[3–5]

Argenteanone A and B from *Aglaia argentea* Lour. prevented the survival of KB cells with IC_{50} values of 7.5μg/mL and 6.5μg/mL, respectively.[6] Cyclopenta[b]benzofurans didesmethyl-rocaglamide inhibit the proliferation of MONO-MAC-6 and MEL-JUSO cell lines cultured *in vitro* with IC_{50} of 0.004μM and 0.013μM dose-dependently.[7]

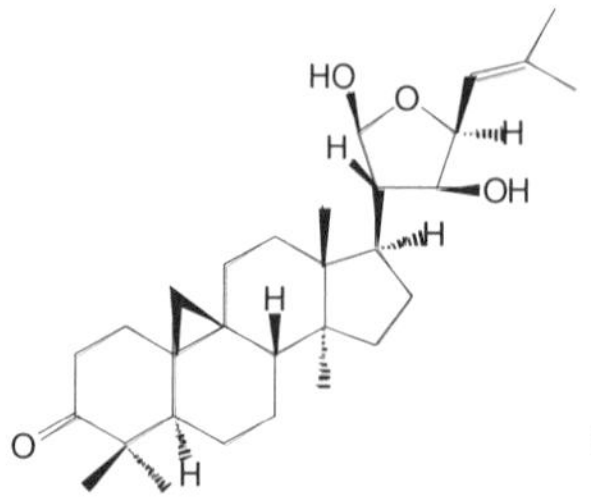

Argenteanone A

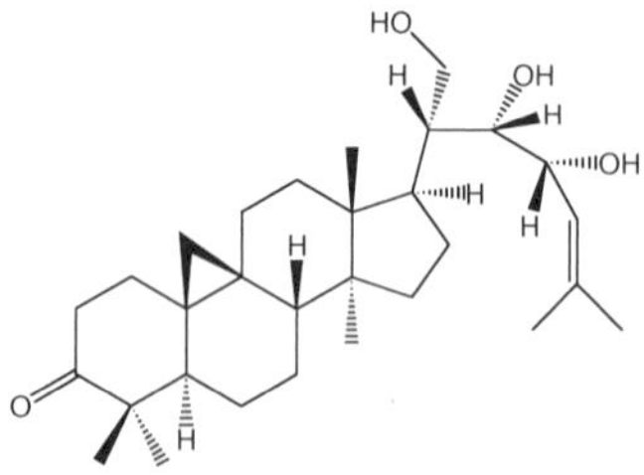

Argenteanone B

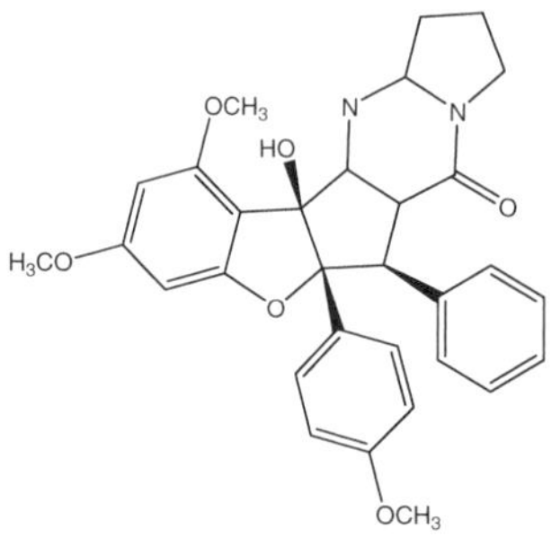

Aglaiastatin

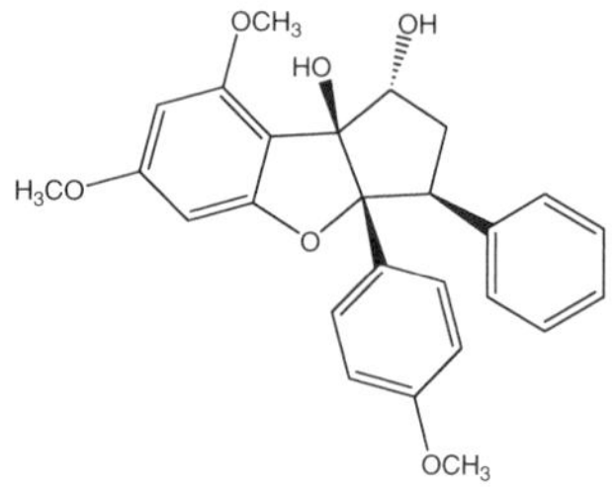

Rocaglaol

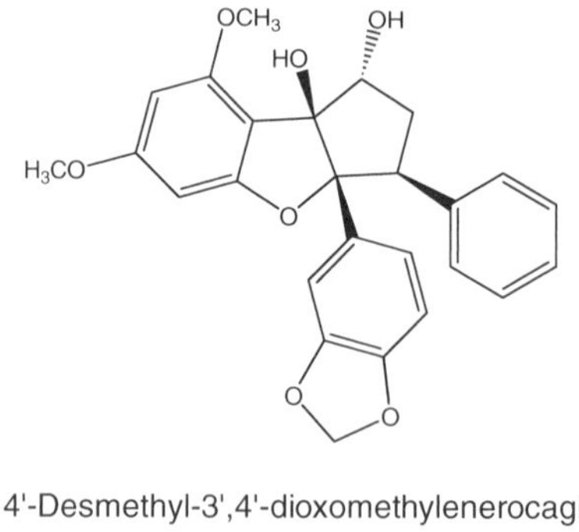

4'-Desmethyl-3',4'-dioxomethylenerocaglaol

Figure 30.4 Cytotoxic principles of *Aglaia*.

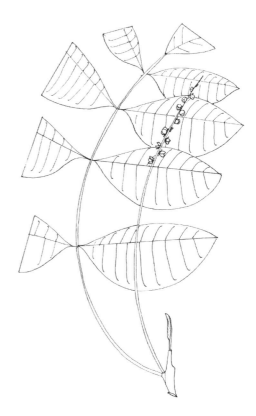

Figure 30.5 *Aphanamixis grandifolia* Bl. [From: FRI
No: 5723. Botanical identification: F. S. P.
Ng, 1967.]

Nanomolar concentrations of aglaiastatin isolated from the leaves of *Aglaia odorata* Lour.
inhibited the growth of K-ras-NRK cells and induced normal morphology in them. It also induced
apoptosis in SW480 and HT29/HI1 carcinoma cells via a p38-mediated stress pathway. In SW480
cells, aglaiastatin stops the cellular cycle in early mitosis.[7–9] What is the antipyretic mechanism of
action of this plant? The question awaits an answer.

30.3 *APHANAMIXIS GRANDIFOLIA* BL.

[From: Greek *aphanos* = invisible and *mixis* = mating, an allusion to the sex organs in the
staminal tube, and from Latin *grandis* = large and *folia* = leaves.]

30.3.1 Botany

Aphanamixis grandifolia Bl. (*Amoora grandifolia*
Bl. *Aglaia aphanamixis* Pellegr., *Amoora grandifolia*
[Bl.] Walp.) is a tree that grows to a height of 12m.
The bark is white, and the inner bark is fibrous and
pink. The folioles are broadly lanceolate and show
about 10 pairs of secondary nerves. The inflorescences are spikes. The flowers are minute and yellowish with an apical pore. The fruits are globose
(Figure 30.5).

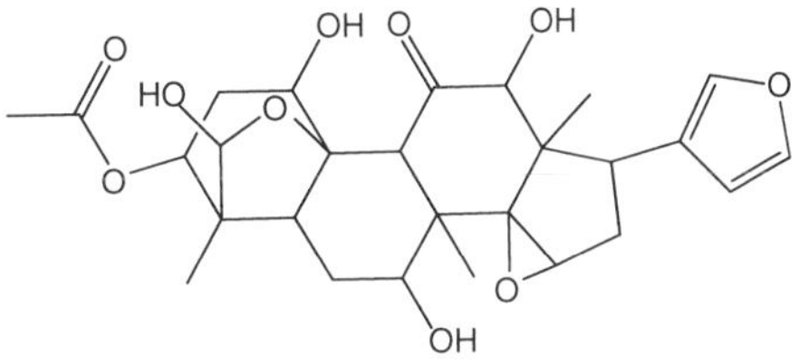

12-Hydroxyamoorastatin

Figure 30.6

30.3.2 Ethnopharmacology

Indonesians drink a decoction of the bark to treat a cold. The seeds are known to contain a series of limonoids, including 12-hydroxyamoorastatin, which has inhibited the growth of the murine P-388 lymphocytic leukemia cell lines.[10]

30.4 *APHANAMIXIS ROHITUKA* (ROXB.) PIERRE

[From: Greek *aphanos* = invisible and *mixis* = mating, an allusion to the sex organs in the staminal tube.]

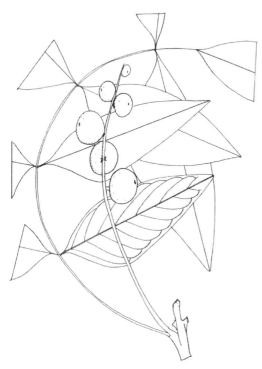

Figure 30.7 *Aphanamixis rohituka* (Roxb.) Pierre. [From: Flora of Malaya. Geographical localization: Logging area in Pahang. FRI No: 28439.]

30.4.1 Botany

Aphanamixis rohituka (Roxb.) Pierre (*Aphanamixis polystachya* [Wall.] Parker, *Aglaia beddomei* [Kosterm.] Jain & Gaur, *Aglaia polystachya* Wall. ex Roxb., *Amoora rohituka* [Roxb.] Wight & Arn., *Aphanamixis sinensis* How & Chen, *Ricinocarpodendron polystachyum* [Wall.] Mabb., *Aphanamixis cumingiana* C. DC., *Amoora aphanamixis* Schult. & Schult. f., and *Ricinocarpodendron sinense*) is a tree that grows to a height of 35m with a girth of 80cm in the rain forests of a geographical area spanning from India to South China to the Solomon Islands. The bark is grayish-brown, scaly, and cracking. The inner bark is red. The leaves are imparipinnate, spiral, and exstipulate. The rachis is up to 1m long and supports up to 15 folioles which are thick and lanceolate, and show about 10 pairs of secondary nerves sunken above. The inflorescences are 60cm yellowish-white spikes. The flowers are minute and are equipped with six yellowish white anthers. The fruits are 4cm in diameter, somehow pumpkin-like, red to pink, on pendulous stalks, and exude a white latex after incision. The seeds are dark brown and glossy, and partially embedded in a red aril (Figure 30.7).

30.4.2 Ethnopharmacology

The vernacular names for *Aphanamixis rohituka* (Roxb.) Pierre include *baddiraj* (Bangladesh), *komalo* (Molucca), *vellakangu* (Tamil), and *elahirilla* (Sanskrit). In Burma, the plant is known as *thit-nee* and the bark provides an astringent remedy. The Taiwanese use the oil expressed from the seeds as medicine.

Jagetia and Venkatesha made the interesting observation that in rodents Ehrlich ascites carcinoma (EAC), exposed to 8 Gy hemi-body gamma radiation, is optimized by 50mg/Kg of ethanolic extract of *Aphanamixis polystachya*.[11]

30.5 *CHISOCHETON PENDULIFLORUS* PLANCH. EX HIERN.

[From: Greek *schizo* = split and *chiton* = tunic, an allusion to the deeply lobed staminal tube, and from Latin *penduliflorus* = with hanging flowers.]

30.5.1 Botany

Chisocheton penduliflorus Planch. ex Hiern. is a tree that grows in the lowland rain forests of Malaysia, Thailand, and Riouw Archipelago up to an altitude of 900m. It is a small tree, no more than 11m in height and 30cm in girth. The bark is blackish. The inner bark is pale fawn. The stems are velvety. The leaves are imparipinnate, spiral, and exstipulate. The rachis is velvety, 40cm long, showing up to eight pairs of folioles, and pubescent below. The folioles are 14–10cm × 5.3–6cm, obovate, pubescent below, and show-ing 11–22 pairs of secondary nerves. The inflo-rescences are 15–23.5cm long with five tiny dull red flowers grouped at the apex. The corolla con-sists of 4–5 petals. The staminal tube has 3–5 apical forked lobes, and the androecium com-prises 3–5 anthers. The nectary disc is obscure and cup-shaped. The ovary is discoid to rounded. The fruits are 3.5cm × 1.2cm and beaked when young, to 5cm long splitting into three valves and contains three glossy seeds which are black with a partial red aril (Figure 30.8).

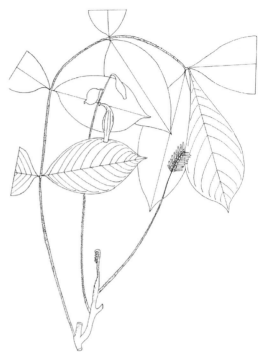

Figure 30.8 *Chisocheton penduliflorus* Planch. ex Hiern. [From: Distributed from The Botanic Gardens, Singapore. Malay Pen-insula. Geographical localization: Koh Mai Forest Reserve, Kedah. April 4, 1938. Botanical identification: M. R. Henderson. Field collector: M. S. Kiah. From FRI No: 28214. Botanical identifi-cation: D. J. Mabberley, July 21, 1987.]

30.5.2 Ethnopharmacology

The fruits are eaten by the Malays as medic-inal food or *ulam*. At present, little evidence is available to support the use of the *Chisocheton* species. The presence of limonoids has been demonstrated by Gunning et al.[12] and Bordoloi et al.[13] 1,2-Dihydro-6-acetoxyazadirone, from the fruits of *Chisocheton paniculatus*, has been shown to be antifungal *in vitro*.[13] The plant is on the verge of extinction.

30.6 *DYSOXYLUM ALLIACEUM* BL.

[From: Greek *dys* = unpleasant and *xylon* = wood, and from Latin *alliaceum* = like onion, owing to the smell of the plant.]

30.6.1 Botany

Dysoxylum alliaceum Bl. (*Dysoxylum thyrsoideum* Griff., *Dysoxylum costulatum* Miq., and *Dysoxylum pulchrum* Ridl.) is a stout tree that grows to a height of 38m with a girth of 2m. The crown is silvery-gray, and the bole is straight and buttressed. The plant grows from lowland rain forests to 800m in the primary rain forests of a zone covering the Andaman Islands to the Solomon

Figure 30.9 *Dysoxylum alliaceum* Bl. [From: Flora of Malaya. FRI No: 4474. Geographical localization: Southeast Kelantan, Ulu S. Aring, near Tarang Village in disturbed lowland undulating forest. Field
collector: T. C. Whitmore, Sept. 22, 1967. Botanical identification: D. J. Mabberley, July 21, 1987.]

Islands. The bark is gray with a few scales. The inner bark is purple–red. The wood has a slight
garlic smell. The leaves are compound, spiral, and exstipulate. The rachis is about 20cm long. The
folioles are 8.1cm × 3–9.3cm × 3.8–10cm × 4.5–8cm × 4.6–5.3cm × 12cm, with 10 pairs of
secondary nerves sunken above. The fruits are gray–green, flushed mauve, hairy, and 3.7cm × 4
cm to 7.5cm in diameter. The sarcotesta is white and 2cm (Figure 30.9).

30.6.2 Ethnopharmacology

The plant is not medicinal and the seeds are known to be poisonous by natives of the countries
where the plant grows. The fruits are not eaten by birds or mammals. The Meliaceae family is a
well-known source of structurally complex, degraded triterpenoids and limonoids. Members of the
genus *Dysoxylum*, however, produce a series of dammarane, glabretal, and apotirucallane triterpenoids, as well as diterpenes which are cytotoxic and nonedible (Figure 30.10). Dysokusones A, B,
and C, from the stems of *Dysoxylum kuskusense,* prevent the survival of HL-60(TB) cells cultured
in vitro with EC_{50} values of 2.25, 6.35, and 2.37µM, respectively. Dysokusone A is cytotoxic against
K-562 and NCI-H522 cells with EC_{50} values of 5.04mM and 4.80mM, respectively.[14] Nymania 3,
a limonoid from *Dysoxylum malabaricum*, is inedible.[15] Dysoxylumic acids A, B, and C, from the
bark of *Dysoxylum hainanense* Merr., show significant toxic activity against *Pieris rapae* L.[16] What
are the cytotoxic principles of *Dysoxylum alliaceum* Bl.?

30.7 *DYSOXYLUM CAULIFLORUM* HIERN.

[From: Greek *dys* = unpleasant and *xylon* = wood, and from Latin *alliaceum* = like onion,
owing to the smell of the plant, and from Latin *cauliflorum* = bearing flowers on older branches.]

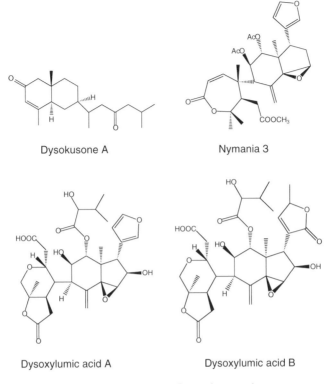

Dysokusone A   Nymania 3

Dysoxylumic acid A   Dysoxylumic acid B

Figure 30.10 Cytotoxic and antifeedant principles of the *Dysoxylum* species.

30.7.1 Botany

Dysoxylum cauliflorum Hiern. is a tree that can reach a height of up to 30m with a girth of 1.5m. Its bole is straight and fluted. The bark is gray and smooth with lenticels, and bosses of defunct inflorescences that are mottled brown with rectangular flakes. The stems are fissured. The inner bark is fawn with a sour smell. The wood is straw-colored. The apical buds of the leaves look like little fists. The leaves are imparipinnate, and up to 60m long × 4mm wide, with up to six pairs of folioles which are subopposite and bullate on both sides. The folioles show reticulate venations and about 9–12 pairs of secondary nerves. The folioles are 11cm × 3–10cm × 4.5–13cm × 5.3–14.5cm × 6.5–10.5cm × 6cm. A few tertiary nerves are visible below. The inflorescences are woody spikes enveloping woody tubercles on the bole and large stems. The flowers are 7mm long, creamy white, fragrant, peppery-scented, and comprise four inflexed petals, a tube which is lobed, forked lobes, eight anthers, a cylindrical disc, a 4-locular ovary, and a protruding stigma. The fruits are globose, 2.5cm × 1.3–4cm in diameter, dehiscent with four valves, with a milky latex. There are 1–4 seeds with an aril covering half of a black testa (Figure 30.11).

30.7.2 Ethnopharmacology

The Malays and Jahuts use a poultice of fruits is used to treat rheumatism. A plaster of boiled roots is applied to treat abdominal pain. The fruits are toxic. One might set the hypothesis that the medicinal properties of the plant are owed to counterirritancy. The fresh fruits are known to abound with dammarane triterpenoids including ocotillone, ocotillol-II, cabralealactone, shoreic acid, and eichlerialactone (Figure 30.12), which are probably cytotoxic and antiplasmodial.[17–19]

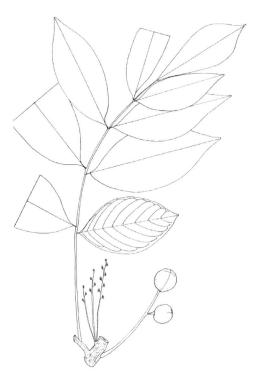

Figure 30.11 *Dysoxylum cauliflorum* Hiern. [From: Flora of Kelantan. Comm. Ex. Hort. Bot. Sing. Geographical localization: Uru River, Kelantan. Alt.: 500m. March 16, 1972. Field collector: M. Shah. No: 2573.]

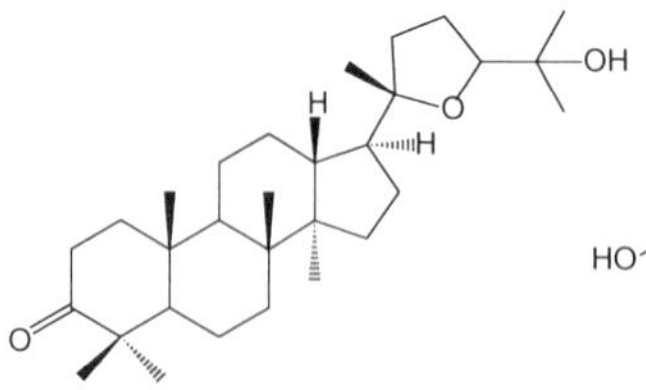

Ocotillone

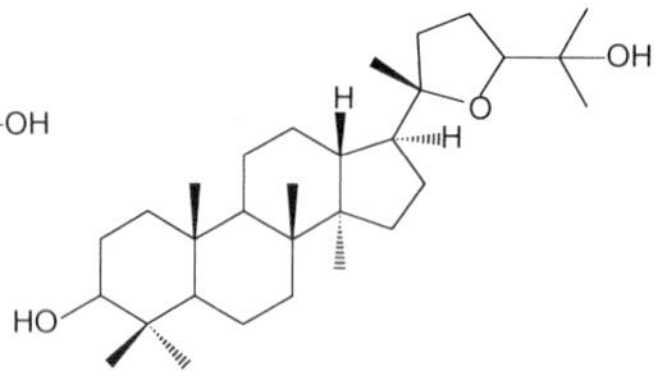

Ocotillol-II

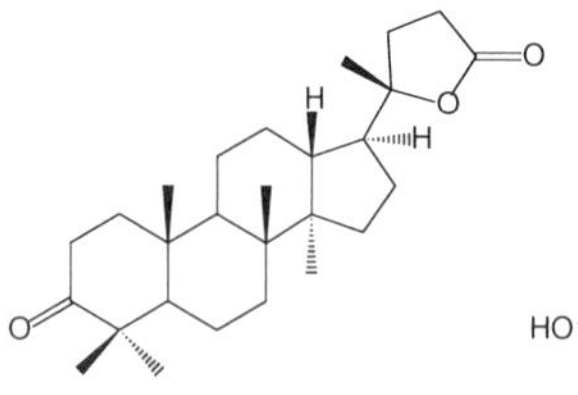

Cabralealactone

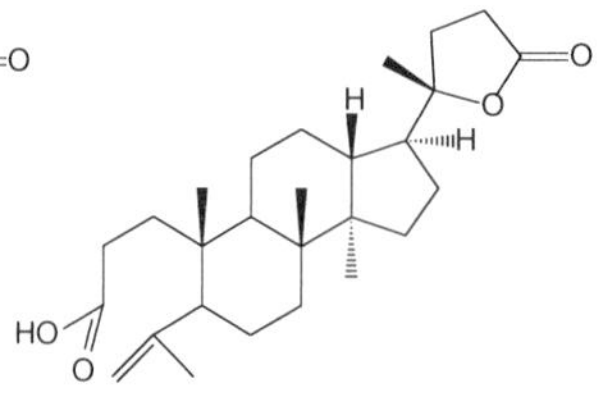

Eichlerialactone

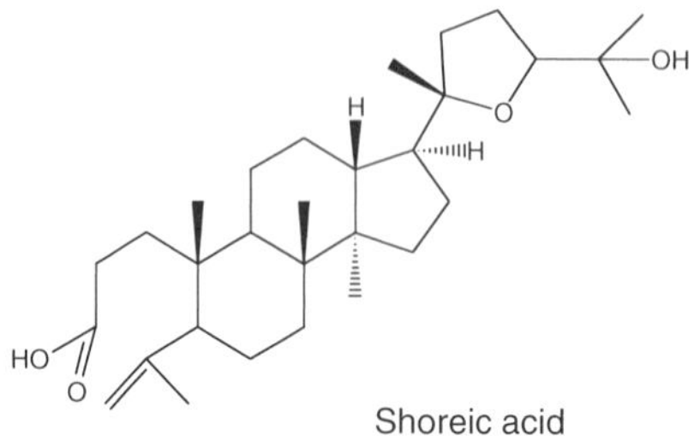

Shoreic acid

Figure 30.12

30.8 *SANDORICUM KOEJAPE* (BURM. F.) MERR.

[From: Moluccan *sandori* = Sandoricum species and from Javanese *kechape* = a name for this tree.]

30.8.1 Botany

Sandoricum koejape (Burm. f.) Merr. (*Sandoricum maingayi* Hiern. and *Sandoricum nervosum* Bl.) is a big tree that grows to a height of 45m with a girth of 2.4m, in the primary rain forests of Southeast Asia up to an altitude of 1200m. The bole is straight and buttressed, the buttresses are up to 5m. The bark is pale pinkish-brown, smooth, and lenticelled. The inner bark is pink. The stems are fissured longitudinally, rough, gray–brown, and lenticelled. The leaves are trifoliolate and hairy, and arranged in terminal spirals which are up to 40cm long. The petiole is 10cm long and somewhat flat at the apex. The petiolules are 5mm long. The rachis is 15cm long and supports three folioles which are 14.6cm × 8–6cm × 8–14.8cm × 11cm and show 10–12 pairs of secondary nerves sunken above. The midrib is broad, flat, and conspicuous above. The flowers are fragrant and in axillary panicles which are 15cm long and drooping. The calyx is truncate and frog-green. The corolla comprises five petals which are yellowish and free. The staminal tube is whitish, cylindrical, ribbed above, and tipped with 10 short appendages. The androecium comprises 10 anthers. The ovary is 4–5-lobed as well as the stigma. The fruits are 3.7cm × 3.6cm, rough, velvety, yellow–red, edible, and with laticiferous drupes (Figure 30.13).

Figure 30.13 *Sandoricum koejape* (Burm. f.) Merr. [From: Flora of Singapore. Comm. Ex. Herb. Hort. Bot. Sing. Geographical localization: Botanic Gardens. March 28, 1933. Field collector: T. Teruya. No: 2330. From FRI No: 39739.]

30.8.2 Ethnopharmacology

The roots of *Sandoricum koejape* (Burm. f.) Merr. are used to treat intestinal disorders throughout Southeast Asia. In the Philippines, the roots are used to promote digestion, and the leaves are used externally to promote sweating; mixed with water they are used as a bath to reduce fever. In Indonesia, the roots are used to treat colic and leukorrhea. In Malaysia, the plant is known as *sentul* and used as a tonic to aid recovery from the exhaustion of childbirth. A paste of the bark is applied to ringworms. The juice squeezed from the leaves is used as a drink to reduce fever, and a decoction of the leaves is used as a drink to stop diarrhea. The Burmese use the roots to stop dysentery. The antipyretic activity of the plant is not substantiated experimentally as of yet, but one might think of a steroid-like mechanism, since 3-oxo-12-oleanen-29-oic acid and katonic acid isolated from the plant protected mice against ear inflammation caused by tetradecanoylphorbol acetate.[16]

Katonic acid inhibits the Epstein–Barr virus early antigen (EBV-EA) activation induced by 12-*O*-tetradecanoylphorbol 13-acetate (TPA); and koetjapic acid alleviates tumor promotion in two-stage mouse skin carcinogenesis induced by 7,12-dimethylbenz(a)anthracene and promoted by TPA.[20] Both 3-oxo-olean-12-en-29-oic acid and katonic acid influence the survival of P-388 cells

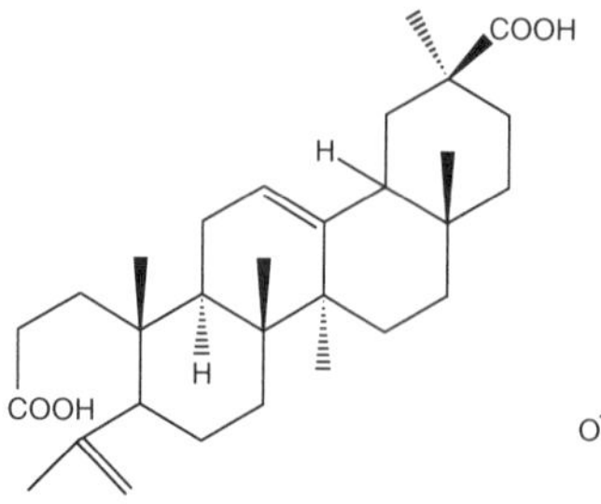

Koetjapic acid 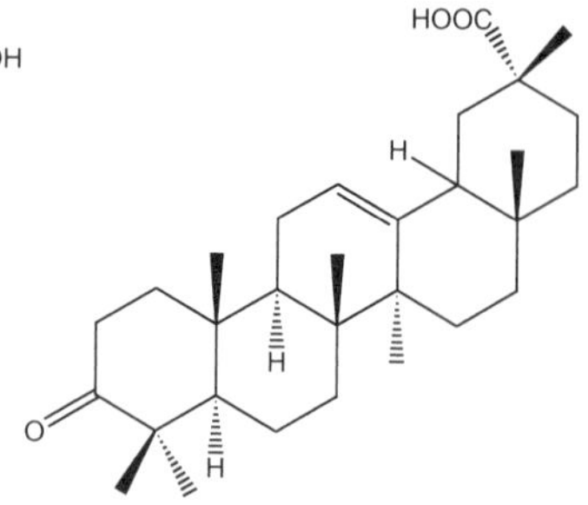   Katonic acid

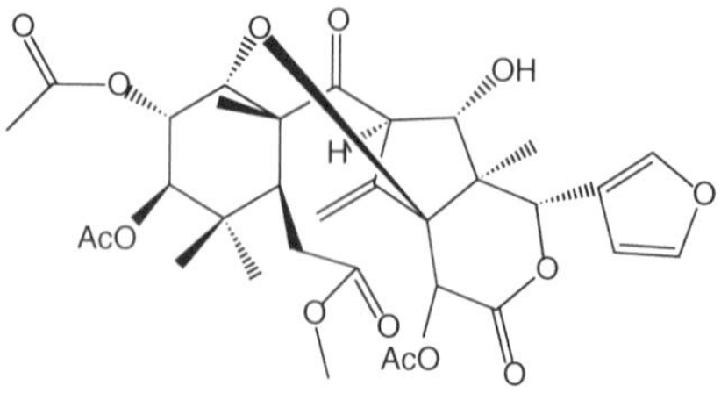

Sandrapin A 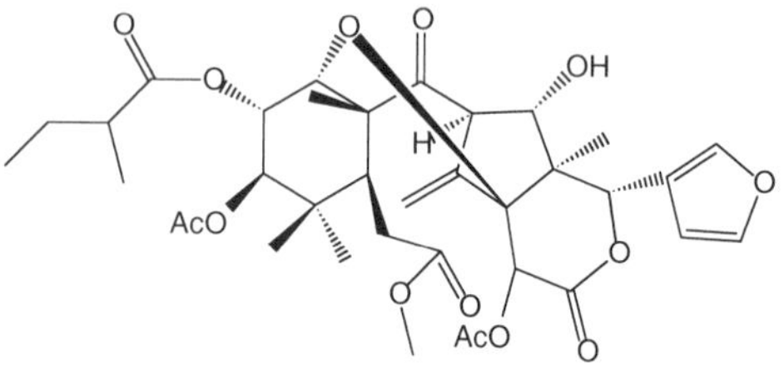   Sandrapin B

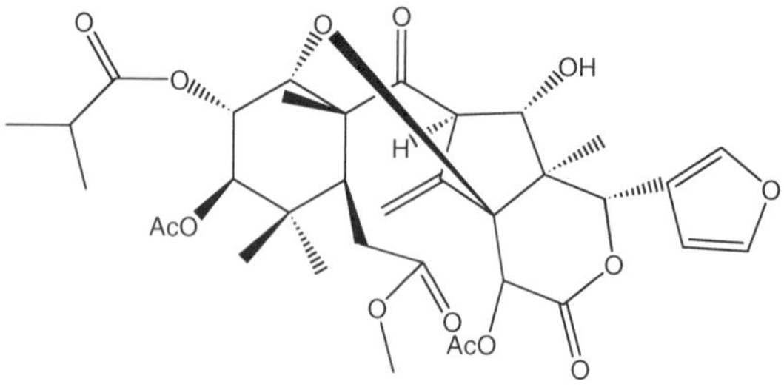

Sandrapin C

Figure 30.14

with ED$_{50}$ values of 0.61μg/mL and 0.11μg/mL, respectively.[21] Note that the leaves of *Sandoricum koejape* (Burm. f.) Merr. abound with trijugin-class limonoids including sandrapins A, B, and C (Figure 30.14), the pharmacological potential of which would be well worth investigating.[22]

The toxicity of the seeds is probably owed to limonoids, such as sandoricin and 6-hydroxysandoricin, which are effective antifeedants.[19–24]

30.9 *TOONA SINENSIS* (JUSS.) ROEM.

[From: Indian *toon* = the name of *Toona* species and from Latin *sinensis* = from China.]

30.9.1 Botany

Toona sinensis (Juss.) Roem. (*Cedrela sinensis* Juss. and *Cedrelas serrata*) is a big tree, 36m tall with a girth of 1.6m, which grows in tropical Asia, from Nepal and China and Taiwan eastward in the highlands of Sumatra and Java. It is cultivated in Europe as a street ornamental. The bole is fissured and shows small bulk, 1m in height and with 30cm buttresses. The bark is fissured and gray. The inner bark is meat-red and laminated, and has a pepperish-garlic smell. The wood is white and valued for making Chinese furniture. The stems are glabrous and lenticelled. The leaves are paripinnate, spiral, and exstipulate. The rachis is 30cm long and holds 10–17 pairs of folioles. The petiolules are 1.2cm long. The blade is serrate, lanceolate, asymmetrical, and apiculate with a 1cm-long tail, and 12.7cm × 3.4–9cm × 2.5–12.2cm × 3–13cm × 3cm. The inflorescences are 70cm- to 1m-long panicles of tiny flowers pendant behind leafy stems. The flowers evoke a powerful sour smell detected 100ft away from the tree. The petals are white, flushed green, and 2–5mm long. The fruits are 2.3cm × 1cm, 5-partite dehiscent fusiform capsules, splitting into five valves around a pentagonal column. The seeds are winged at one end (Figure 30.15).

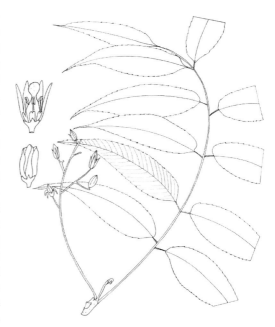

Figure 30.15 *Toona sinensis* (Juss.) Roem. [From: Flora of the Malay Peninsula. Geographical localization: Path to Ayer Baka, Cameron Highlands. Alt.: 4500ft. June 6, 1933. Field collector: E. F. Symington. No: 31033. Botanical identification: J. M. Edmonds, Feb. 25, 2000.]

30.9.2 Ethnopharmacology

The vernacular names of the plant include Chinese Cedar, Chinese Mahogany, Red Toon; *chuen tien shu* (Chinese), *surian* (Malay), and *surian bawang* (Indonesian). In China, the bark is boiled in water to make a drink which is taken to treat red spots on skin. The root bark affords a remedy for gynecological troubles, including irregular menses. The powdered roots are used to cool and to promote urination. The young leaves are eaten to stimulate digestion. The gynecological properties of the plant are unexplored as of yet, but Poon et al. made the interesting observation that an extract of the plant significantly inhibited both basal and human chorionic gonadotropin (hCG)-stimulated testosterone productions in the rodent.[25]

The evidence currently available shows that the plant has an antiproliferative effect on human lung cancer cells, and improves the secretion of insulin in diabetic rats.[26–28] One might consider the hypothesis that a series of triterpenoids supports this activity. Note that methyl gallate (Figure 30.16) from *Toona sinensis* (Juss.) Roem. protects DNA in canine Cocker Spaniel kidney cell line (MDCK) cells against hydrogen peroxide-induced oxidative stress.[29]

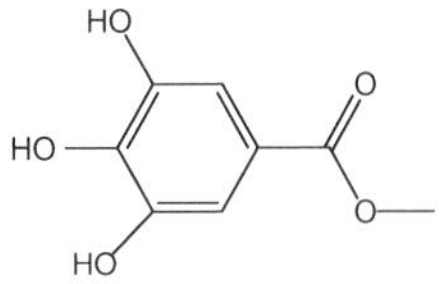

Methyl gallate

Figure 30.16

30.10 *TOONA SURENI* (BL.) MERR.

[From: Indian *toon* = the name of *Toona* species and from Malay *surian Toona sureni* (Bl.) Merr.]

30.10.1 Botany

Toona sureni (Bl.) Merr. (*Toona febrifuga* Roem., *Cedrela febrifuga* Bl., *Cedrela toona* Roxb. ex Rottl., and *Cedrela sureni* [Bl.] Burk.) is a tree that grows to a height of 30m with a girth of

2m in the primary rain forests of tropical Asia, India, China, and Papua New Guinea, up to an altitude of 1200m. The bole is buttressed, the buttresses are up to 3m high and 2m in girth. The bark has elongated scales, fissured regularly longitudinally, and grayish. The inner bark is fibrous, evokes a pleasant cedar fragrance, is orange–red or pink, and turns rapidly orange–brown. The sapwood is yellowish-white. The wood is sweet scented and peppery. The leaves are paripinnate with 4–5 pairs of folioles. The rachis is velvety and 38–50cm long. The petiolules are 4mm long. The folioles are 17cm × 5–14cm × 4–12.8cm × 3.5–13cm × 4.5cm, and show 11 pairs of secondary nerves. The flowers are 4mm long on 2mm-long pedicels, powerfully and obnoxiously scented, detectable 100 feet away, white, and in 40cm-long panicles. The corolla is white or flushed pink and 3mm long. The fruits are 1.5cm × 8mm rusty, lenticelled capsules opening to release several flat seeds winged at both ends (Figure 30.17).

Figure 30.17 *Toona sureni* (Bl.) Merr. [From: Flora of the Malay Peninsula. Forest Department. Geographical localization: Telom Valley, riverside of Loi River. Alt.: 3000ft. Aug. 11, 1934. No: 34163. Field collector: R. C. Barnard. Botanical identification: J. M. Edmonds, Feb. 25, 2000.]

30.10.2   Ethnopharmacology

The vernacular name is *Suryan* (Malay), *surian wangi*. In Cambodia, Laos, Vietnam, and Indonesia the bark is used as a tonic to lower fever and to assuage rheumatic pains. In Bali, the leaf tips are applied to swellings. A significant advance in the pharmacological exploration of *Toona sureni* (Bl.) Merr. has been provided by the work of Takahashi et al.[30] They showed that an extract of the plant prevents the survival of *Leishmania* cultured *in vitro*, probably due to its content of triterpenoids.[31]

30.11 *TRICHILIA CONNAROIDES* (WIGHT & ARN.) BENTVELZEN

[From: Greek *tricha* = three parts, referring to the capsule splitting into three, and from Latin *connaroides* = resembling *Connarus*.]

30.11.1   Botany

Trichilia connaroides (Wight & Arn.) Bentvelzen (*Heynea trijuga* Roxb. ex Sims., *Heynea trijuga* Roxb. ex Sims. var. *multijuga* C. DC., and *Walsura tenuifolia* Ridl.) is a handsome tree that grows in the lowland rain forests in a geographical zone covering India to the Philippines and Borneo including Vietnam and South China. The plant grows up to 15m high and 65cm in diameter. The crown is spreading. The bark is greenish-brown with white patches. The inner bark is whitish. The stems are black, 2mm in diameter, and longitudinally fissured and lenticelled. The leaves are pinnate, light green, spiral, hairy, and exstipulate, comprising 5–6 pairs of folioles which are 10–11.5–8.5cm × 4–3.5–3.4cm, and asymmetrical at base. The petiolules are 1cm long. There are 12 pairs of nerves below. The rachis is 32cm and somewhat swollen and constricted at the nodes. The inflorescences

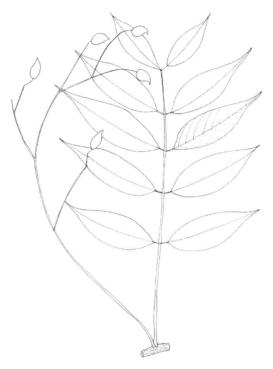

Figure 30.18 *Trichilia connaroides* (Wight & Arn.) Bentvelzen. [From: Flora of Malaya. Geographical localization: Johor, Kota Tinggi.]

are 40cm long corymbose cymes which are 50cm across. The flowers are small, delicate, and pure white. The androecium includes 10 bright yellow anthers. The ovary is 2-celled. The fruits are pinkish capsules, 2cm × 1.2cm, bivalved with a single seed in a white aril (Figure 30.18).

30.11.2   Ethnopharmacology

In the Hainan Islands and in Malaysia, the leaves are boiled in water to make a drink used to treat cholera. To date the pharmacological and especially the antibacterial properties of the plant are unexplored. These properties would be worth investigating in further experimental studies. Note that *Trichilia connaroides* (Wight & Arn.) Bentvelzen is most likely antibacterial, parasiticidal, and cytotoxic.[32–34] Sanogo et al.[35] showed that an aqueous extract of the roots of *Trichilia roka* Chiov. inhibits the thermogenic effects of yeast in rats when given orally at doses of 0.25g/Kg, 0.5g/Kg, and 1.0g/Kg, supporting thereby the antipyretic effect of Meliaceae. What are the principles involved here and what is the precise pharmacological mechanism? Answering these questions may lead to the discovery of new antipyretic agents.

30.12 *XYLOCARPUS GRANATUM* KOENIG.

[From: Greek *xylon* = wood and *karpon* = fruit, and from Latin *granatum* = like pomegranate, referring to the fruits.]

30.12.1   Botany

Xylocarpus granatum Koenig. (*Carapa moluccensis* sensu Ridl., *Carapa obovata* Bl., *Carapa granatum* [Koening] Alston, and *Xylocarpus minor* Ridl.) is a tree that grows to a height of 15m

Figure 30.19 *Xylocarpus granatum* Koenig. [From: Herbarium of the University of Wisconsin, Madison, 1958. Flora of Malay Peninsula. Forest Department. Geographical localization: Trong Forest Reserve, Perak. Nov. 11, 1940. No: 45399. Botanical identification: G. K. Noamesi, 1958.]

with a girth of 2.7m in the mangroves from East Africa to Tonga. The bole is crooked and develops snake-like buttresses without pneumatophores. The bark is smooth. The rachis is 8cm long. The folioles are elliptical, 10.5–11.3–18.4cm × 4.3cm, and show nine pairs of secondary nerves below. The folioles are rounded at the apex. The fruits are globose, 8.8cm × 8.5cm, hard, and smooth (Figure 30.19).

30.12.2   Ethnopharmacology

In Indonesia, the roots are used to treat cholera and the bark is used to stop dysentery throughout Southeast Asia, including Malaysia, where the plant is called *Nyireh bunga* (Malay). The Filipinos eat the fruits to treat diarrhea. Indonesians use the fruits as a tonic and externally to soothe inflammation. Tannins which abound in this mangrove tree might account for the reported effectiveness of *Xylocarpus moluccensis* (Lamk.) Roem against diarrhea in rodents poisoned with castor oil and magnesium sulfate.[36] One might also propose tannins as imparting to the plant its antiinflammatory property. Other principles include a series of limonoids including xyloccensin I–V.[37–42] As of yet, the effect of xyloccensin on cancer cells, neurones, viruses, bacteria, and parasites are unknown (Figure 30.20).

30.13 *XYLOCARPUS MOLUCCENSIS* (LAMK.) ROEM.

[From: Greek *xylon* = wood and *karpon* = fruit, and from Latin *moluccensis* = from the Moluccas.]

30.13.1   Botany

Xylocarpus moluccensis (Lamk.) Roem. (*Carapa moluccensis* Lamk. and *Carapa obovata sensu* Ridl. non Bl.) is a tree that grows to a height of 18m with a girth of 2m, in the mangroves of Southeast Asia to North Australia. The stems are glabrous, lenticelled, and 5mm in diameter. The folioles are broadly elliptical, pointed at the apex, 7.6cm × 2.2–9cm × 4cm, showing nine pairs of

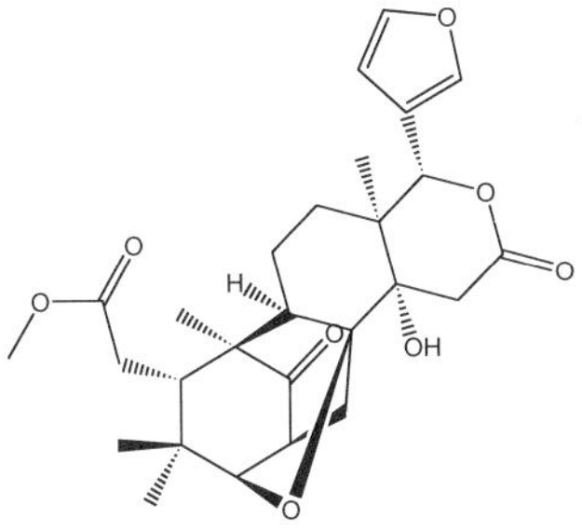

Xyloccensin K 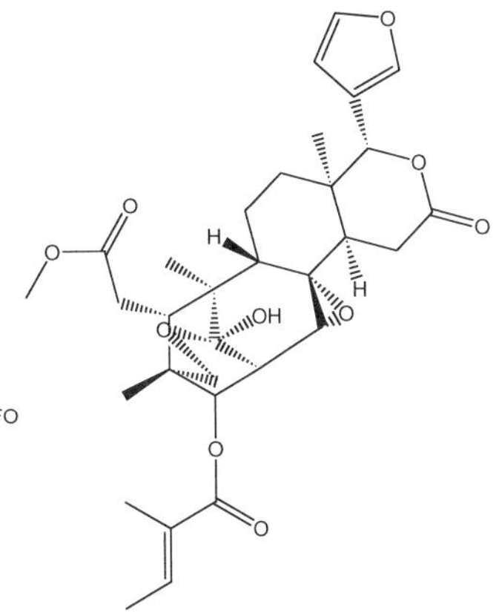   Xyloccensin L

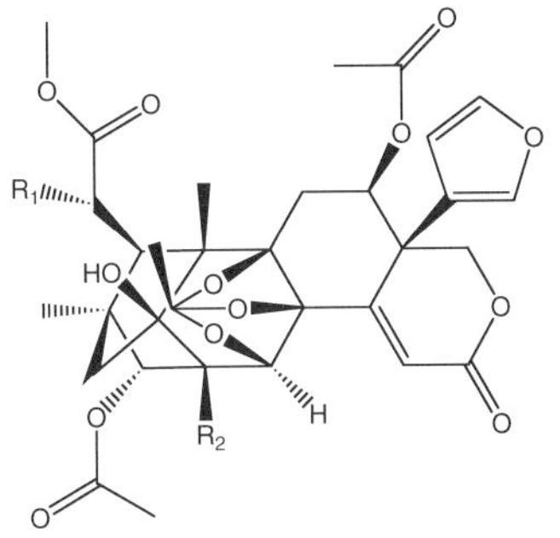

Xyloccensin Q, R_1=OCOCH$_3$ R_2=OH

Xyloccensin R, R_1=OH R_2=OH

XyloccensinS,R_1=OH R_2=OCOCH$_3$

XyloccensinT,R_1=OH R_2=H

XyloccensinU,R_1=H R_2=OH

XyloccensinV,R_1=H R_2=OCOCH$_3$

Figure 30.20 Phragmalin-type limonoids xyloccensin from *Xylocarpus granatum*.

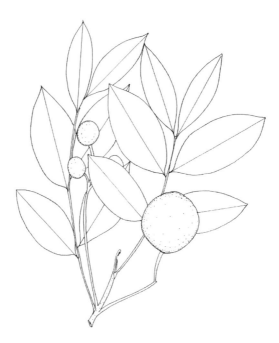

Figure 30.21 *Xylocarpus moluccensis* (Lamk.) Roem. [From: Singapore Field No: 18527. Distributed from the Botanic Gardens, Singapore. Geographical localization: Tioman Island, Pulau Tulai Island. Alt.: Sea level. May 27, 1927. Field collector: M. R. Henderson.]

secondary nerves. The rachis is 15–20cm long × 2mm. The petiolules are 4mm. The fruits are 4.9cm in diameter, green, woody, and contain 5–20 large tetrahedral seeds (Figure 30.21).

30.13.2   Ethnopharmacology

In Indonesia, the roots are used to treat cholera and the bark is used to stop dysentery throughout Southeast Asia, including Malaysia and Indonesia where the plant is called *nyireh batu* (Malay), *paradong jantan* (Malay, Indonesian). The Filipinos eat the fruits to treat diarrhea. Indonesians use the fruits as a tonic and externally to soothe inflammation. The antidiarrheal and antibacterial properties of the plant are confirmed.[36] Taylor identified xyloccensins G, H, and I in the plant.[43] Are these pharmacologically active? One wonders.

REFERENCES

1. Siddiqui, B. S., Afshan, F., Gulzar, T., and Hanif, M. 2004. Tetracyclic triterpenoids from the leaves of *Azadirachta indica*. *Phytochemistry*, 65, 2363.
2. Boeke, S. J., Boersma, M. G., Alink, G. M., van Loon, J. J. A., van Huis, A., Dicke, M., Rietjens, I. M. C. 2004. Safety evaluation of neem (*Azadirachta indica*) derived pesticides. *J. Ethnopharmacol.*, 94, 25.
3. Greger, H., Pacher, T., Brem, B., Bacher, M., and Hofer, O. 2001. Insecticidal flavaglines and other compounds from Fijian *Aglaia* species. *Phytochemistry*, 57, 57.
4. Schneider, C., Bohnenstengel, F. I., Nugroho, B. W., Wray, V., Witte, L., Hung, P. D., Kiet, L. C., and Proksch, P. 2000. Insecticidal rocaglamide derivatives from *Aglaia spectabilis* (Meliaceae). *Phytochemistry*, 54, 731.
5. Hort, J. C., Nugroho, B. W., Bohnenstengel, F. I., Wray, V., Witte, L., Hung, P. D., Kiet, L. C., Sumaryono, W., and Proksch, P. 1999. New insecticidal rocaglamide derivatives from flowers of *Aglaia duperreana* (Meliaceae). *Phytochemistry*, 52, 837.

6. Omobuwajo, O. R., Martin, M. T., Perromat, G., and Païs, M. 1996. Cytotoxic cycloartanes from *Aglaia argentea*. *Phytochemistry*, 41, 1325.

7. Hausott, B., Greger, H., and Marian, B. 2004. Flavaglines: a group of efficient growth inhibitors block cell cycle progression and induce apoptosis in colorectal cancer cells. *Int. J. Cancer*, 109, 933.

8. Wu, T. S., Liou, M. J., Kuoh, C. S., Teng, C. M., Nagao, T., and Lee, K. H. 1997. Cytotoxic and antiplatelet aggregation principles from *Aglaia elliptifolia*. *J. Nat. Prod.*, 60, 606.

9. Bohnenstengel, F. I., Steube, K. G., Meyer, C., Nugroho, B. W., Hung, P. D., Kiet, L. C., and Proksch, P. 1999. Structure activity relationships of antiproliferative rocaglamide derivatives from *Aglaia* species *(*Meliaceae*)*. *Z. Naturforsch*, 54, 55.

10. Polonsky, J., Varon, Z., Marazano, C., Arnoux, B., Pettit, G. R., Schmid, J. M., Ochi, M., and Kotsuki, H. 1979. The structure of amoorastatone and the cytotoxic limonoid 12-hydroxyamoorastatin. *Experientia*, 35, 987.

11. Jagetia, G. C. and Venkatesha, V. A. 2005. Enhancement of radiation effect by *Aphanamixis polystachya* in mice transplanted with Ehrlich ascites carcinoma. *Biol. Pharm. Bull.*, 28, 69.

12. Gunning, P. J., Lloyd, B., Jeffs, L. B., Isman, M. B., and Towers, G. H. N. 1994. Two limonoids from *Chisocheton microcarpus*. *Phytochemistry*, 36, 1245.

13. Bordoloi, M., Bhogeswar, S. B., Mathur, R. K., and Goswami, B. N. 1993. A meliacin from *Chisocheton paniculatus*. *Phytochemistry*, 34, 583.

14. Fujioka, T., Yamamoto, M., Kashiwada, Y., Fujii, H., Mihashi, K., Ikeshiro, Y., Chen, I. S., and Lee, K. H. 1998. Cytotoxic diterpenes from the stem of *Dysoxylum kuskusense*. *Bioorg. Med. Chem. Lett.*, 8, 3479.

15. Govindachari, T. R., Suresh, G., Kumari, G. N. K., Rajamannar, T., and Partho, P. D. 1999. Nymania-3: a bioactive triterpenoid from *Dysoxylum malabaricum*. *Fitoterapia*, 70, 1, 83.

16. Luo, X. D., Wu, S. H., Wu, D. G., Ma, Y. B., and Qi, S. H. 2002. Novel antifeeding limonoids from *Dysoxylum hainanense*. *Tetrahedron*, 58, 7797.

17. Benosman, A., Richomme, P., Roussakis, C., Hadi, A. H., and Bruneton, J. 2000. Effects of triterpenes from the stem bark of *Dysoxylum cauliflorum* on a non-small-cell bronchopulmonary carcinoma cell line (NSCLC-N6). *Anticancer Res.*, 20, 1855.

18. Huang, R., Harrison, L. J., and Sim, K. Y. 1999. A triterpenoid with a novel abeo-dammarane skeleton from *Dysoxylum cauliflorum*. *Tetrahedron Lett.*, 40, 1607.

19. Horgen, F. D., Edrada, R. A., de los Reyes, G., Agcaoili, F., Madulid, D. A., Wongpanich, V., Angerhofer, C. K., Pezzuto, J. M., Soejarto, D. D., and Farnsworth, N. R. 2001. Biological screening of rain forest plot trees from Palawan Island (Philippines). *Phytomedicine*, 8, 71.

20. Rasadah, M. A., Khozirah, S., Aznie, A. A., and Nik, M. M. 2004. Anti-inflammatory agents from *Sandoricum koetjape* Merr. *Phytomedicine*, 11, 261.

21. Kaneda, N., Pezzuto, J. M., Kinghorn, A. D., Farnsworth, N. R., Santisuk, T., Tuchinda, P., Udchachon, J., and Reutrakul, V. 1992. Plant anticancer agents, L. cytotoxic triterpenes from *Sandoricum koetjape* stems. *J. Nat. Prod.*, 55, 654.

22. Ismail, I. S., Ito, H., Mukainaka, T., Higashihara, H., Enjo, F., Tokuda, H., Nishino, H., and Yoshida, T. 2003. Ichthyotoxic and anticarcinogenic effects of triterpenoids from *Sandoricum koetjape* bark. *Biol. Pharm. Bull.*, 26, 1351.

23. Powell, R. G., Mikolajczak, K. L., Zilkowski, B. W., Mantus, E. K., Cherry, D., and Clardy, J. 1991. Limonoid antifeedants from seed of *Sandoricum koetjape*. *J. Nat. Prod.*, 54, 241.

24. Ismail, I. S., Ito, H., Hatano, T., Taniguchi, S., and Yoshida, T. 2003. Modified limonoids from the leaves of *Sandoricum koetjape*. *Phytochemistry*, 64, 1345.

25. Poon, S. L., Leu, S. F., Hsu, H. K., Liu, M. Y., and Huang, B. M. 2005. Regulatory mechanism of *Toona sinensis* on mouse leydig cell steroidogenesis, *Life Sci.*, 76, 1473.

26. Chang, H. S., Hung, W. C., Huang, M. S., and Hsu, H. K. 2002. Extract from the leaves of *Toona sinensis* Roem. exerts potent antiproliferative effect on human lung cancer cells. *Am. J. Chin. Med.*, 30, 307.

27. Yu, J. Y. L. 2002. *Toona sinensis* extract affects gene expression of GLUT4 GLU00 glucose transporter in adipose tissue of alloxan induced diabetic rats. *Proc. 5th Cong. Int. Diabet. Soc.*

28. Yang, Y. C., Hsu, H. K., Hwang, J. H., and Hong, S. J. 2003. Enhancement of glucose uptake in 3T3-L1 adipocytes by *Toona sinensis* leaf extract Kaoshiung. *J. Med. Sci.*, 9, 327.

29. Hsieh, T. J., Liu, T. Z., Chia, Y. C., Chern, C. L., Lu, F. J., Chuang, M. C., Mau, S. Y., Chen, S. H., Syu, Y. H., and Chen, C. H. 2004. Protective effect of methyl gallate from *Toona sinensis* (Meliaceae) against hydrogen peroxide-induced oxidative stress and DNA damage in MDCK cells. *Food Chem. Toxicol.*, 42, 843.

30. Takahashi, M., Fuchino, H., Satake, M., Agatsuma, Y., and Sekita, S. 2004. *In vitro* screening of leishmanicidal activity in Myanmar timber extracts. *Biol. Pharm. Bull.*, 27, 921.

31. Kraus, W. and Kypke, K. 1979. Surenone and surenin, two novel tetranortriterpenoids from *Toona sureni* [blume] Merrill. *Tetrahedron Lett.*, 20, 2715.

32. Aladesanmi, A. J. and Odediran, S. A. 2000. Antimicrobial activity of *Trichilia heudelotti* leaves. *Fitoterapia*, 71, 179.

33. Sparg, S. G., van Staden, J., and Jäger, A. K. 2000. Efficiency of traditionally used South African plants against schistosomiasis. *J. Ethnopharmacol.*, 73, 209.

34. Germanò, M. P., D'Angelo, V., Sanogo, R., Catania, S., Alma, R., De Pasquale, R., and Bisignano, G. 2005. Hepatoprotective and antibacterial effects of extracts from *Trichilia emetica* Vahl. (Meliaceae). *J. Ethnopharmacol.*, 96, 227.

35. Sanogo, R., Germanò, M. P., D'Angelo, V., Forestieri, A. M., Ragusa, S., and Rapisarda, A. 2001.*Trichilia roka* Chiov. (Meliaceae): pharmacognostic researches. *J. Farmaco*, 56, 357.

36. Uddin, S. J., Shilpi, J. A., Alam, S. M. S., Alamgir, M., Rahman, M. T., and Sarker, S. D. 2005. Antidiarrheal activity of the methanol extract of the barks of *Xylocarpus moluccensis* in castor oil and magnesium sulfate-induced diarrhea models in mice. *J. Ethnopharmacol.*, in press.

37. Alvi, K. A., Crews, P., Aalbersberg, B., and Prasad, R. 1991. Limonoids from the Fijian medicinal plant *dabi* (*xylocarpus*). *Tetrahedron*, 47, 8943.

38. Kokpol, U., Chavasiri, W., Tip-Pyang, S., Veerachato, G., Zhao, F., Simpson, J., and Weavers, R. T. 1996. A limonoid from *Xylocarpus granatum*. *Phytochem.*, 41, 903.

39. Wu, J., Zhang, S., Xiao, Q., Li, Q., Huang, J., Long, L., and Huang, L. 2004. Xyloccensin L, a novel limonoid from *Xylocarpus granatum*, *Tetrahedron Lett.*, 45, 591.

40. Wu, J., Xiao, Q., Zhang, S., Li, X., Xiao, Z., Ding, H., and Li, Q. 2005. Xyloccensins Q–V, six new 8,9,30-phragmalin ortho ester antifeedants from the Chinese mangrove *Xylocarpus granatum*. *Tetrahedron*, 61, 8382.

41. Cui, J., Deng, Z., Li, J., Fu, H., Proksch, P., and Lin, W. 2005. Phragmalin-type limonoids from the mangrove plant *Xylocarpus granatum*. *Phytochemistry*, in press.

42. Wu, J., Xiao, Q., Huang, J., Xiao, Z., Qi, S., Li, Q., and Zhang, S. 2004. Xyloccensins O and P, unique 8,9,30-phragmalin ortho esters from *Xylocarpus granatum*, *Org. Lett.*, 6, 1841.

43. Taylor, D. A. H. 1983. Limonoid extractives from *Xylocarpus moluccensis*. *Phytochemistry*, 22, 1297.

CHAPTER 31

Medicinal Plants Classified in the Family Rutaceae

31.1 GENERAL CONCEPT

The family Rutaceae (A. L. de Jussieu, 1789) consists of 150 genera and 1500 species of prickly treelets, shrubs, and herbs known to produce limonoids, essential oils, flavonoids (hesperidin), coumarins, and several sorts of alkaloids including most notably carbazole and acridone. In the field, Rutaceae are easily recognized by three main botanical features: the leaves are compound and conspicuously dotted with translucent oil cells; the flowers are pure white, ephemeral, and endowed with a conspicuous oily stigma; and the fruits which are baccate or succulent (hesperidia) or capsular (Figure 31.1).

Several fruit trees are provided by this taxon: *Citrus limon* (L.) Burm. f. (Lemon), *Citrus aurantium* L. (Sour Orange), *Citrus sinensis* (L.) Osbeck (Sweet Orange), and *Citrus aurantifolia* (Chaistm.) Swingle (Lime). The oil obtained by mechanical means from the fresh peel of the fresh orange *Citrus sinensis* (Orange oil, *Oleum Aurantii, British Pharmaceutical Codex*, 1963) has been used as a flavoring agent and in perfumery. Bergamot oil (*Oleum Bergamottae, British Pharmaceutical Codex*, 1949), obtained by expression from the fresh peel of the fruit of *Citrus bergamia*, has been used by the perfumery industry in preparations for the hair (Cologne Spirit or *Spiritus Coloniensis*). Lemon oil (*Oleum Limonis, British Pharmaceutical Codex*, 1963), which is obtained by expression of fresh lemon peel (*Citrus limon, Citrus*

Figure 31.1 Botanical hallmarks of Rutaceae. **(See color insert following page 168.)**

limonia, Citrus medica) is carminative and used as a flavoring agent. The dry peel of *Citrus aurantium* (*Aurantii Cortex Siccatus, British Pharmacopoeia*, 1963) has been used as a flavoring agent and for its bitter and carminative properties. Of therapeutic importance are *Pilocarpus jaborandi* Holmes, *Ruta graveolens* L., *Agathosma betulina*, *Peganum harmala*, *Zanthoxylum americanum,* and *Zanthoxylum clavaherculis. Pilocarpus jaborandi* Holmes contains an imidazole alkaloid, pilocarpine, which is occasionally used to treat glaucoma. An infusion of *Ruta graveolens* L. (Common Rue, Herb of Grace) has been used to promote menses. The oil of rue has been known to stop spasms, to promote menses, and to produce skin irritation (Rue, *British Pharmaceutical Codex*, 1934). The dried leaves of *Agathosma betulina* (*Barosma betulina* or *buchu*) were used to promote urination. The dried seeds of *Peganum harmala* (Wild Rue was used in India to expel intestinal worms and to induce drowsiness. The dried bark of *Zanthoxylum americanum* (Northern Prickly Ash) and *Zanthoxylum clavaherculis* (Southern Prickly Ash) have been used to stop flatulence (*Zanthoxylum*, toothache bark, *Xanthoxylum, British Pharmaceutical Codex*, 1934). The leaves of *Murraya koenigii* (curry leaves) are used to flavor Asian food.

An interesting feature of the plants classified within the family Rutaceae is that they elaborate several sorts of unusual cytotoxic, antimicrobial, neuroactive, and musculotropic quinoline, quinazoline, and acridone alkaloids derived from anthranilic acid.[1] Examples of such alkaloids are acronycine and rutaecarpine from *Acronychia baueri* Scott and *Euodia rutaecarpa* Hook. f., respectively (Figure 31.2).

Acronycine is cytotoxic and has undergone clinical trials whereas rutaecarpine displays a uterotonic property. Allocryptopine, an alkaloid from *Fagara coca*, and *Zanthoxylum brachycanthum* are more effective than quinidine in controlling atrial arrhythmia. Investigating the family Rutaceae for pharmacologically active natural products should be a fruitful task.

Skin injuries respond to furanocoumarins and especially bergapten after sun exposure through biochemical processes which are still obscure. Bergamot oil, expressed from *Citrus aurantium* L. ssp. *bergamia* (Wight & Arn.) Engler, causes a similar effect after contact and sunlight exposure with blisters and vesicles. The photodynamic sensitizing properties of furanocoumarins have been used to treat psoriasis, but risks of gastrointestinal disorders, photosensitization, and cancer are there. Note that a growing body of evidence suggests that coumarins of Rutaceae are antitumoral. Approximately 50 species of plants classified within the family Rutaceae are of medicinal value in the Pacific Rim. Note that many of them are used to treat diseases of the respiratory tract, infections, to reduce fever, and to promote digestion, often on account of their essential oils.

31.2 *AEGLE MARMELOS* CORREA

[After Aegle, one of the hesperides of the nymphs who cared for the golden apples of immortality.]

31.2.1 Botany

Aegle marmelos Correa is a fruit tree that grows wild in India, Sri Lanka, Burma, Thailand, Cambodia, Laos, and Vietnam, up to 1000m in altitude. The plant is cultivated throughout the Pacific Rim. The tree grows to a height of 12m and a diameter of 1.2m. The stems are lenticelled and show conspicuous woody thorns. The leaves are alternate, exstipulate, and trifoliolate. The petiole is 4.2–6.7cm long. The folioles are 8.5cm × 4cm – 7.5cm × 3.3cm – 8.2cm × 4.2cm – 6.5cm × 3.2cm – 5.7cm × 3.4cm, and show 5–8 pairs of secondary nerves. The margin is crenate and the apex is notched. The flowers are greenish-white, fragrant, in axillary panicles, with many stamens. The gynaecium is 8–20-celled. The fruits are yellowish-green, 10.5cm in diameter, woody, globose, and smooth. The seeds are numerous and woolly (Figure 31.3).

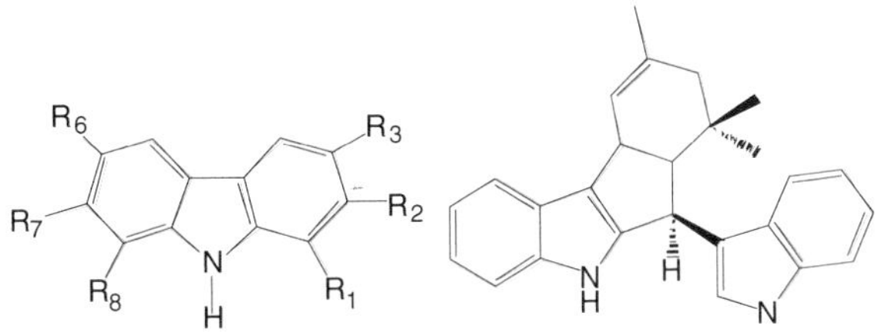

Indole alkaloids

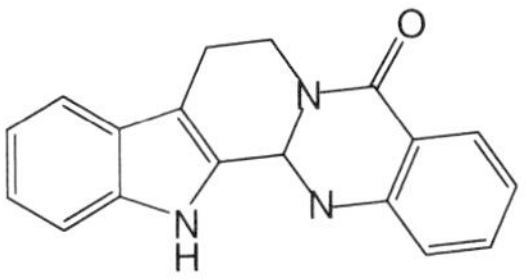

Rutaecarpine

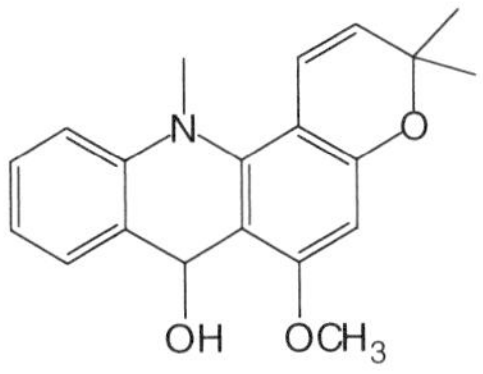

Acronycine

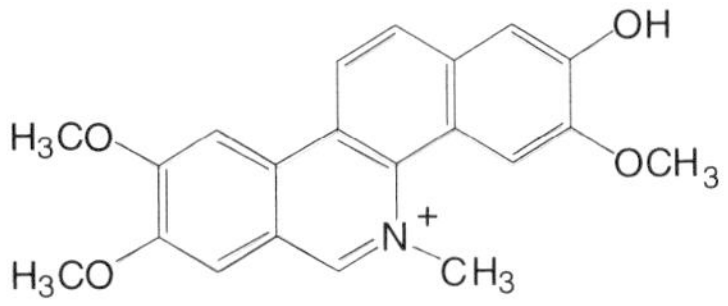

Fagaronine

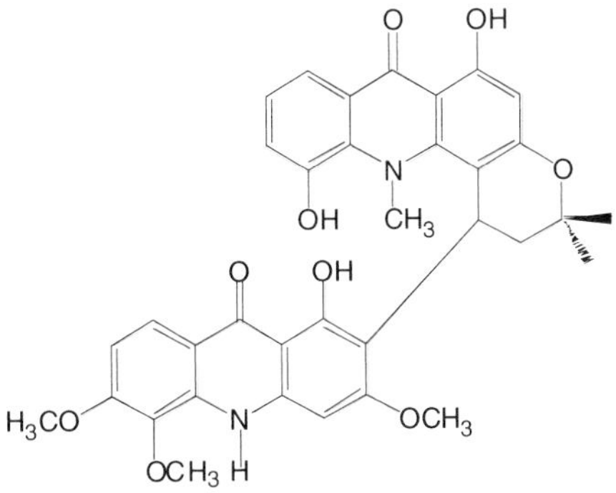

Buntanbismine

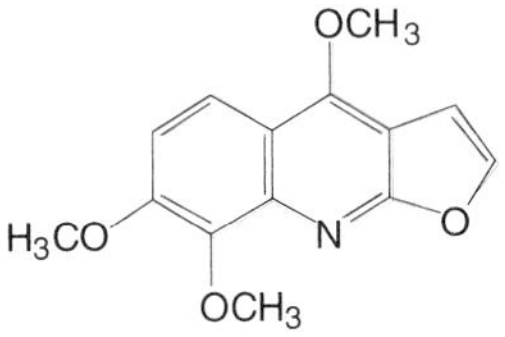

Skimmianine

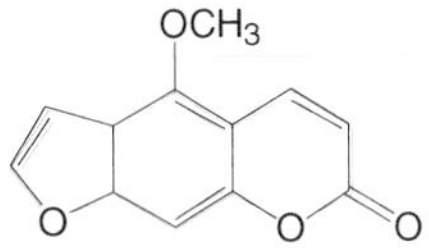

Coumarin

Figure 31.2 Examples of bioactive natural products from the family Rutaceae.

Figure 31.3 *Aegle marmelos* Correa. [From: Flora of Malaya. FRI No: 27344. Geographical localization: Teluk Bahang, Penang, in compound of the home of an Indian family. Botanical identification: F. S. P. Ng, May 11, 1983. Field collector: F. S. P. Ng, June 23, 1982.]

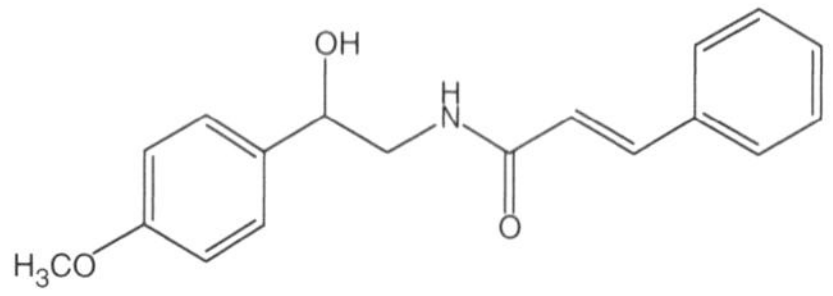

31.2.2 Ethnopharmacology

The fruits of *Aegle marmelos* Correa, or Bengal Quince, Indian Bael, Wood Apple, Golden Apple, Bek Fruit, Stone Apple, *bael* (Hindi), *bilwa* (Sanskrit), *kuvalum* (Tamil), and *buah mentaga* (Malay), offer a remedy for dysentery throughout a geographical area including India, Burma, and Indonesia. In Burma, the juice squeezed from the young leaves is instilled into the eyes to treat ophthalmia. The Indonesians eat the ripe fruits to promote digestion and to soothe inflammation of the rectum; the young leaves are used externally to soothe inflammation, sores, and to deflate swellings, and are also eaten to induce abortion. The roots are infused in hot water to make a drink that is said to calm palpitation of the heart. In Malaysia, a decoction is used as a drink to stop vomiting. In Cambodia, the fruits are used to treat tuberculosis and liver dysfunction. In India, the plant is regarded as sacred and the leaves are used for the worship of Shiva. In Sri Lanka, the plant is used to treat diabetes.

The antiinflammatory property of the plant has been substantiated by Arul et al.[2] Extracts of the leaves caused significant inhibition of the carrageenan-induced paw edema, cotton-pellet granuloma, paw-licking, and hyperpyrexia in rodents. What is the active principle? Might it be related to the series of hydroxyamide alkaloids of which *N*-2-[4-(3′,3′-dimethylallyloxy)phenyl]ethyl cinnamide, *N*-2-hydroxy-2-[4-(3′,3′-dimethylallyloxy)phenyl]ethyl cinnamide, *N*-4-methoxystyryl cinnamide, *N*-2-hydroxy-2-(4-hydroxyphenyl)ethyl cinnamide, aegeline (Figure 31.4), and marmeline, which are known to abound in the plant?

Aegeline

Figure 31.4

Aqueous decoctions of *Aegle marmelos* Correa lowered the fasting blood glucose level and improved glucose tolerance in rodents, confirming thereby the antidiabetic property of the plant.[3,4] Sarath et al.[5] showed that rats fed with the plant produced hepatic lesions which included central vein abnormalities. The antituberculosic property of the plant is unexplored as of yet, but essential oil isolated from the leaves inhibited concentration as well as time dependence of the germination of spores of several strains of fungi.[6]

With regard to the antidiarrheal property of the plant, Shoba and Thomas[7] showed that a methanolic extract of *Aegle marmelos* Correa protected rodents against castor oil-induced diarrhea with reduction of both the induction time of diarrhea and the total weight of the feces. Is the smooth muscle involved here?[8]

Aegle marmelos Correa given at a dose of 1.00g/Kg lowered the serum levels of T_3, hence has some potential for the regulation of hyperthyroidism.[9]

A remarkable advance in the pharmacology of *Aegle marmelos* Correa has been provided by the work of Lampronti et al.[10] They showed that extracts of the plant inhibit the *in vitro* proliferation of the leukemia K562 cell line, on account of butyl-*p*-tolyl sulfide, 6-methyl-4-chromanone, and butylated hydroxyanisole, which displayed levels of activity comparable to those of cisplatin, chromomycin, cytosine arabinoside, and 5-fluorouracil.

In summary, the evidence available thus far strongly indicates that *Aegle marmelos* Correa has manifold pharmacological properties, but much less is known about the principle involved. Note that aegeline and congeners have a chemical structure related to arylethanolamines, which are notably pharmacophores of beta-adrenergic agents well known for their cardiovascular effects and which have some effects in the control of glycemia.[11–13]

31.3 *ATALANTIA MONOPHYLLA* DC.

[After Atalanta of the golden apples in Greek mythology and from Greek *mono* = single, and *phullon* = leaves.]

31.3.1 Botany

A physical description of *Atalantia monophylla* DC. (*Atalantia spinosa* Tanaka) includes a very spiny shrub or treelet up to 6m tall that is common on rocky and sandy coasts and in dry open country throughout Southeast Asia. The blade is 9.4cm × 6 cm. The fruit pedicel is 1.5cm long, and the fruit is 2cm in diameter. The petiole is 1.4cm long. There are 11 pairs of secondary nerves below, 5.5cm × 5cm – 4.9cm × 3.7cm – 7.3cm × 5cm –9.4cm × 6cm – 5.5cm × 5cm. The leaves are notched at the tip. The flowers are 1.5cm long. The calyx splits onto two irregular lobes. The stamens are joined into a tube (Figure 31.5).

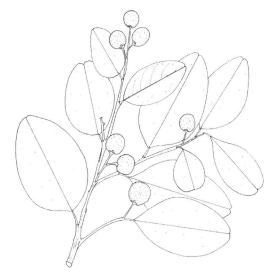

Figure 31.5 *Atalantia monophylla* DC. [From: National University of Singapore, Dept. of Botany Herbarium (SINU). Flora of Pulau Sibu, Johor, West Malaysia. Honours Class Field trip collections. 1991/1992. Botanical identification: K. S. Chua, H. J. Samsuri, H. T. W. Tan, and I. M. Turner, July 30, 1991.]

31.3.2 Ethnopharmacology

In Cambodia, Laos, Vietnam, and Malaysia, the leaves of the plant are used to treat lung disorders, probably because of their essential oils which irritate the bronchial mucosa and stimulate the movement of bronchial villosities. The plant elaborates an interesting series of acridone alkaloids such as atalaphylline and atalaphylline 3,5-dimethyl ether, the cytotoxic properties of which would be worth investigating.[14–17] As a matter of fact, acridone derivatives inhibit the enzymatic activity of telomerase, which is responsible for the maintenance of telomere length in more than 80% of all tumors and is not expressed in normal somatic cells.[16] In addition, acridone derivatives are able to influence DNA topoisomerase II and protein kinase C, hence there is some activity on replication of Human Immunodeficiency Virus (HIV) at the transcriptional level.[18] Are acridones the cytotoxic and antiretroviral principles of *Atalantia monophylla* DC.?

31.4 *ATALANTIA ROXBURGHIANA* HOOK. F.

[After Atalanta of the golden apples in Greek mythology and after William Roxburgh (1751–1815), botanist.]

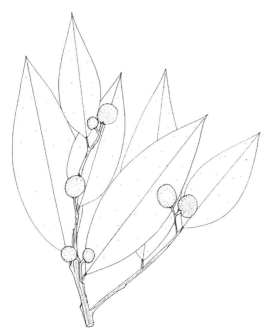

Figure 31.6 *Atalantia roxburghiana* Hook. f. [From: Forest Department, Malaya. Botanical identification: B. C. Stone, June 18, 1970. Field collector: S. Ahmad. Geographical localization: Rawang, Batu Tako.]

31.4.1 Botany

Atalantia roxburghiana Hook. f. (*Atalantia kwangtungensis* Merr.) is a treelet that grows to a height of 6m in open country and on the limestone hills of Cambodia, Laos, Vietnam, Malaysia, and Thailand. The bole is crooked, smooth, and the bark is dark green. The stems are spiny. The leaves are simple, spiral, and exstipulate. The crushed leaves have a strong citrus odor. The blade is dotted with oil cells, acute, sharp, or minutely blunt-tipped. The flowers are white and minute, with a 4-lobed calyx and stamens which are free. The secondary nerves are inconspicuous. The fruits are pale yellow (Figure 31.6).

31.4.2 Ethnopharmacology

In Cambodia, Laos, Vietnam, and Malaysia, the leaves of the plant are used to treat lung disorders, probably on account of their essential oils. The plant is called *limau hutan* (Malay). It has not yet been explored for pharmacology. One can reasonably expect the isolation of cytotoxic and/or anti-HIV acridone alkaloids from it.

31.5 *CITRUS HYSTRIX* DC.

[From: Latin *citrus* = limon and Greek *hystrix* = porcupine.]

31.5.1 Botany

Citrus hystrix DC. (*Citrus hystrix* H. Perrier, *Fortunella sagittifolia* K. M. Feng and P. I. Mao) is a treelet that grows up to 2m high. The bark is smooth, greenish-yellow, and the inner bark is creamy white. The young stems, petioles, and gynaecium are light green and pubescent. The leaves are simple, spiral, and exstipulate. The petiole is winged. The wings are as broad and similar to the blade. The blade is dotted with oil cells, wavy, glossy, and 2.5cm × 2.2cm – 3.4cm – 2.7cm × 1.9cm. The fruits are bumpy, green, and strongly aromatic (Figure 31.7).

31.5.2 Ethnopharmacology

The vernacular names of the plant include Wild Lime, Thai Lime, Kaffir Lime, *shauk-nu* (Burmese), *suan gan* (Chinese), *kok mak khi hout* (Laos), *makrut* (Thai), and *truc* (Vietnamese).

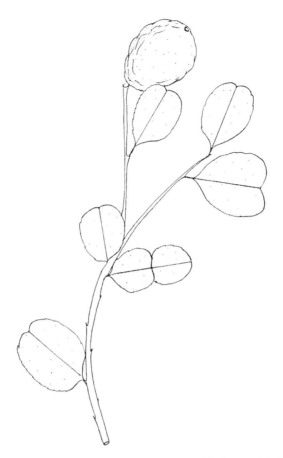

Figure 31.7 *Citrus hystrix* DC. [From: Flora of Malaya. FRI No: 38024. Geographical localization: Terengganu, Besut.]

Citrus hystrix DC. is the lemon swanggi of Rumphius. The Indonesians use the juice expressed from the fruits as a tonic ingredient. In Thailand, the fruit is commonly used as a flavor or condiment in *tom yum*. Tiwawech et al.[19] made the interesting observation that *Citrus hystrix* DC. promotes the hepatocarcinogenicity of 2-amino-3,8-dimethylimidazo(4,5)quinoxaline.

31.6 *CITRUS MITIS* BLCO.

[From: Latin *citrus* = limon and *mitis* = gentle.]

31.6.1 Botany

Citrus mitis Blco. (*Citrus microcarpa* Bunge) is a common ornamental plant native to China, thought to have been taken in early times to Indonesia and the Philippines. The plant grows to a height of 3m and is commonly cultivated. The crown is spreading, the bole is thorny, the bark is smooth, the inner bark is whitish, and the wood is white. The leaves are simple, spiral, and exstipulate. The petiole is 1.5–1.7cm long. The blade is 8.1 × 2.4cm – 7.7cm × 2.4cm – 7.7cm × 1.8cm – 7.8cm × 2.1cm – 11cm × 3.6cm, lanceolate, minutely notched at the apex, crenate, and has seven pairs of secondary nerves which are indistinct. The fruits are yellow to orange, sour, and measure 2cm × 3.5cm (Figure 31.8).

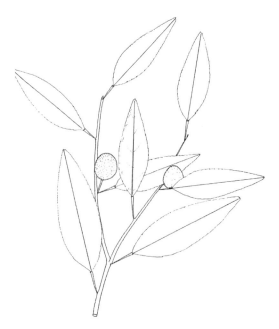

Figure 31.8 *Citrus mitis* Blco. [Flora of Malaya. FRI No: 370973. Oct. 22, 1992. Geographical localization: Kelantan, Kuala Kra.]

31.6.2 Ethnopharmacology

The plant is known as calamondin or Panama orange, or *limau kesturi* (Malay). In the Philippines, the juice expressed from the fruit is applied externally to calm insect bites, to heal buboes, to promote the growth of hair, to cool, to treat cough, to perfume, and to soothe inflammation. The pharmacological properties of this plant are unexplored as of yet.

31.7 *CLAUSENA EXCAVATA* BURM. F.

[After Clausen, a botanist known to Burmann, and from Latin *cavus* = hole.]

31.7.1 Botany

Clausena excavata Burm. f. (*Clausena lunata* Hay., *Clausena moningerae* Merr., and *Lawsonia falcata* Lour.) is a treelet that grows in an area that runs from India to the Philippines. The plant is commonly wild or cultivated in lowland, forest edges, villages, and open country. The leaves are spiral, exstipulate, and compound with 5–31 pairs of folioles which are asymmetrical, serrate, and measure 2.4cm × 1cm – 3.1cm × 1cm – 5cm × 1.7cm – 4cm × 1.7cm. The folioles show numerous cell oils and nine pairs of secondary nerves. The rachis is 29cm long. The internodes are 1.5cm long. The flowers are very small and arranged in 15cm-long panicles. The fruits are 5mm in diameter on 7mm-long pedicels (Figure 31.9).

31.7.2 Ethnopharmacology

In Burma, the plant is used as a remedy for stomach troubles. In Cambodia, Laos, and Vietnam, the plant is used to invigorate, to promote menses and digestion, and to treat paralysis. In Taiwan, a decoction of the roots is used as a drink to promote sweating. The leaves are used to kill vermin.

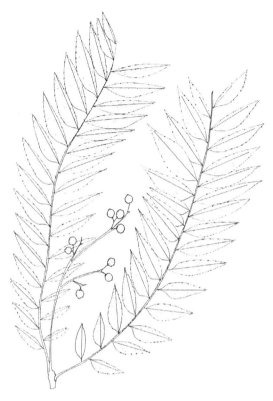

Figure 31.9 *Clausena excavata* Burm. f. [From: Flora of Malaya. FRI No: 0424. Geographical localization: Hedge of a rest house, Patani River, Kedah. Field collector: T. C. Whitmore, June 13, 1966. Botanical identification: 1968.]

The Malays use the roots to heal sores, the leaves to mitigate headaches, and to invigorate after the exhaustion of childbirth. In Indonesia, the juice squeezed from the leaves is used to reduce fever and to expel worms from the intestines.

The plant produces a limonoid, clausenolide-1-ethyl ether, which exhibited HIV-1 inhibitory activity, as well as coumarins, including dentatin and nor-dentatin, which were cytotoxic in syncitium assay.[21]

Dentatin, nor-dentatin, clausenidin, and the alkaloids 3-formyl-carbazole, mukonal, 3-methoxycarbonylcarbazole, 2-hydroxy-3-formyl-7-methoxycarbazole, and clauszoline J eliminated *Mycobacteria* with MIC values ranging from 50–200μg/mL (Figure 31.10). 3-Formylcarbazole, mukonal, 3-methoxycarbonylcarbazole, and 2-hydroxy-3-formyl-7-methoxycarbazole showed antifungal activity with IC_{50} values of 13.6, 29.3, 9.5, and 2.8μg/mL, respectively.[21,22] Carbazole alkaloids of the *Clausena* species and Rutaceae in general are related in structure to the antitumor alkaloid ellipticine and hold some potential as cytotoxic and antiviral agents.[23]

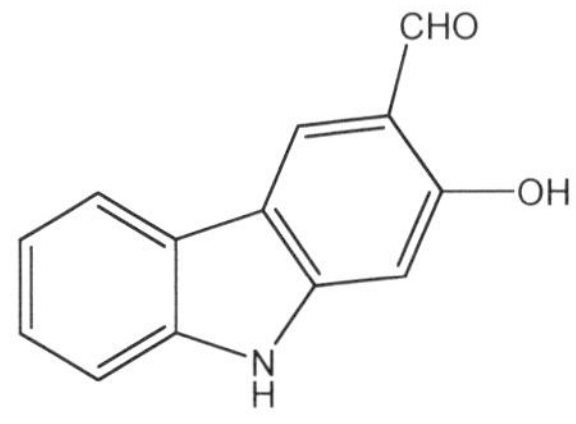

Mukonal

Figure 31.10

31.8 *CLAUSENA LANSIUM* (LOUR.) SKEELLS

[After Clausen, a botanist known to Burmann.]

31.8.1   Botany

Clausena lansium (Lour.) Skeells (*Clausena wampi* [Blco.] Oliv.) is a tree that grows to a height of 9m with a girth of 15cm. The leaves are compound, spiral, and exstipulate. The rachis is 15cm long with 3–4 pairs of folioles. The petiolules are 1.1cm long. The folioles are asymmetrical, 8cm × 4.3cm – 7.5cm × 4cm – 5.7cm × 3.8cm, and show 10 pairs of secondary nerves. The blade is minutely tipped at the apex. The gynaecium is gland-dotted and hairy. The fruits are a ripe golden yellow fruit, sour, with a yellow furry skin, and whitish pulp surrounding several greenish-black seeds (Figure 31.11).

31.8.2   Ethnopharmacology

The plant is known as *wong pee*, *wampee*, *huang p'i tz*, and *huang p'i ku* (Chinese). The Chinese eat the fruit as an antidote for litchi excess and believe that the fruit must be eaten on a full stomach. In the Pentsao, the fruits are mentioned as a stomachic, cooling, and anthelmintic remedy. The plant is known to elaborate

Figure 31.11 *Clausena lansium* (Lour.) Skeells. [From: Flora of Malay Peninsula. Geographical localization: Near railway station, Kepong, Selangor. Oct. 30, 1947. Field collector: W. T. Moy. Botanical identification: B. C. Stone, 1970.]

a series of coumarins, including chalepensin, chalepin, gravelliferone, and angustifolin, which might be involved in the antiinflammatory action. Carbazole alkaloids, including 3-formyl-6-methoxy-carbazole, methyl 6-methoxycarbazole-3-carboxylate, 3-formyl-1,6-dimethoxycarbazole, and 2,7-dihydroxy-3-formyl-1-(3'-methyl-2'-butenyl)carbazole, are perhaps involved in the anthelmintic action mentioned above (Figure 31.12).[24,25]

Of particular interest is a cyclic amide, clausenamide, first isolated by Yang et al.,[26] which exhibited nootropic potency at 10mg/Kg, 50–100 times more than piracetam (Nootropil®), a drug used for the improvement of memory.[27,28] An interesting development from this observation would be to study the *Clausena* species and Rutaceae for clausenamide and congeners for neuropharmacological potential.

31.9 *EUODIA ELLERYANA* F. MUELL.

[From: Greek *eu* = good and *odion* = smell, and after Ellery, Robert Lewis John (1827–1908), astronomer and public servant.]

31.9.1   Botany

Euodia elleryana (F. Muell.) (*Melicope elleryana* [F. Muell.] T. G. Hartley) is a riparian tree that grows to a height of 30m in coastal riverine rain forests and streambanks of Australia and Papua New Guinea. The bole is 17m long and 40cm in diameter. The bark is brown, cracked, and lenticelled. The inner bark is yellow. The stems are squarish, lenticelled, 4mm in diameter with 4.5cm-long internodes. The nodes are broad. The leaves are decussate, exstipulate, and 3-foliolate. The petiole is 4.5–5cm long and somewhat flat and triangular. The petiolules are 5–8 mm – 1.2cm long and grooved above. The folioles are 10.1cm × 6.5cm – 7.5cm × 5cm – 9cm

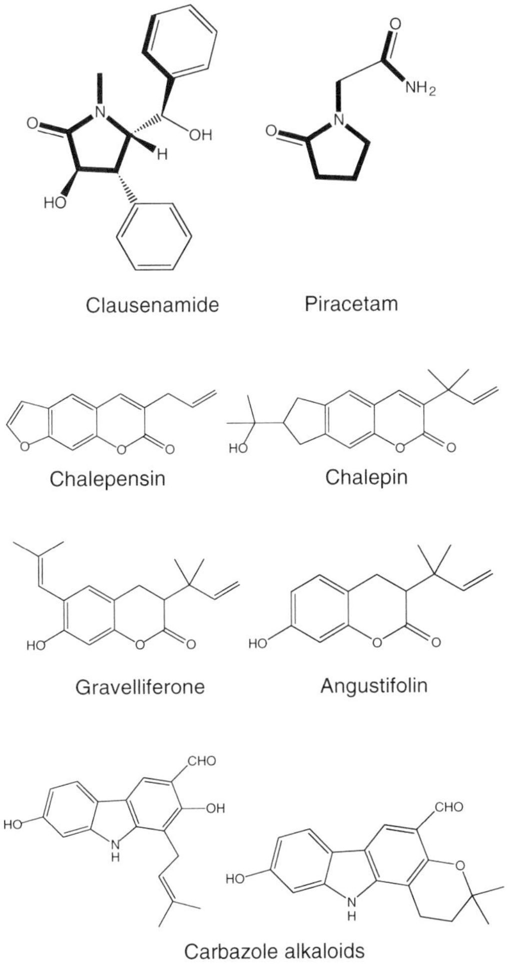

Clausenamide   Piracetam

Chalepensin   Chalepin

Gravelliferone   Angustifolin

Carbazole alkaloids

Figure 31.12 Chemical constituents of *Clausena lansium* (Lour.) Skeells.

× 4.4cm, with 10 pairs of secondary nerves, and with internerves sunken above and raised below. The midrib is sunken above. The inflorescences are axillary cymes which are 2.5cm long. The flowers are pink–purple. The fruits are green, dotted with oil cells, 1.2cm in diameter, and dehiscent on 1.3cm-long pedicels. A vestigial calyx is present at the base of the fruits (Figure 31.13).

31.9.2 Ethnopharmacology

The sap of *Euodia elleryana* F. Muell var. *teragona* (K. Sch.) W.D. Francis, or Pink *Euodia*, is applied in Papua New Guinea to promote the healing of sores, where the young green fruits are known as poisonous. Vernacular names include *iliek* and *ilik* (Mooi). The healing property is probably owed to antibacterial effects. Khan et al.[29] showed that methanol extracts of various parts of the plant inhibited the growth of a broad spectrum of bacteria. While the antibacterial principle here is unknown, one could reasonably expect that evodiamine and rutaecarpine-like quinazoline alkaloids are involved. Evodiamine, extracted from several species of *Euodia* fruits, inhibited cell proliferation and migration of several types of cancer cell lines, including leukemia Human Caucasian acute lymphoblastic leukaemia cell-line (CCRF-CEM) cells in a concentration-dependent manner with an IC_{50} of 0.57μM via apoptosis following microtubular cytoskeleton abrogation

Figure 31.13 *Euodia elleryana* F. Muell. [From: BOZWESEN, Nederlands Nieuw Guinea. Botanical identification: Chr. Versteegh and C. Kalkman. Geographical identification: Netherlands New Guinea, Div. West New Guinea, Warsamson River, about 25Km east of Sorong. In primary forest. Alt.: About 60m. Sandy, clayey soil.]

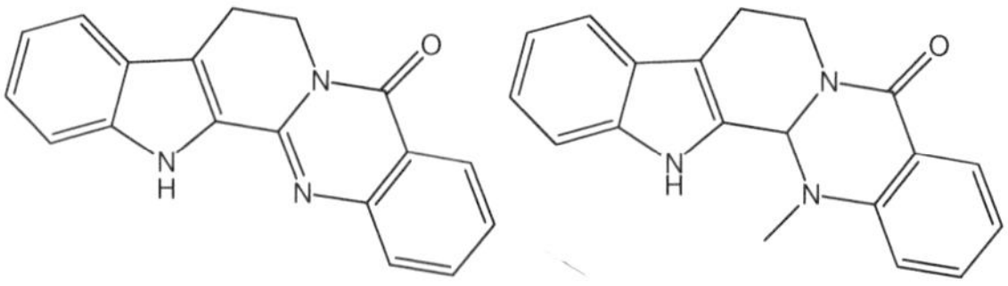

                    Rutaecarpine   Evodiamine

Figure 31.14 Cytotoxic alkaloids from the *Euodia* species.

(Figure 31.14).[30,31] It would be interesting to learn whether further study of *Euodia elleryana* F. Muell. discloses any cytotoxic alkaloid. It is likely.

31.10 *LUNASIA AMARA* BLCO.

[After Latin *amaritudo* = bitterness.]

31.10.1   Botany

Lunasia amara Blco. (*Lunasia costulata* Miq.) is a slender shrub, 2–8m high, which grows in the primary rain forests of the Philippines and Celebes. The sapwood is pale orange. The bark is gray and smooth. The inner bark is green. The stems are grayish-brown and minutely scaly. The leaves are simple, spathulate, papery, glossy above, crenate, wavy, and with yellowish nervations. The midrib is flat above. The petiole is grayish-green, minutely scaly, 3.5–6cm long, and slender. The blade is 26cm × 8.5cm – 21.5cm × 7.2cm, thinly coriaceous, with 17–20 pairs of secondary nerves and scalariform tertiary nerves, both of which are prominent below. The base is cordate to auriculate, the margin is wavy, and the apex is acuminate. The inflorescences are

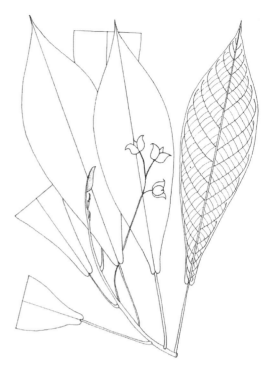

Figure 31.15 *Lunasia amara* Blco. [From: Philippines Plant Inventory. PPI. Flora of the Philippines, Joint project of the Philippines National Museum, Manila and B. P. Bishop Museum, Honolulu. Supported by NSF/USAID. Field collectors: F. J. Reynoso et al. PPI No: 7489. Geographical localization: Island Visayas, Province: Western Samar, Brgy, Tinama, Paranas. Found on the ridge. Oct. 10, 1992.]

axillary spikes or racemes that are 13.5cm long. The fruits are 3-lobed, 1.1cm × 2.2cm, and each lobe is 5mm in diameter. The flower buds are cream-colored. The flowers are khaki to yellow. The fruits are green, covered with a few light brown scales, dusty white or powdered, dehiscent and 3-seeded (Figure 31.15).

31.10.2 Ethnopharmacology

In Indonesia, the leaves and bark are boiled in water to make a lotion which is used to soothe inflammation. Filipinos use the bark to counteract snake poisoning and to mitigate stomachache. The pharmacological properties of *Lunasia amara* Blco. are unexplored as of yet. Note that the *Lunasia* species elaborate a series of quinoline alkaloids, such as lunacridine, which would be worth investigating for pharmacology (Figure 31.16).[32]

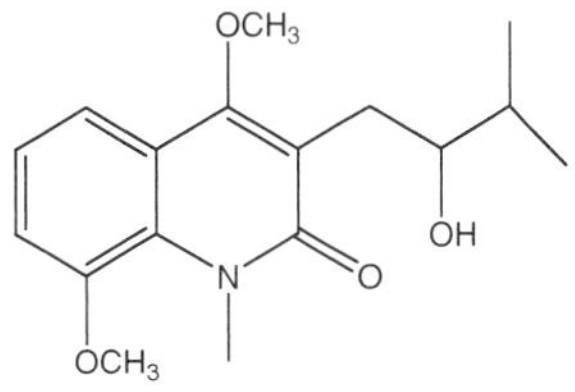

Lunacridine

Figure 31.16

31.11 *MICROMELUM MINUTUM* (FORST. F.) W. & A.

[After Greek *micro* = very small and *melon* = apple, and from Latin *minutum* = very small.]

31.11.1 Botany

Micromelum minutum (Forst. f.) W. & A. (*Micromelum pubescens* Bl.) is a shrub that grows to a height of 3m in Southeast Asia and the Pacific Islands. The bark is pale yellowish-brown. The inner bark is pale yellow. The stems are terete, glabrous, and 2mm in diameter. The leaves are compound, spiral, and exstipulate. The rachis shows 7–9 pairs of folioles which are asymmetrical, membranaceous, and 9.8cm × 3.5cm – 10.4cm × 3.7cm – 6.5cm × 2.4cm – 8cm × 2cm.

Figure 31.17 *Micromelum minutum* (Forst. f.) W. &.A.

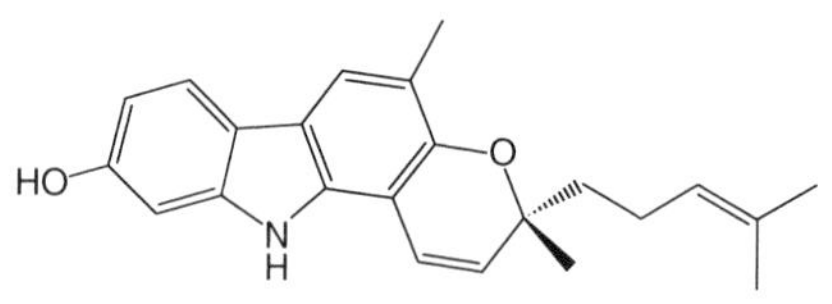

Mahanine

Figure 31.18 Mahanine, a cytotoxic alkaloid from the *Micromelum* species.

The petiolules are 5mm long. The blade shows several oil cells, and nine pairs of secondary nerves. The margin of the foliole is wavy, the base is asymmetrical, and the apex is acuminate. The inflorescences are panicles of little greenish-white flowers which are stellate, and strongly and sweetly scented. The fruits are elliptic, 7mm × 5mm, yellowish, turning red, and marked at the apex by a vestigial disc (Figure 31.17).

31.11.2   Ethnopharmacology

In Cambodia, Laos, and Vietnam, the leaves are used to calm itchiness and to promote menses. In Malaysia, the roots are boiled and pounded to produce a paste which is applied to treat ague. In the Philippines, the leaves are used to mitigate headaches. Vernacular names include *jahaka* and *ling wangi*. A remarkable advance in the pharmacology of *Micromelum minutum* (Forst f.) W. & A. has been provided by the work of Roy et al.[33] They showed that mahanine, a carbazole alkaloid, induces apoptosis in the human myeloid cancer cell HL-60 at 10μM via activation of caspase-3 through a mitochondrial-dependent pathway (Figure 31.18). Another significant contribution to the pharmacology of this plant has been provided by the work of Ma et al.[34] Using anti-*Mycobacterium tuberculosis* bioassay-directed fractionation, they isolated a series of carbazole alkaloids, including micromeline, lansine, 3-formylcarbazole, and 3-formyl-6-methoxycarbazole, with MIC values between 14.3μg/mL and 42.3μg/mL, as well as an unusual lactone derivative of oleic acid, (–)-Z-9-octadecene-4-olide, which showed potent *in vitro* anti-TB activity against the H37R strain, with an MIC value of 1.5μg/mL, and the Erdman strain of *Mycobacterium tuberculosis* in a J774 mouse macrophage model, with an IC_{90} value of 5.6μg/mL.

31.12 *PARAMIGNYA ANDAMANICA* TANAKA

[From: Greek *paramignunai* = to mix in and from Latin *andamanica* = from the Andaman Islands in the Bay of Bengal.]

31.12.1   Botany

Paramignya andamanica Tanaka (*Paramignya armata* Oliv. var *andamanica* King, and *Paramignya andamanica* [Ling] Tan) is a climber that grows in the primary rain forests of Cambodia, Laos, Vietnam, Thailand, and Borneo. The stems are 5mm in diameter, velvety at the apex, with regular internodes which are 1–2cm long. A few tiny recurved thorns are present along the stem. The leaves are simple, exstipulate, and spiral. The petiole is velvety when young, twisted,

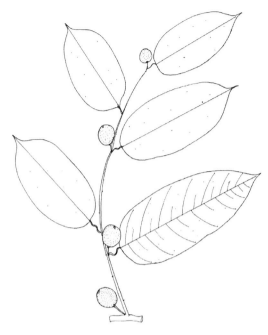

Figure 31.19 *Paramignya andamanica* Tanaka. [From: Flora of North Borneo. Distributed from the Herbarium of the Forest Department, Sandakan, North Borneo District: Kinabatangan. Locality Bumbulud Summit, Gomentong Caves, Hill, 20 miles south of Sandakan. Alt.: 725ft above sea level. July 22, 1954. Field collector: G. H. S. Wood. SAN A 4642. Botanical identification: Singapore, June 1955.]

channeled above, and 6mm–1.2cm long. The blade is oblong, lanceolate, dotted with oil cells, thinly coriaceous, 13cm × 5cm – 10cm × 4cm – 11.2cm × 4cm, with seven pairs of secondary nerves plus internerves and a fine network of tertiary nerves. The base is round, and the apex is shortly acuminate. The fruits are 1.7cm in diameter, globose, on 7mm-long pedicels from axillary tubercles (Figure 31.19).

31.12.2 Ethnopharmacology

In Cambodia, Laos, and Vietnam, the leaves and fruits are boiled in water to make a drink used to treat cough and bronchitis. The pharmacological potential of this plant and the genus *Paramignya* are to date unexplored.

31.13 *TODDALIA ASIATICA* (L.) LAMK.

[From: Latin *asiatica* = from Asia.]

31.13.1 Botany

Toddalia asiatica (L.) Lamk. (*Toddalia aculeata* Pers.) is a massive climber up to 20m long that grows in the primary rain forests of India, Ceylon, Southeast Asia, and South China. The plant is also found in East Africa and Mauritius. The main stems are 10–15cm in diameter. The young stems are spiny, 4–8mm in diameter, and fissured longitudinally. The leaves are spiral, exstipulate, and 3-foliolate. The petioles are 2.5–4cm long. The folioles are glossy, very dark green above, 6cm × 2.5cm – 7.3cm × 3cm – 6.5cm × 3cm – 7cm × 3cm – 6cm × 2.1cm, gland-

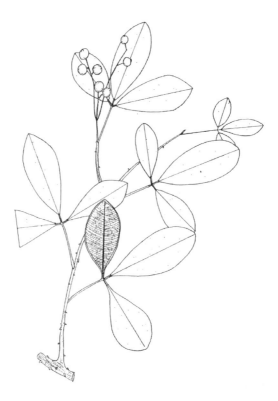

Figure 31.20 *Toddalia asiatica* (L.) Lamk.

dotted, with several pairs of secondary nerves. Intramarginal nerves are present. The flowers are yellow with a gray gynaecium. The fruits are 7mm × 1cm (Figure 31.20).

31.13.2 Ethnopharmacology

In the Philippines, the roots are used to treat diarrhea and to relieve the exhaustion of malarial fever, while the leaves are used to promote digestion. In Cambodia, Laos, and Vietnam, the leaves are eaten to treat lung troubles. The roots are used to invigorate. The Javanese drink a decoction of the bark as a bitter tonic or an antipyretic. In Burma and Taiwan, the fruits are used to treat rheumatism and intestinal troubles. *Toddalia* (*British Pharmaceutical Codex*, 1911) lists the dried root bark of *Toddalia asiatica* (L.) Lamk.

Using rodents, Hao et al.[35] showed that alkaloids of *Toddalia asiatica* reduce the auricle swelling caused by xylol, joint swelling caused by agar, and leucocyte migration brought on by sodium carboxymethyl cellulose, and also exhibit analgesic effects. The plant contains isopimpinellin, which has cardiovascular effects.[36]

A new antiplasmodial coumarin, 5,7-dimethoxy-8-(3′-hydroxy-3′methyl-1′-butene)-coumarin, has been isolated from the roots of *Toddalia asiatica* (L.) Lamk. This finding supports the traditional use of this plant for the treatment of malaria.[37]

The main constituent of the plant is toddalolactone, the pharmacological potential of which is unexplored.

While the antipyretic property of the plant has not yet been substantiated, accumulated experimental evidence indicates that the plant abounds with inhibitors of platelet aggregation.[38] Such compounds include chelerythrine, dictamine, 4-methoxy-1-methyl-2-quinolone, haplopine, γ-fagarine 2,6-dimethoxy-*p*-benzoquinone, and braylin (Figure 31.21). The work showed complete inhibition of the aggregation of platelets at 100mg/mL induced by arachidonic acid *in vitro*. Is the

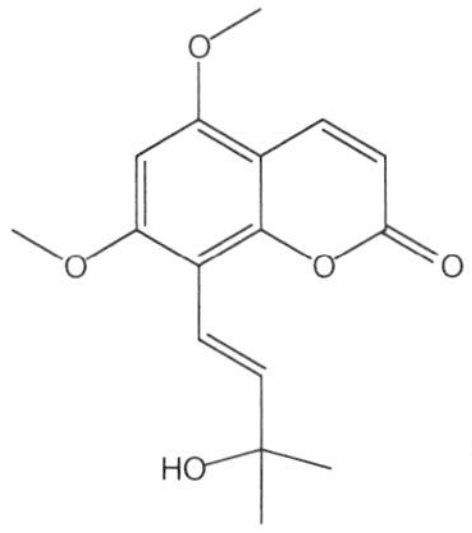

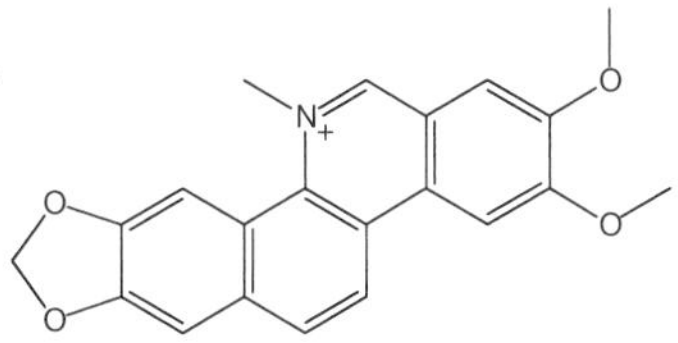

5,7-Dimethoxy-8-(3'-hydroxy-3'methyl-1'-butene)-coumarin   Nitidine

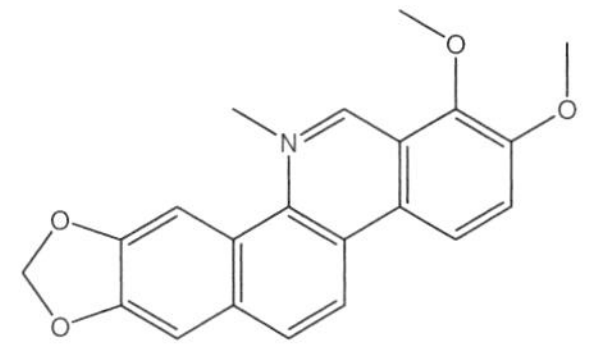

Chelerythrine 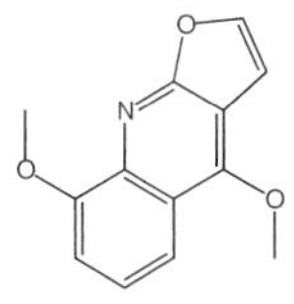   γ-Fagarine

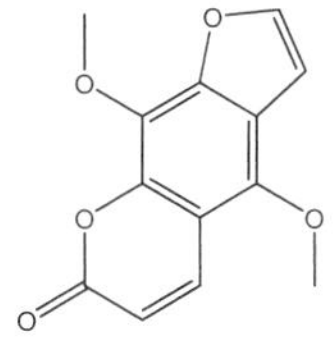

Isopimpinellin 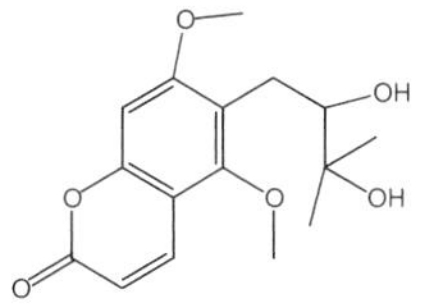   Toddalolactone

Figure 31.21

antipyretic and antiinflammatory property of *Toddalia asiatica* (L.) Lamk. related to the inhibition of eicosanoic acid? Probably yes.

31.14 *ZANTHOXYLUM AVICENNAE* (LAMK.) DC.

[From: Greek *zanthos* = yellow and *xylon* = wood, and after Avicenna or Ibn Sina, Arabian physician and philosopher (980–1037).]

31.14.1   Botany

Zanthoxylum avicennae (Lamk.) DC. (*Zanthoxylum clava-herculis sensu* Lour., also wrongly spelled as *Zanthoxyllum* or *Xanthoxylum*) is a woody climber that grows to a length of 8m, in Vietnam, Cambodia, and Laos. It is spiny along the bole and stems. The stems are terete and 5mm in diameter. The leaves are spiral, exstipulate, and compound. The rachis is squarish, 12cm long, and shows 10–11 pairs of folioles which are round and crenate. The blade of the foliole is without secondary nerves, and measures 1.5cm × 6 mm – 2.2cm × 3mm, on 2mm-long petiolules. The

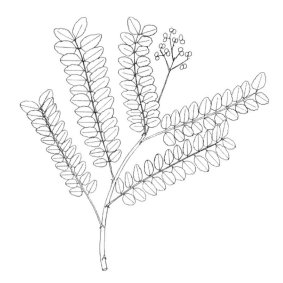

Figure 31.22 *Zanthoxylum avicennae.* [From: Plants of Borneo. Geographical localization: Sabah, Ranau District, Takutan Village, 5Km. Dec. 18, 1995.]

inflorescences are 8cm-long panicles. The fruits are 6mm in diameter, dark green, and smell like lemons (Figure 31.22).

31.14.2   Ethnopharmacology

In Cambodia, Laos, and Vietnam, the bark is used as a bitter tonic. Filipinos boil the stems in water to make a drink that is used to promote digestion and to counteract snake poisoning. The plant's pharmacology is as of yet unexplored, but Fish et al.[39] isolated a series of alkaloids including chelerythrine, which inhibits the aggregation of platelets.

The genus *Zanthoxylum* has attracted a great deal of interest on account of its ability to elaborate a broad series of benzo[c]phenanthridine alkaloids including chelerythrine and nitidine, quinoline alkaloids including skimmianine and aporphine such as liriodenine.[40–42] Benzo[c]phenanthridine alkaloids such as nitidine are of particular interest as they are able to inhibit the enzymatic activity of topoisomerase, hence its anticancer potential.[43–45]

Of recent interest is the work of Cheng et al.[46] Using the root bark of *Zanthoxylum ailanthoides*, they isolated a series of alkaloids which inhibited the survival of HIV in H9 lymphocyte cells cultured *in vitro*, including γ-fagarine and (+)-tembamide which showed anti-HIV activities with EC_{50} values inferior to 0.1μg/mL, as well as sesquiterpenes, 10β-methoxymuurolan-4-en-3-one, and 10α-methoxycadinan-4-en-3-one (Figure 31.23).

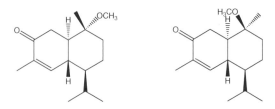

10α-Methoxycadinan-4-en-3-one   10β-Methoxymuurolan-4-en-3-one

Figure 31.23

31.15 *ZANTHOXYLUM MYRIACANTHUM* WALL. EX HK. F.

[From: Greek *zanthos* = yellow and *xylon* = wood, and Latin *myriacanthum* = many thorns.]

31.15.1   Botany

Zanthoxylum myriacanthum Wall. ex Hk. f. is a tree that grows up to 27m in height and 23cm in diameter. The bark is fawn and smooth with big spines on a rounded podia. The stems are terete, hollowed, and covered with numerous spines. The leaves are spiral, exstipulate, and compound. The rachis is 20–36cm and 3mm in diameter. The folioles are subopposite, 7cm × 2 cm – 7.8cm × 2.7cm – 8cm × 3.3cm – 7.5cm × 4cm – 6.7cm × 3.8cm, aromatic, glossy and leathery above, with 7–12 pairs of secondary nerves that are sunken above and raised below, with several oil cells. The margin is minutely serrate and the apex is minutely notched. The inflorescence are cymoses and 12cm long. The flowers are minute, 3mm × 2mm. The fruits are in groups of two or three globose folioles which open to show a glossy seed (Figure 31.24).

31.15.2   Ethnopharmacology

The seeds are burned and the smoke inhaled to heal syphilitic ulceration of the nose. The little evidence available so far indicates that the plant abounds with benzophenanthridine alkaloids such as nitidine and dihydronitidine, which are worth investigating.[47,48]

Figure 31.24 *Zanthoxylum myriacanthum* Wall. ex Hk. f. [From: FRIM. Geographical localization: Selangor, Ulu Gombak Forest Reserve, Genting. Hill mixed dipterocarp forest. Alt.: 700m. FRI No: 39055. Field collectors: Chua et al., Oct. 15, 1992. Botanical identification: L. Chua, Dec. 6, 1996.]

REFERENCES

1. Michael, J. P. 2004. Quinoline, quinazoline and acridone alkaloids. *Nat. Prod. Res.*, 21, 650.
2. Arul, V., Miyazaki, S., and Dhananjayan, R. 2005. Studies on the anti-inflammatory, antipyretic and analgesic properties of the leaves of *Aegle marmelos* Corr. *J. Ethnopharmacol.*, 96, 163.
3. Kamalakkannan, N. and Prince, S. M. 2003. Hypoglycemic effect of water extracts of *Aegle marmelos* fruits in streptozotocin diabetic rats. *J. Ethnopharmacol.*, 87, 207.
4. Karunanayake, E. H., Welihinda, J., Sirimanne, S. R., and Adorai, G. S. 1984. Oral hypoglycaemic activity of some medicinal plants of Sri Lanka. *J. Ethnopharmacol.*, 11, 223.
5. Sarath, N., Arseculeratne, A. A., Gunatilaka, L., and Panabokke, R. G. 1985. Studies on medicinal plants of Sri Lanka. Part 14: Toxicity of some traditional medicinal herbs. *J. Ethnopharmacol.*, 13, 323.
6. Rana, B. K., Singh, U. P., and Taneja, V. 1997. Antifungal activity and kinetics of inhibition by essential oil isolated from leaves of *Aegle marmelos*. *J. Ethnopharmacol.*, 57, 29.

7. Shoba, F. G. and Thomas, M. 2001. Study of antidiarrheal activity of four medicinal plants in castor-oil induced diarrhea. *J. Ethnopharmacol.*, 76, 73.

8. Arul, V., Miyazaki, S., and Dhananjayan, R. 2004. Mechanisms of the contractile effect of the alcoholic extract of *Aegle marmelos* Corr. on isolated guinea pig ileum and tracheal chain. *Phytomedicine*, 11, 679.

9. Kar, A., Panda, S., and Bharti, S. 2002. Relative efficacy of three medicinal plant extracts in the alteration of thyroid hormone concentrations in male mice. *J. Ethnopharmacol.*, 81, 281.

10. Lampronti, I., Martello, D., Bianchi, N., Borgatti, M., Lambertini, E., Piva, R., Jabbar, S., Choudhuri, M. S., Khan, M. T., and Gambari, R. 2003. *In vitro* antiproliferative effects on human tumor cell lines of extracts from the Bangladeshi medicinal plant *Aegle marmelos* Correa. *Phytomedicine*, 10, 300.

11. Grimaldi, A., Bennet, P., Delas, B., Chapelon, C., Lebargy, F., Sobel, J., and Bosquet, F. 1985. Betablockers and hypoglycemia. *Ann. Med. Interne. (Paris)*, 136, 186.

12. Delarue, J., Schneiter, P., Cayeux, C., Haesler, E., Jequier, E., and Tappy, L. 1997. Effects of adrenergic blockade on hepatic glucose production during ethanol administration. *Clin. Physiol.*, 17, 509.

13. Szelachowska, M. and Kinalska, I. 1990. Effect of beta-adrenergic blockade on cortisol secretion in insulin-induced hypoglycemia in animals with experimental diabetes mellitus. *Pol. Tyg. Lek.*, 45, 376.

14. Govindachari, T. R., Viswanathan, N., Pai, B. R., Ramachandran, V. N., and Subramaniam, P. S. 1970. Alkaloids of *Atalantia monophylla* Correa. *Tetrahedron*, 26, 2905.

15. Basa, S. C. 1975. Atalaphyllinine, a new acridone base from *Atalantia monophylla*. *Phytochemistry*, 14, 835.

16. Bahar, M. H., Shringarpure, J. D., Kulkarni, G. H., and Sabata, B. K. 1982. Structure and synthesis of atalaphylline and related alkaloids. *Phytochemistry*, 21, 2729.

17. Gururaj, H., Kulkarni, G. H., and Sabata, B. K. 1981. An acridone alkaloid from the root bark of *Atalantia monophylla*. *Phytochemistry*, 20, 867.

18. Harrison, R. J., Reszka, A. P., Haider, S. M., Romagnoli, B., Morrell, J., Read, M. A., Gowan, S. M., Incles, C. M., Kelland, L. R., and Neidle, S. 2004. Evaluation of by disubstituted acridone derivatives as telomerase inhibitors: the importance of G-quadruplex binding. *Bioorg. Med. Chem. Lett.*, 14, 5845.

19. Tiwawech, D., Hirose, M., Futakuchi, M., Lin, C., Thamavit, W., Ito, N., and Shirai, T. 2000. Enhancing effects of Thai edible plants on 2-amino-3,8-dimethylimidazo(4,5-f)quinoxaline-hepatocarcinogenesis in a rat medium-term bioassay, *Cancer Lett.*, 158, 195.

20. Sunthitikawinsakul, A., Kongkathip, N., Kongkathip, B., Phonnakhu, S., Daly, J. W., Spande, T. F., Nimit, Y., Napaswat, C., Kasisit, J., and Yoosook, C. 2003. Anti-HIV-1 limonoid: first isolation from *Clausena excavata*. *Phytother. Res.*, 17, 1101.

21. Sunthitikawinsakul, A., Kongkathip, N., Kongkathip, B., Phonnakhu, S., Daly, J. W., Spande, T. F., Nimit, Y., and Rochanaruangrai, S. 2003. Coumarins and carbazoles from *Clausena excavata* exhibited antimycobacterial and antifungal activities. *Planta Med.*, 69, 155.

22. Sunthitikawinsakul, A., Kongkathip, N., Kongkathip, B., Phonnakhu, S., Daly, J. W., Spande, T. F., Nimit, Y., and Rochanaruangrai, S. 2003. Coumarins and carbazoles from *Clausena excavata* exhibited antimycobacterial and antifungal activities. *Planta Med.*, 69, 155.

23. Hirata, K., Ito, C., Furukawa, H., Itoigawa, M., Cosentino, L. M., and Lee, K. H. 1999. Substituted 7H-pyrido[4,3-c]carbazoles with potent anti-HIV activity. *Biorg. Med. Chem. Lett.*, 9, 119.

24. Li, W. S., McChesney, J. D., and El-Feraly, F. S. 1991. Carbazole alkaloids from *Clausena lansium*. *Phytochemistry*, 30, 343.

25. Kumar, V., Vallipuram, K., Adebajo, A. C., and Reisch, J. 1995. 2,7-Dihydroxy-3-formyl-1-(3′-methyl-2′-butenyl)carbazole from *Clausena lansium*. *Phytochemistry*, 40, 1563.

26. Yang, M. H., Chen, Y. Y., and Huang, L. 1988. Three novel cyclic amides from *Clausena lansium*. *Phytochemistry*, 27, 445.

27. Zhang, J. T., Duan, W., Jiang, X. Y., Liu, S. L., and Zhao, M. R. 2000. Effect of (−)clausenamide on impairment of memory and apoptosis. *Neurobiol. Aging*, 21, Suppl., 1, 243.

28. Tang, K. and Zhang, J. T. 2004. Mechanism of (−)clausenamide induced calcium transient in primary culture of rat cortical neurons. *Life Sci.*, 74, 1427.

29. Khan, M. R., Kihara, M., and Omoloso, A. D. 2000. Antimicrobial activity of *Evodia elleryana*. *Fitoterapia*, 71, 72.

30. Huang, Y. C., Guh, J. H., and Teng, C. M. 2004. Induction of mitotic arrest and apoptosis by evodiamine in human leukemic T-lymphocytes. *Life Sci.*, 75, 35.

31. Zhang, Y., Zhang, Q. H., Wu, L. J., Tashiro, S., Onodera, S., and Ikejima, T. 2004. Atypical apoptosis in L929 cells induced by evodiamine isolated from *Evodia rutaecarpa*. *J. Asian Nat. Prod. Res.*, 6, 19.

32. Bowman, R. M., Gray, G. A., and Grundon, M. F. 1973. Quinoline alkaloids. XV. Reactions of a quinoline isoprenyl epoxide with hydride reagents. Asymmetric synthesis and stereochemistry of lunacridine and related Lunasia alkaloids. *J. Chem. Soc.*, 10, 1051.

33. Roy, M. K., Thalang, V. N., Trakoontivakorn, G., and Nakahara, K. 2004. Mechanism of mahanine-induced apoptosis in human leukemia cells (HL-60). *Biochem. Pharmacol.*, 67, 41.

34. Ma, C., Case, R. J., Wang, Y., Zhang, H. J., Tan, G. T., Van Hung, N., Cuong, N. M., Franzblau, S. G., Soejarto, D. D., Fong, H. H., and Pauli, G. F. 2005. Anti-tuberculosis constituents from the stem bark of *Micromelum hirsutum*. *Planta Med.*, 71, 261.

35. Hao, X. Y., Peng, L., Ye, L., Huang, N. H., and Shen, Y. M. 2004. A study on anti-inflammatory and analgesic effects of alkaloids of *Toddalia asiatica*. *Zhong Xi Yi Jie He Xue Bao*, 2, 450.

36. Guo, S., Li, S., Peng, Z., and Ren, X. 1998. Isolation and identification of active constituent of *Toddalia asiatica* in cardiovascular system. *Zhong Yao Cai*, 21, 515.

37. Oketch-Rabah, H. A., Mwangi, J. W., Lisgarten, J., and Mberu, E. K. 2000. A new antiplasmodial coumarin from *Toddalia asiatica* roots. *Fitoter.*, 71, 636.

38. Tsai, I. L., Wun, M. F., Teng, C. M., Ishikawa, T., and Chen, I. S. 1998. Anti-platelet aggregation constituents from formosan *Toddalia asiatica*. *Phytochemistry*, 48, 1377.

39. Fish, F., Gray, A. I., and Waterman, P. G. 1975. Coumarin, alkaloid and flavonoid constituents from the root and stem barks of *Zanthoxylum avicennae*. *Phytochemistry*, 14, 841.

40. Ng, K. M., Gray, A. G., and Waterman, P. G. 1987. Benzophenanthridine alkaloids from the stem bark of a *Zanthoxylum* species. *Phytochemistry*, 26, 325.

41. Yang, Y. P., Cheng, M. J., Teng, C. M., Chang, Y. L., Tsai, I. L., and Chen, I. S. 2002. Chemical and anti-platelet constituents from Formosan *Zanthoxylum simulans*. *Phytochemistry*, 61, 567.

42. Martin, M. T., Rasoanaivo, L. H., and Raharisololalao, A. 2005. Phenanthridine alkaloids from *Zanthoxylum madagascariense*. *Fitoterapia*, 76, 590.

43. Bongui, J.B., Blanckaert, A., Elomri, A., and Seguin, E. 2005. Constituents of *Zanthoxylum heitzii* (Rutaceae), *Biochem. System. Ecol.*, 33, 845.

44. Li, D., Zhao, B., Sim, S. P., Li, T. K., Liu, A., Liu, L. F., and LaVoie, E. J. 2003. 2,3-Dimethoxy-benzo[i]phenanthridines: topoisomerase I-targeting anticancer agents. *Bioorg. Med. Chem.*, 11, 521.

45. Holden, J. A., Wall, M. E., Wani, M. C., and Manikumar, G. 1999. Human DNA topoisomerase I: quantitative analysis of the effects of camptothecin analogs and the benzophenanthridine alkaloids nitidine and 6-ethoxydihydronitidine on DNA topoisomerase I-induced DNA strand breakage. *Arch. Biochem. Biophys.*, 370, 66.

46. Cheng, M. J., Lee, K. H., Tsai, I. L., and Chen, I. S. 2005. Two new sesquiterpenoids and anti-HIV principles from the root bark of *Zanthoxylum ailanthoides*. *Bioorg. Med. Chem.*, in press.

47. Waterman, P. G. 1975. Alkaloids from the root bark of *Zanthoxylum myriacanthum*. *Phytochemistry*, 14, 2530.

48. Sukari, M. A., Salim, W. S. W., Ibrahim, N. H., Rahmani, M., Aimi, N., and Kitajima, M. 1999. Phenantridine alkaloids from *Zanthoxylum myriacanthum*. *Fitoterapia*, 70, 197.

CHAPTER 32

Medicinal Plants Classified in the Family Loganiaceae

32.1 GENERAL CONCEPT

The family Loganiaceae (Martius, 1827 nom. conserv., the Logania Family) consists of approximately 20 genera and 500 species of tropical trees, shrubs, and climbers that commonly produce iridoids and monoterpenoid indole alkaloids that are formed by the condensation of tryptamine and secologanin (an iridoid). Looking for Loganiaceae in the field might not be a very easy task; it is advised to look for trees or dichotomous climbers with opposite simple leaves, interpetiolar stipules, no latex, and tubular flowers which are whitish with five lobes, a 2-celled gynaecium, and fruits which are always superior capsules, berries, or drupes.

Classical examples of pharmaceutical products of Loganiaceous origin are the dried ripe seeds of *Strychnos nux-vomica* L., a plant of India, Ceylon, Thailand, Cambodia, Laos, Vietnam, and North Malaysia. *Nux vomica* (*British Pharmacopoeia*, 1963) and *Strychnos ignatii* (Ignatia, *British Pharmaceutical Codex*, 1934) consist of the dried ripe seeds containing not less than 1.2% of strychnine (Figure 32.1). It was used as a bitter and as an ingredient of purgative pills and tablets. Strychnine (*British Pharmaceutical Codex*, 1959) was formerly used to stimulate blood circulation during surgical shock, but its use is now more limited to invigorating breathing

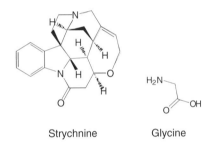

Strychnine   Glycine

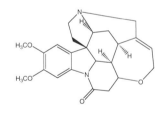

Brucine

Figure 32.1 Examples of bioactive alkaloids from the family Loganiaceae.

during poisoning. Strychnine given in a small dose to humans and animals binds to the glycinergic receptor and enhances the motor response of the spinal reflex. Large doses cause tremors and slight twitching of the limbs, followed by sudden convulsions of all muscles. The body becomes arched backwards in hyperextension with the legs and arms extended and the feet turned inward. The facial muscles produce a characteristic grinning expression known as *risus sardonicus*. Death from medullary paralysis usually follows the second or fifth seizure. The convulsions are mediated spinally and believed to result from a blockade of inhibitory glycinergic sites. Antidotes for strychnine poisoning are short-acting barbiturates and muscle-relaxing drugs. The seeds of *Strychnos nux-vomica* L. are used to treat eye diseases because strychnine instilled locally increases the ability to discriminate colors and intensities of illumination, particularly in the area of the blue visual field. *Strychnos nux-vomica* L. was once used to treat amblyopia.

The dried rhizome and roots of *Gelsenium sempervirens* (*Gelsenium*, *British Pharmaceutical Codex*, 1963) contain no less than 0.32% of gelsemine, which has been used as a tincture to treat migraine (Gelsenium Tincture, *British Pharmaceutical Codex*, 1963). Note that *Gelsemium sempervirens* (L.) Ait. f. (Evening Trumpet Flower) is a common ornamental garden plant in North America. Another example of medicinal Loganiaceae is *Gelsemium nitidum* (American Yellow Jasmine), the roots of which are occasionally used to reduce headache.

While the genus *Strychnos* has attracted a great deal of interest, very little is known about the pharmacological potential of the remaining genera, a gap that is worth investigating further.[1-10] An exciting reserve of potentially active Loganiaceae is in the medicinal plants of the Pacific Rim, where about 20 species are used to invigorate, to counteract putrefaction, to treat eye diseases, and to expel worms from the intestines. Among these are *Fagraea auriculata* Jack, *Fagraea blumei* G. Don., *Fagraea obovata* (non Wall.) King, and *Neuburgia corynocarpa* (A. Gray) Leenh.

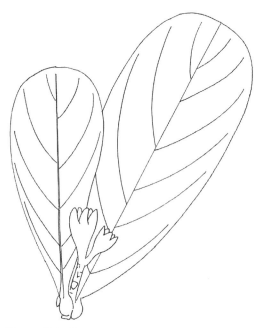

Figure 32.2 *Fagraea auriculata* Jack. [From: Singapore Field No: 36752. Distributed from The Botanic Gardens, Singapore. Geographical localization: Pengkalan Raja, Pontian. L. Johor. July 3, 1939. Alt.: Sea level. Field collector: I. Ngadiman and M. R. Henderson in peat forest.]

32.2 *FAGRAEA AURICULATA* JACK

[After J. T. Fagraeus (1729–1747), a Swedish naturalist, and from Latin *auriculata* = with ears.]

32.2.1 Botany

Fagraea auriculata Jack is a climber or small tree that grows from sea level to 1200m in Burma, Thailand, Cambodia, Laos, Vietnam, and throughout the Malay Archipelago. The leaves are opposite, simple, and stipulate. The petiole shows auricles at its base. The blade is 24cm × 10cm, obovate, fleshy, and has 5–8 pairs of secondary nerves that are not prominent below. The flowers are among the largest in the group of flowering plants, 30cm long and 30cm wide when fresh. The sepals are 2–7.5cm × 1.5cm long and almost free. The corolla tube is up to 15cm long (Figure 32.2).

32.2.2 Ethnopharmacology

The Indonesians of Sumatra use the bark to heal ulcers. The pharmacological properties of this plant are as of yet unknown.

32.3 *FAGRAEA BLUMEI* G. DON.

[After J. T. Fagraeus (1729–1747), a Swedish naturalist, and after Carl Ludwig Blume (1789–1862), a German–Dutch botanist.]

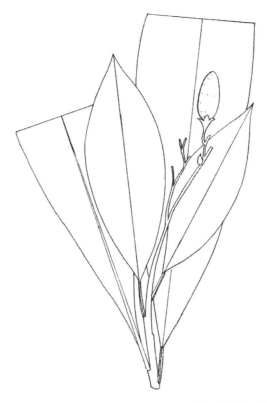

Figure 32.3 *Fagraea blumei* G. Don. [From: W. J. J. O. de Wilde and B. E. E. de Wilde-Duyfjes, Aug. 13, 1978. Geographical localization: Ketambe Mountain and vicinity, 8–15Km, southwest from the mouth of Lau Ketambe, c. 40Km northwest of Kubatjame. Alt.: 2400m, Indonesia.]

32.3.1 Botany

Fagraea blumei G. Don. (*Fragraea vaginata* King & Gamble and *Fagraea obovata* Wall.) is a river bank tree from Malaysia, Borneo, Java, and the Philippines that grows to a height of 15m with a girth of 35cm. The leaves are simple, opposite, and stipulate. The petiole is 2cm long and without auricles. The blade is elliptic, pointed at the base, and the apex is 29cm × 6.5cm – 12cm × 4cm, with clearly visible secondary nerves. The inflorescences are 4.9cm long, warty, and lenticelled. The flowers are tubular, cream-colored, less than 6cm long, and funnel-shaped. The fruits are green and glossy, ovoid to fusiform, and 3cm long on a persistent calyx (Figure 32.3).

32.3.2 Ethnopharmacology

In Indonesia, the leaves are used to reduce fever and to ease headaches. The pharmacological properties in this plant are unexplored. Note, however, that it is known to elaborate a series of lignans, of which pinoresinol showed analgesic and spasmolytic properties in rodents.[11]

32.4 *FAGRAEA OBOVATA* (NON WALL.) KING

[After J. T. Fagraeus (1729–1747), a Swedish naturalist, and from Latin *obovata* = obovate, referring to the leaves.]

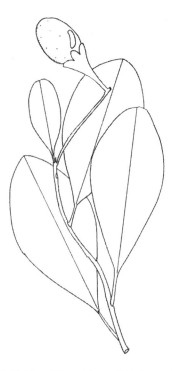

Figure 32.4 *Fagraea obovata* (non Wall.) King. [From: Flora of Thailand. Royal Forest Department. No: 26274. Geographical localization: Central: Pisamlok, Kaeng Sopa. May 2, 1961. No: 128.]

32.4.1 Botany

Fagraea obovata (non Wall.) King is a tree in Thailand and Malaysia that grows to a height of 12m with a girth of 92cm near streams in the rain forests of. The leaves are simple, opposite, and stipulate. The blade is 11cm × 5.4cm – 12cm × 5cm – 11.5cm × 5cm, thick, and obovate. The fruits are 3.5cm in diameter, ovoid, green, glossy, and seated on a persistent calyx which is 2cm × 3cm (Figure 32.4).

32.4.2 Ethnopharmacology

In Indonesia, the leaves are used to reduce fever and to ease headaches. The pharmacological properties of this plant are unexplored. One could investigate the plant for analgesic or antiinflammatory principles.

32.5 *NEUBURGIA CORYNOCARPA* (A. GRAY) LEENH.

[From: Latin *coryno* = club-like and *carpus* = fruits, referring to the shape of the fruits.]

32.5.1 Botany

Neuburgia corynocarpa (A. Gray) Leenh. (*Couthovia britannica* Kaneh. & Hatus, *Couthovia corynocarpa* A. Gray, *Couthovia novobritannica* Kaneh. & Hatus, and *Couthovia seemannii* A. Gray) is a timber tree that grows from Papua New Guinea to the Solomon Islands. It grows to a height of 9m. The leaves are simple, opposite, and stipulate. The stipules are interpetiolar and dissected. The blade is dark green, glossy, 11cm × 6cm – 13.5cm × 6.3cm – 9cm × 5cm,

Figure 32.5 *Neuburgia corynocarpa* (A. Gray) Leenh. [From: Logging area. Heravat, New Britain. Alt.: 150ft. Field collector: A. G. Floyd, Dec. 28, 1954. Botanical identification: P. W. Leenhouts.]

elliptic–lanceolate, with about 7–9 pairs of secondary nerves, and a yellow midrib. The inflorescences are 7.5cm-long cymes. The fruits are ovoid, kidney-shaped, pointed at the apex, white, and succulent (Figure 32.5).

32.5.2 Ethnopharmacology

In the Solomon Islands, the bark is used externally to treat skin diseases. The pharmacological properties of this plant are unexplored as of yet. Papooses call the plant *teittaka* and *chiew*.

REFERENCES

1. Philippe, G., Angenot, L., De Mol, P., Goffin, E., Hayette, M. P., Tits, M., and Frédérich, M. 2005. *In vitro* screening of some *Strychnos* species for antiplasmodial activity. *J. Ethnopharmacol.*, 97, 535.
2. Yin, W., Wang, T. S., Yin, F. Z., and Cai, B. C. 2003. Analgesic and anti-inflammatory properties of brucine and brucine N-oxide extracted from seeds of *Strychnos nux-vomica*. *J. Ethnopharmacol.*, 88, 205.
3. Thongphasuk, P., Suttisri, R., Bavovada, R., and Verpoorte, R. 2003. Alkaloids and a pimarane diterpenoid from *Strychnos vanprukii*. *Phytochemistry*, 64, 897.
4. Houghton, P. J., Mensah, A. Y., Iessa, N., and Hong, L. Y. 2003. Terpenoids in Buddleja: relevance to chemosystematics, chemical ecology and biological activity. *Phytochemistry*, 64, 385.
5. Philippe, G., De Mol, P., Zèches-Hanrot, M., Nuzillard, J. M., Tits, M. H., Luc Angenot, L., and Michel Frédérich, M. 2003. Indolomonoterpenic alkaloids from *Strychnos icaja* roots. *Phytochemistry*, 62, 623.
6. Melo, Mde F., Thomas, G., and Mukherjee, R. 1988. Antidiarrheal activity of bisnordihydrotoxiferine isolated from the root bark of *Strychnos trinervis* (Vell.) Mart. (Loganiaceae). *J. Pharm. Pharmacol.*, 40, 79.
7. Thepenier, P., Jacquier, M. J., Massiot, G., Le Men-Olivier, L., and Delaude, C. 1988. Alkaloids from *Strychnos staudtii*. *Phytochemistry*, 27, 657.
8. Brasseur, T. and Angenot, L. 1988. Six flavonol glycosides from leaves of *Strychnos variabilis*. *Phytochemistry*, 27, 1487.
9. Quetin-Leclercq, J. and Angenot, L. 1988. 10-Hydroxy-Nb-methyl-corynantheol, a new quaternary alkaloid from the stem bark of *Strychnos usambarensis*. *Phytochemistry*, 27, 1923.

10. Massiot, G., Massoussa, B., Jacquier, M. J., Thépénier, P., Le Men-Olivier, L., Delaude, C., and Verpoorte, R. 1988. Alkaloids from roots of *Strychnos matopensis*. *Phytochemistry*, 27, 3293.
11. Okuyama, E., Suzumura, K., and Yamazaki, M. 1995. Pharmacologically active components of Todopon Puok (*Fagraea racemosa*), a medicinal plant from Borneo. *Chem. Pharm. Bull. (Tokyo)*, 43, 2200.

CHAPTER 33

Medicinal Plants Classified in the Family Gentianaceae

33.1 GENERAL CONCEPT

The family Gentianaceae (A. L. de Jussieu, 1789 nom. conserv., the Gentian Family) consists of approximately 75 genera and about 1000 species of annual or perennial herbs, mainly belonging to the genus *Gentiana,* which includes about 400 species. The leaves of Gentianaceae are simple, mostly opposite, without stipules, often connate at the base or connected by a transverse line. The flowers are perfect, actinomorphic, showy, tubular and brightly colored, and 4–12-lobed, which are contorted or very rarely imbricate. The gynoecium consists of two carpels united to form a superior, and one locular ovary which develops into a capsule containing numerous seeds.

A large number of Gentianaceae are bitter, but have been used in Western medicine to promote appetite. These include *Sabatia angularis* (American Centaury), *Centaurium erythraea* Rafn. (European Centaury), the dried fermented rhizome and root of *Gentiana lutea* L. (Yellow Gentian) (Gentian, *British Pharmacopoeia*, 1963), *Gentiana catesbaei, Gentiana macrophylla, Gentiana punctata*, and *Gentiana purpurea*. The dried flowering tops of the Common Centaury *Centaurium minus (Centaurium umbellatum* and *Erythraea centaurium*) and other species of *Centaurium* (Petite Centaurée, *French Pharmacopoeia*, 1965) have been used as a bitter in the form of a liquid extract (1 in 1 dose 2–4mL) and infusion (1 in 20 dose 30–60mL). The bitterness of Gentianaceae is imparted by a series of iridoid glycosides such as gentiopicroside, swiertiamarin, and amarogentin (Figure 33.1). Besides iridoids, there is a massive body of evidence to demonstrate that the Gentianaceae Family is a vast source of xanthones and aglycones, the pharmacological properties of which need to be investigated for monoamine oxidase (MAO) inhibition, antiviral activity, antiinflammation, and antiplatelet aggregation. One such compound is norathyriol, which relaxes the thoracic aorta in the rat mainly by suppressing the Ca^{2+} influx through both voltage-dependent and receptor-operated calcium channels.[1] Chalcone dimers and flavonoids from *Gentiana lutea* exhibited interesting levels of enzymatic inhibition against MAO B with an IC_{50} value of 2.9µM for 2-methoxy-3-(1,1′-dimethylallyl)-6α,10α-dihydrobenzo(1.2c)chroman-6-one.[2] However, as of yet, only a few active principles have been isolated from this family. About 20 plants classified within the family Gentianaceae are used in the Pacific Rim to invigorate, to reduce fever, to stimulate appetite and urination, to relieve the bowels from constipation, and to counteract putrefaction of the skin.

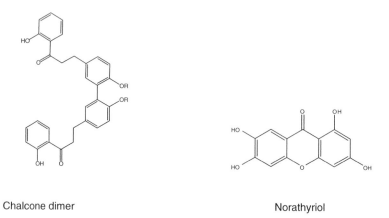

Gentiopicroside   Swiertiamarin   Amarogentin

2-Methoxy-3-(1,1'-dimethylallyl)-6α, 10α -dihydrobenzochroman-6-one

Chalcone dimer   Norathyriol

Figure 33.1 Examples of bioactive natural products from the family Gentianaceae.

33.2 *NYMPHOIDES INDICA* (THWAITES) KUNTZE

[From: Latin *indica* = from India.]

33.2.1 Botany

Nymphoides indica (Thwaites) Kuntze (*Limnanthemum indicum* [L.] Griseb., *Menyanthes indica* L., *Limnanthemum esquirolii* H. Léveil, and *Nymphoides humboldtiana* Kuntze) is an aquatic herb found in pools, pans, marshes, and rivers throughout southern Taiwan, Cambodia, India, Indonesia, Japan, Korea, Malaysia, Burma, Nepal, Sri Lanka, Vietnam, Australia, the Pacific Islands, China, tropical Africa, New Zealand, and India at an altitude of 1600m.

The plant, which is grown for ornamental purposes, grows from a rhizome and looks like a member of the family Nympheaceae. The stems are fleshy, terete, and unbranched. The leaves are simple, exstipulate, and alternate, which is why it is often incorporated in the family Menyanthaceae. The petiole is 1–2cm. The blade is broadly ovate to subcordate, 3cm × 18cm, thinly coriaceous, and abaxially densely glandular. The base is cordate, the margin is entire, and the blade shows a net of palmate nervations. The flowers are white with a yellow center, 7mm–12cm long, with five lobes which are ovate–elliptical, and densely fimbriate-pilose. The anthers are sagitate, 2–2.2mm long. The ovary is cylindrical and the stigma lobes are triangular. The fruits are elliptic capsules which are 3–5mm long (Figure 33.2).

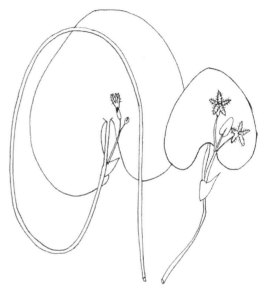

Figure 33.2 *Nymphoides indica* (Thwaites) Kuntze. [From: University of Illinois at Chicago. Plants of New Guinea. Geographical localization: Irian Jaya District of Manokuvari, Subdistrict Anggi, Village Iraiweri, Hamlet, Irai, Danau Anggi Gigi, on the lake. Alt.: 1800m. Field collector: E. A. Widjaja. No: 4274. June 8, 1991. Collected under the sponsorship of the U.S. National Cancer Institute. NCI sample: U44 Z2621 K.]

33.2.2 Ethnopharmacology

In Papua New Guinea, the plant is known as *misuanto* (Hatarn), *auggi meritz* (Sougb), and used to promote pregnancy. In Vietnam, the plant is used to reduce fever, to invigorate, as a carminative, and as an antiscorbutic. To date the pharmacological properties of *Nymphoides indica* (Thwaites) Kuntze, and in a broader sense the genus *Nymphoides,* are as of yet unexplored. Note that the presence of gynecological and antipyretic properties suggest some interferences with the eicosanoic acid pathways.

33.3 *SWERTIA JAVANICA* BL.

33.3.1 Botany

Swertia javanica Bl. is a woody herb that grows up to 1.5m tall and grows in marshes and by rivers and lakes. The stems are woody, 4–8mm in diameter, terete, and regularly marked by leaf

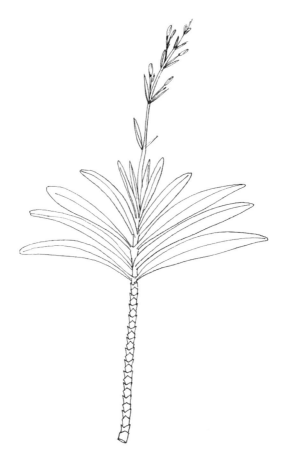

Figure 33.3 *Swertia javanica* Bl. [From: W. J. J. O. de Wilde and B. E. E. de Wilde-Duyfjes 16305. Geographical localization: Leuser Mountain Nature Reserve Aceh, North Sumatra, April 10, 1975. Return Camp 8–9. Climbing Leuser Mountain west top, from Penosan via Putjuk Angasan, c. 25Km, southwest of Blang Kedjeren, c. Alt.: 3200m. Exposed along stream in open swamp.]

scars. The leaves are simple, opposite, and exstipulate. The blade is linear, 5–15cm × 4–6 mm. The inflorescences are pseudoterminal cymes which are about 15cm long, and show a few 5-lobed flowers (Figure 33.3).

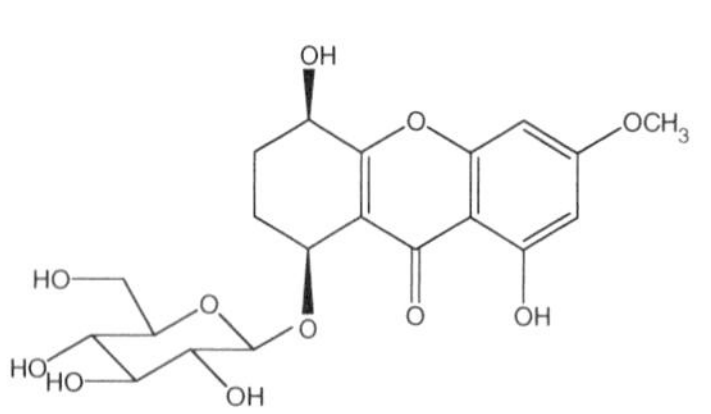

Tetrahydroswertianolin

Figure 33.4

33.3.2 Ethnopharmacology

The plant is bitter and used to promote digestion in Indonesia. The pharmacological properties of *Swertia javanica* Bl. are to date unexplored. Note that genus *Swertia* has received a great deal of interest on account of its ability to elaborate xanthones, which exhibit a broad spectrum of pharmacological activities including antidepressant, antileukemic, antitumor, antitubercular, choleretic, diuretic, antimicrobial, antifungal, antiinflammatory, antiviral, cardiotonic, and hypoglycemic.[3–5]

Although xanthones in this genus have been the focus of attention by pharmacologists, very little work has been done with regard to the iridoids which abound in the genus. Of particular interest is tetrahydroswiertianolin (Figure 33.4) from *Swertia japonica*, which protected rodents against hepatic apoptosis induced by intraperitoneal injection of D-galactosamine (700mg/Kg) and lipopolysaccharide (10mg/Kg) via blockade of tumor necrosis factor (TNF)-α production at the transcriptional level.[6]

REFERENCES

1. Ko, F. N., Lin, C. N., Liou, S. S., Huang, T. F., and Teng, C. M. 1991. Vasorelaxation of rat thoracic aorta caused by norathyriol isolated from Gentianaceae. *Eur. J. Pharmacol.*, 192, 133.
2. Haraguchi, H., Tanaka, Y., Kabbash, A., Fujioka, T., Ishizu, T., and Yagi, A. 2004. Monoamine oxidase inhibitors from *Gentiana lutea*. *Phytochemistry*, 65, 2255.
3. Neerja, P., Jain, D. C., and Bhakuni, R. S. 2000. Phytochemicals from genus *Swertia* and their biological activities. *Ind. J. Chem.*, 39B, 565.
4. Peres, M., Tanus, J. N., and de Fernando, O. F. 2000. Tetraoxygenated naturally occurring xanthones. *Phytochemistry*, 55, 683.
5. Jiang, D. J, Zhu, H. Q., Tan, G. S., Liu, S. Q., Xu, K. P., and Li, Y. J. 2004. Demethylbellidifolin preserves endothelial function by reduction of the endogenous nitric oxide synthase inhibitor level. *J. Ethnopharmacol.*, 93, 295.
6. Hase, K., Xiong, Q., Basnet, P., Namba, T., and Kadota, S. 1999. Inhibitory effect of tetrahydroswertianolin on tumor necrosis factor-α-dependent hepatic apoptosis in mice. *Biochem. Pharmacol.*, 57, 1431.

Medicinal Plants Classified in the Family Apocynaceae

34.1 GENERAL CONCEPT

The family Apocynaceae (A. L. de Jussieu, 1789 nom. conserv., the Dogbane Family) consists of approximately 250 genera and 2000 species of tropical trees, shrubs, woody climbers, and herbs. The cardinal botanical features are an exudation of an abundant milky latex; simple, exstipulate, opposite or whorled leaves; and showy, often pure white, salver-shaped and slightly fragrant flowers with five contorted lobes, with fruits in pairs (Figure 34.1).

The most remarkable characteristic is the plant's ability to elaborate a series of dimers of monoterpenoid indole alkaloids including, notably, vinblastine and vincristine, from *Catharanthus roseus* G. (Figure 34.2). Vinblastine (Velbe®) is particularly useful in treating Hodgkin's disease whereas vincristine sulphate

Figure 34.1 Botanical hallmarks of Apocynaceae. **(See color insert following page 168.)**

(Oncovi®) is used to treat acute leukemia in children. A classical example of Apocynaceae is *Rauwolfia serpentina* (L.) Benth. ex Kurz. The root of this plant has been used in India for a very long time in the treatment of insomnia and certain forms of insanity. In Western medicine, *Rauwolfia serpentina* tablets and powdered *Rauwolfia serpentina* (*U.S. National Formulary*, 1965) consisting of the dried, finely powdered roots have been used to treat hypertension and migraine.

Rauwolfia serpentina (L.) Benth. ex Kurz contains reserpine, an indole alkaloid which blocks the adrenergic transmission by depleting norepinephrine from sympathetic neurons. Ibogaine is a psychostimulating alkaloid from *Tabernanthe iboga*, which protects the *N*-methyl-aspartate neuron receptors against the excessive release of excitatory amino acids, and represents, therefore, a potential therapeutic agent for the treatment of Alzheimer's disease, Huntington's chorea, and other brain conditions. An additional interesting feature of Apocynaceae is the production of cardiotonic glycosides, steroidal alkaloids, and iridoids.

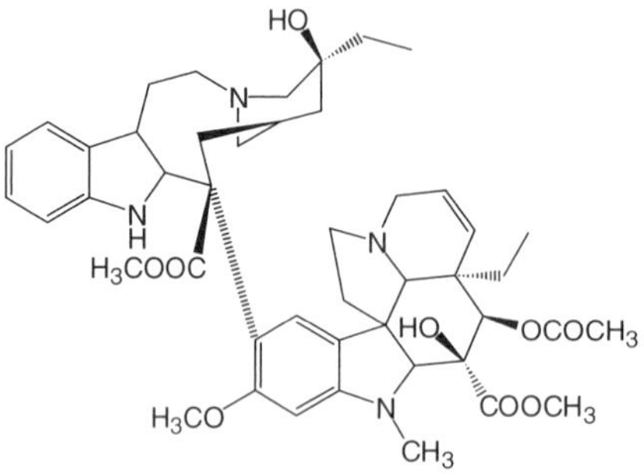

Vinblastine

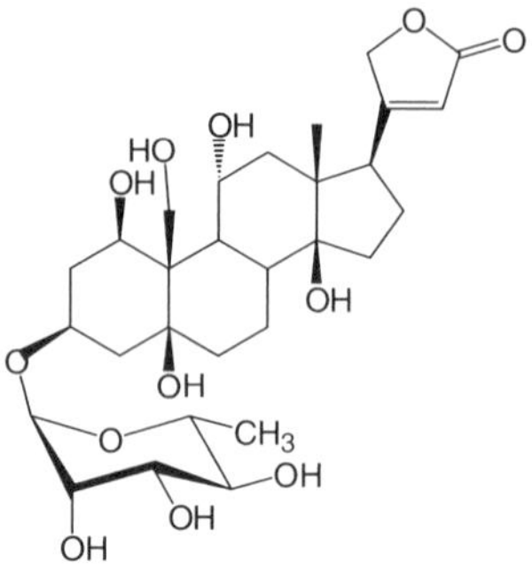

Ouabain

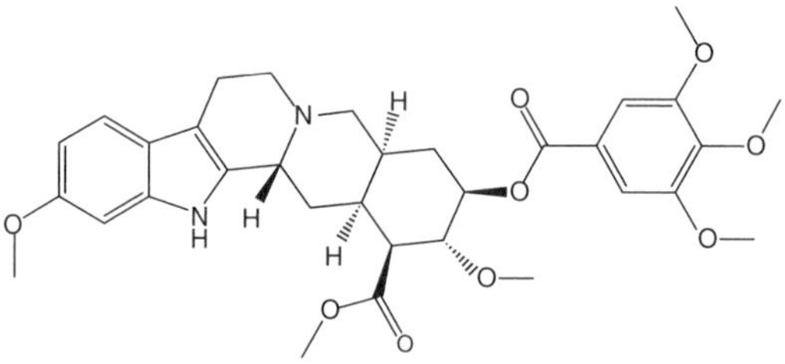

Reserpine

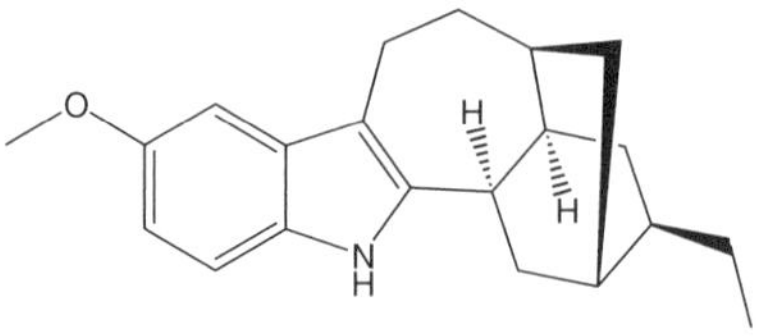

Ibogaine

Figure 34.2 Examples of bioactive natural products from the family Apocynaceae.

Acokanthera, Adenium, Cerbera, Nerium, Strophanthus, and *Thevetia* species abound with cardiotonic glycosides, of which ouabain (Ouabain, *British Pharmacopoeia,* 1958), obtained from the seeds of *Strophanthus gratus* or from the wood of *Acokanthera schimperi* or *Acokanthera ouabaio,* is used to treat acute congestive heart failure. *Funtumia, Holarrhena, Kibatalia,* and *Malouetia* species contain steroidal alkaloids of relative therapeutic value. A preparation made from the bark of *Holarrhena antidysenterica* (Roxb.) Wall is *Holarrhena* (*British Pharmaceutical Codex,* 1949).

The traditional systems of medicine of the Pacific Rim use about 80 species of plants classified within the family Apocynaceae. Most of them are used to treat gastrointestinal ailments, to reduce fever and pains, and to treat diabetes and infectious diseases.

34.2 *ALSTONIA ANGUSTIFOLIA* WALL. EX A. DC.

[After C. Alston (1685–1760), a Scottish botanist, and from Latin *angustifolia* = narrow leaf.]

34.2.1 Botany

Alstonia angustifolia Wall. ex A. DC. is a tree that grows to a height of 25m in the seasonal swamps of Malaysia, Singapore, Sumatra, and Borneo. The bole is cracked and the bark is brown. The crown is conical. The inner bark is yellowish without latex. The sapwood is light brown and the wood yellow. A white latex is present at the stems. The leaves are simple, exstipulate, and in whorls of three. The petiole is 1.5cm long. The blade is elliptical to lanceolate, 6cm × 2.5cm – 11.5cm × 2.3cm – 16cm × 2.1cm, and shows 15–20 pairs of secondary nerves which are distant and arising at an acute angle to the midrib. The calyx, pedicel, and peduncle are covered with gray–yellow hairs more or less persisting in the fruits. The petals are broadly rounded and up to 2.5mm long and tomentose. The tube is 3–3.5mm long and tomentose outside. The fruits are pairs of follicles which are 25–70cm × 3mm, and contain numerous hairy seeds at the apices and are pointed at one end (Figure 34.3).

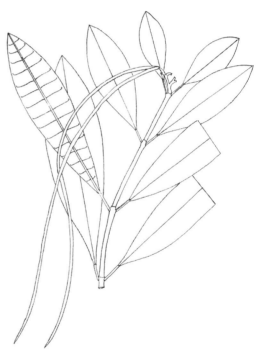

Figure 34.3 *Alstonia angustifolia* Wall. ex A. DC. [From: Flora of Malaysia. Geographical localization: Pahang, Fraser's Hill. Hill for-

34.2.2 Ethnopharmacology

The Malays call the plant *pelai penipu paya.* They heat and oil the leaves then apply them to the spleen to break malarial fever. The pharmacological properties of the plant are unexplored but worth investigating, since Kam and Choo isolated a series of alkaloids including alstolactone, affinisine oxindole, lagumicine, N(4)-demethylalstonerine, N(4)-demethylalstonerinal, and 10-methoxycathafoline N(4)-oxide (Figure 34.4).[1]

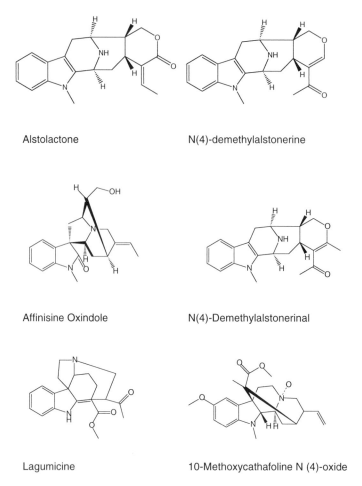

Figure 34.4 Indole alkaloids from *Alstonia angustifolia* Wall. ex A. DC.

34.3 *ALSTONIA MACROPHYLLA* WALL. EX G. DON.

[After C. Alston (1685–1760), a Scottish botanist, and from Greek *makro* = large and *phullon* = leaf.]

34.3.1 Botany

Alstonia macrophylla Wall. ex G. Don. (*Alstonia pangkorensis* King & Gamble) is a tree that grows to a height of 30m with a girth of 2.1m in the low hill rain forests of Southeast Asia. The bole is straight, with low flutes and small buttresses. The bark is blackish-brown, smooth, grid-cracked, with small square adherent scales. The inner bark is cream with broken, orange–yellow laminations, and without latex. The wood is yellowish-brown. The leaves are simple, exstipulate, and in whorls of four. The petiole is 2–3cm long. The blade is elliptical, 15–25cm × 9cm and shows 20–24 pairs of secondary nerves which are distant and arising at an acute angle to the midrib. The blade is finely velvety below. The calyx, pedicel, and peduncle are covered with yellowish hairs persisting in the fruits. The petals are broadly rounded up to 3.7–5.7 mm, glabrous with ciliate margins, and the tube is 4.5–6mm long and glabrous. The fruits are pairs of follicles which are 30cm × 3mm and contain numerous hairy seeds at the apices and are pointed at one end (Figure 34.5).

34.3.2 Ethnopharmacology

In the Philippines, the bark is used to treat fever, fatigue, irregular menses, liver disease, dysentery, malaria, diabetes, and to expel worms from the intestines. The leaves are heated, piled, and applied to sprains, bruises, and dislocated joints. The plant is known as *pulai* or *penipu bukit* in Malaysia and *chuharoi* in the Nicobar Islands. In India, the leaves and stem bark are made into a drink to assuage stomachache and to counteract putrefaction of urine. Externally it is used as a treatment for the skin. The leaves are heated and oiled and applied to sprains and dislocated joints. The antiseptic property of the plant has been confirmed by Chattopadhyay et al.[2] They tested the polar extract of the leaves of *Alstonia macrophylla* against various strains of bacteria cultured *in vitro* and found potent activity against Gram-positive *Staphylococcus aureus*, *Staphylococcus saprophyticus*, and *Streptococcus faecalis*; and Gram-negative *Escherichia coli*, *Proteus mirabilis*, and dermatophyte moulds *Trichophyton rubrum*, *Trichophyton mentagrophytes* var. *mentagrophytes*, and *Microsporum gypseum* with minimum inhibitory concentration values ranging from 64–1000mg/mL for bacteria and 32–128mg/mL for dermatophytes.

The antipyretic property is not substantiated yet, but a methanolic extract from the leaves at a concentration of 200mg/Kg and 400mg/Kg, protected rodents against carrageenan and dextran-induced hind paw edema with a level of potencies comparable to that of the standard drug indomethacin (10mg/Kg).[3]

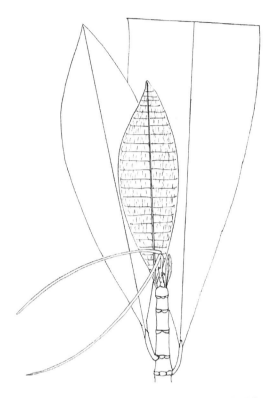

Figure 34.5 *Alstonia macrophylla* Wall. ex A. DC. [From: Flora of Malaya. Kep. Field No: 20543. Geographical localization: Pinang Island, Pinang Hill Moniots Road. Secondary Forest. Alt.: 1800ft. Field collector: T. C. Whitemore, Oct. 22, 1971. Botanical indentification: Loh, June 13.

34.4 *ALSTONIA SPECTABILIS* R. BR.

[After C. Alston (1685–1760), a Scottish botanist, and Latin *spectabilis* = spectacular, showy.]

34.4.1 Botany

Alstonia spectabilis R. Br. (*Alstonia villosa* Bl.) is a tree that grows up to a height of 8m. It is found in a geographical zone covering the Philippines, Java, Queensland, and east to the Solomon Islands. The bole is straight and coarsely fluted. The bark is smooth. White latex is present. The leaves are simple, exstipulate, and whorled. The petiole is 1.3cm long and channeled above. The blade is elliptic, lanceolate, thinly coriaceous, 14.2cm × 3.8cm – 18cm × 5.5cm, and shows 22 pairs of secondary nerves which are conspicuously spaced. The flowers are salver-shaped, membranous, and cream (Figure 34.6).

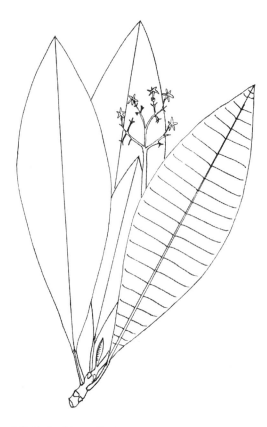

Figure 34.6 *Alstonia spectabilis* R. Br. [From: Ex Queensland Herbarium. Field collectors: P. I. Forester et al. (PIF 6516). Geographical localization: Bwinning Point, Weipa, Red Soil Rain Forest. 12°35′ S, 141°54′ E. Alt.: 3m. Collection No: Q 508294.]

34.4.2 Ethnopharmacology

In the Moluccas, the bark is used to treat stomach and intestinal discomfort. The pharmacological properties of this plant have not yet been explored. An interesting development would be to assess the plant for serotonine-reuptake inhibition.

34.5 *ALSTONIA SPATULATA* BL.

[After C. Alston (1685–1760), a Scottish botanist, and Latin *spatulata* = like a spatula.]

34.5.1 Botany

Alstonia spatulata Bl. is a tree that grows to a height of 12–24m with a girth of 1.2–1.6m. The bole is 6m tall. The crown is pagoda-like. The bark is smooth and grayish. The inner bark is yellow, and shows an abundant milky latex. The sapwood is pale orange. The leaves are simple, sessile, whorled, and exstipulate. The blade is spathulate and thick, 6.5cm × 2.7cm – 5cm × 2.3cm, and shows about 24 pairs of secondary nerves crowded, at right angles to the midrib. The inflorescences are loose, few flowered, and the pedicels are up to 1.3cm long. The flowers are salver-shaped, 5-lobed, the lobes contorted, 7mm – 1.2cm long. The fruits are pairs of green–blue, with 13.5cm × 2mm follicles (Figure 34.7).

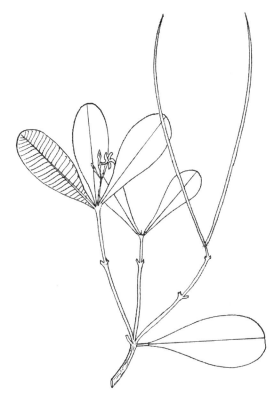

Figure 34.7 *Alstonia spatulata* Bl. [From: Harvard University Herbarium. No: 126179. Flora of Sabah, Herbarium of the Forest Department, Sandakan. Geographical localization: Madahan Forest Reserve, District: Papar. Alt.: 30m. Field collector: A. G. Amir Sigun. Feb. 21, 1991. Botanical identification: D. Middleton.]

34.5.2 Ethnopharmacology

In Cambodia, Laos, Vietnam, and Malaysia, the latex from *Alstonia spatulata* Bl. is applied externally to sores and diseased skin. The bark is used to lower fever and to expel worms from the intestines. The Malays call the plant *pulai basong*. The plant has never been studied for its pharmacology. What is the potential of this plant in dermatology?

34.6 *CARISSA CARANDAS* L.

[From: Indian *karaunda* = name of the plant.]

34.6.1 Botany

Carissa carandas L. (*Arduina carandas* [Linnaeus] K. Schumann, *Damna-canthus esquirolii* H. Léveillé) is a spiny treelet that grows up to 5m tall and is native to India and cultivated in Taiwan, India, Indonesia, Malaysia, Burma, Sri Lanka, Thailand, and the Pacific Islands. Its fruits, which can be eaten raw, are also made into jelly, or used for pies. The stems are up to 5cm long and show numerous spines which are woody, simple or forked. The leaves are decussate, simple, and exstipulate. The petiole is 5mm long. The blade is light green, oblong, broadly ovate to oblong, 3cm – 7cm × 1.5cm – 4.5cm. The base of the blade is broadly cuneate to rounded, and the apex

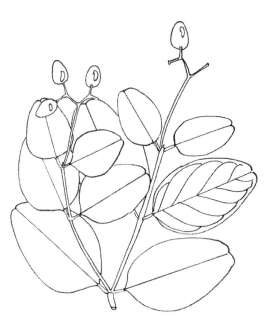

Figure 34.8 *Carissa carandas* L. [From: Flora of Malay Peninsula. Forest Department. Geographical localization: Kamoung Bahru, cultivated. No: 7450. Field collector: J. Omar. Botanical identification: M. R. Henderson.]

is shortly apiculate. The blade shows 5–8 pairs of secondary nerves. The inflorescences are terminal, usually 3-flowered cymes that are up to 2.5cm long. The flowers are fragrant. The calyx has five lobes which are 2.5–7mm long, with many basal glands inside. The corolla is white or pale rose, the corolla tube is about 2cm long, puberulent inside, and develops five linear contorted 1cm-long acute lobes. The fruits are reddish-purple, 3cm × 2.3cm – 6cm × 3.7cm – 4.3cm × 3cm, 1.5cm – 2.5cm × 1cm – 2cm, ellipsoid to grape-like (Figure 34.8).

34.6.2 Ethnopharmacology

Carissa carandas L. is known as Bengal Currant or Christ's Thorn, Karanda, *kerenda* (Malay), *karaunda* (India), *nam phrom* or *namdaeng* (Thailand), *caramba* (Philippines), *kalakai* (Tamil), and *ci huang guo* (Chinese). In Cambodia, Laos, and Vietnam, the fruits are eaten to treat liver dysfunction, to break fever, and to counteract the putrefaction of blood. The roots are bitter and used to promote digestion. The juice expressed from the roots is applied externally to calm itching. The plant is known to produce pentacyclic triterpenoids, including carissin and lignans.[4] Vohra and De reported some levels of cardioactivity from this plant.[5] A remarkable advance in the pharmacology of *Carissa* species has been provided by the work of Lindsay et al.[6] They isolated from the wood of *Carissa lanceolata* R. Br. a series of quinones with antibacterial activity — carindone, carissone, and dehydrocarissone (Figure 34.9). Dehydrocarissone inhibited the growth of both *Staphylococcus aureus* and *Escherichia coli* with Minimum Inhibition Concentration (MIC) values inferior to 0.5mg/mL and about 2mg/mL against the Gram-negative bacillus *Pseudomona aeruginosa*. Taylor et al.[7] made the interesting observation that a plant classified in the genus *Carissa* inhibits the survival of Herpes Simplex Virus (HSV), Sindbis virus, and poliovirus.[8] Are carindone, carissone, and dehydrocarissone involved here? Does *Carissa carandas* L. have antiviral principles?

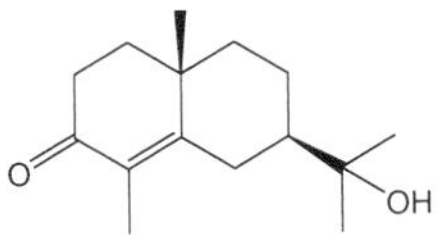

Carissone

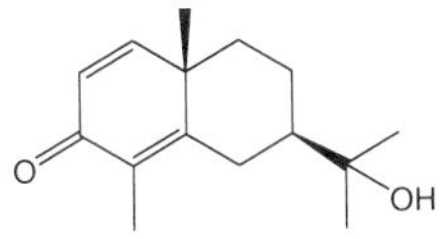

Dehydrocarissone

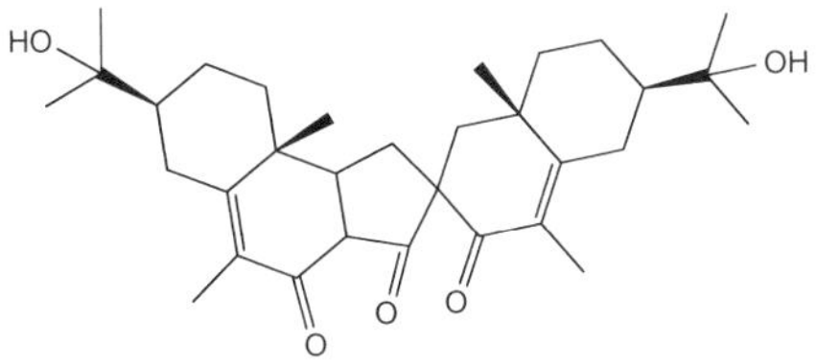

Carindone

Figure 34.9 Bioactive constituents from the genus *Carissa:* carissone, dehydrocarissone, carindone.

Note that the plant has probably some level of antidiabetic activity since oral administration of ethanolic extracts of leaves from *Carissa edulis* lowered blood glucose both in normal and streptozotocin (STZ) diabetic rats.[8]

34.7 *EPIGYNUM MAINGAYI* HOOK. F.

[From: Greek *epi* = around and *gyne* = female, and after Alexander Carroll Maingay (1836–1869), physician and botanist.]

34.7.1 Botany

Epigynum maingayi Hook. f. is a small woody climber that grows in the rain forests of Malaysia. The stems are lenticelled and exude white latex after incision. The leaves are simple, opposite, and exstipulate. The petiole is 4mm long. The blade is 7.5cm × 4cm – 8.7cm × 4cm – 4.4cm × 2.3cm – 1cm × 2cm and shows 11 pairs of secondary nerves. The inflorescences are cymose and terminal. The flowers are salver-shaped, 5-lobed, and the lobes are contorted. The fruits consist of pairs of follicles which are terete and contain several oblong, flat, comose seeds (Figure 34.10).

Figure 34.10 *Epigynum maingayi* Hook. f. [From: Flora of Malaya, Phytochemical Survey of Malaysia Herbarium. Geographical localization: Jerantut, Pahang. Field collectors: L. E. Teo and F. Remy. Det.: F. Remy. Duplicata: Kew, Leyd, Kepong. Nov. 11, 1991.]

34.7.2 Ethnopharmacology

The Malays use the plant to promote the secretion of milk. Note that pregnane saponins are known to occur in the genus.[9]

34.8 *ERVATAMIA SPHAEROCARPA* BL.

[From: Latin *sphaerocarpa* = globose fruits.]

34.8.1 Botany

Ervatamia sphaerocarpa Bl. is a tree that grows to a height of 15m. The plant is found in the primary rain forests of Thailand, Malaysia, and Sumatra. The stems are terete, lenticelled, and exude a white latex after incision. The leaves are simple, opposite, and exstipulate. The petiole is 1cm long and channeled above. The blade is elliptic, ovate, 13cm × 7cm – 8.5cm × 4.3cm – 9cm × 3cm, and shows six pairs of secondary nerves which are well spaced. The fruits are pairs of follicles which are 4cm × 3cm, beaked, and yellowish (Figure 34.11).

34.8.2 Ethnopharmacology

The plant is used after childbirth as a protective remedy. It also is an external remedy for venereal infections in Malaysia and South Thailand. To date, the pharmacological potential of this

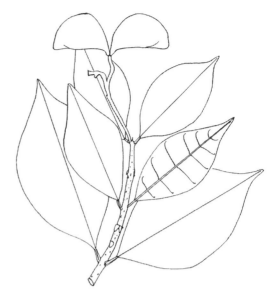

Figure 34.11 *Ervatamia sphaerocarpa* Bl. [From: Flora of Malaya. Comm. Ex. Herb. Hort. Bot. Sing. Geographical localization: Jungle path to Bujang Mountain, Malacca, Chenderiang Peak. Alt.: 2000ft. Nov. 2, 1975. No: MS 3372. Field collector: M. Shah.]

plant has not been explored. An interesting development would be to confirm its antibacterial activity and to identify the active constituents. Alkaloids are expected.

34.9 *HOLARRHENA CURTISII* KING & GAMBLE

[After William Curtis (1746–1799), botanist.]

34.9.1 Botany

Holarrhena curtisii King & Gamble (*Holarrhena crassifolia* Pierre, *Holarrhena densiflora* Ridl., and *Holarrhena pulcherima* Ridl.) is a tree that grows in Vietnam, Cambodia, Laos, Thailand, and Malaysia. The leaves are simple, decussate, and exstipulate. The petiole is 5mm long and grooved above. The blade is spathulate, thick, and velvety below, 3.5cm × 2.5cm – 6.5cm × 3.4cm – 9cm × 3.4cm, and shows 8–16 pairs of secondary nerves as well as a few tertiary nerves below. The inflorescences are long-stalked terminal cymes. The flowers are white and tubular with five contorted lobes. The fruits are pairs of follicles which are 18.5–25cm × 5mm (Figure 34.12).

34.9.2 Ethnopharmacology

In Cambodia, Laos, and Vietnam, the plant is used to treat dysentery. The pharmacological

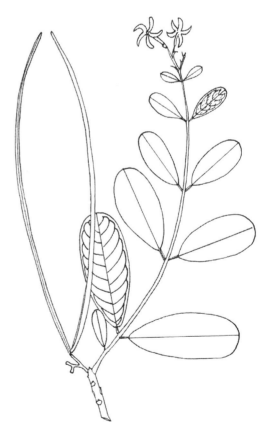

Figure 34.12 *Holarrhena curtisii* King & Gamble. [From: Flora of Malay Peninsula. Forest Department. Geographical localization: Padang Mengkudu, Tangga Hill Forest Reserve, Kedah. No: 56965. Field collector and botanical identification: C. F. Symington.]

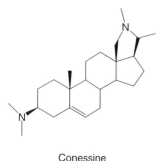

Conessine

Figure 34.13

properties of the plant are unknown. The plant is known, however, to abound in steroidal alkaloids which probably impart the antidysenteric property.[10] Voravuthikunchai et al.[11] showed that *Holarrhena antidysenterica* (Roxb.) Wall. prevents the survival of the Gram-negative bacillus *Escherichia coli* probably because of its steroidal alkaloid content.[12] *Holarrhena* (*British Pharmaceutical Codex*, 1949) consists of the dried bark from the stem and roots of *Holarrhena antidysenterica* (Roxb.) Wall. containing not less than 2% of total alkaloids including the steroidal alkaloid conessine (Figure 34.13). Conessine hydrobromide (*French Pharmacopoeia*, 1965) has been used to treat amebic dysentery, but it has been dropped on account of severe neuropsychopathic effects. Steroidal alkaloids given orally at a dose of 200–800mg/Kg protected rodents against diarrhea induced by castor oil. It would be interesting to assess the properties of *Holarrhena curtisii* King & Gamble against the Gram-negative bacteria, *Entameba histolytica* and *Giardia lambia*.[13]

34.10 *KIBATALIA ARBOREA* (BL.) G. DON.

[From: Sundanese *ki batali* = name of *Kibatalia* species and from Latin *arborea* = tree-like.]

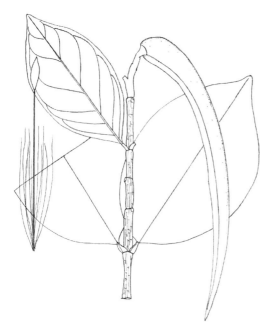

Figure 34.14 *Kibatalia arborea* (Bl.) G. Don. [From: Forest Research Institute, Kepong, Malaysia. Kep. 80947. Botanical identification: P. S. Ashton, August 1966. Geographical localization.: Near reservoir, FRF Kepong hillside.]

34.10.1   Botany

Kibatalia arborea (Bl.) G. Don. (*Kickxia arborea* Bl., *Hasseltia arborea* Bl., *Kickxia arborea* [Bl.] Bl.) is a tree that grows to a height of 36m with a girth of 2.7m in the primary rain forests of Malaysia, Sumatra, and Java. The crown is small. The bole is straight and nonbuttressed. The bark is grayish-black and smooth. The inner bark is granular, white and orange, and with a milky latex which can be used to make rubber. The stems are thick, fissured, lenticelled, articulated, and measuring up to 8mm in diameter. The leaves are simple, decussate, and exstipulate. The petiole is about 1cm long and channeled above. The blade is ovate, elliptical, 20cm × 10cm, and showing 6–10 pairs of secondary nerves visible below only. The fruits are impressively massive, woody, up to 80cm-long pairs of follicles containing several linear hairy seeds up to 15cm long, somewhat like *Strophanthus* seeds (Figure 34.14).

34.10.2   Ethnopharmacology

The Malays and Indonesians call this plant *jelutong pipit*. In Indonesia small doses of the latex are ingested to expel worms from the intestines. The pharmacological potential properties of this plant have not yet been explored. One can reasonably expect steroidal alkaloids as active principles for the anthelminthic property mentioned above.[14]

34.11 *KOPSIA LARUTENSIS* KING & GAMBLE

[After J. Kops (1765–1849), first editor of the *Dutch Flora Batavia*.]

34.11.1   Botany

Kopsia larutensis King & Gamble is a laticiferous treelet that grows up to 8m tall. It is found in the primary rain forests of Malaysia. The stems are squarish and swollen at the nodes. The leaves are simple, decussate, and exstipulate. The petiole is 1cm long and channeled above. The blade is 15cm × 3.5cm – 19cm × 6cm, elliptical, membranaceous, pointed at both ends, with the apex showing below a little brownish gland. The blade has a few secondary nerves and the midrib is sunken above. The inflorescences are 3cm-long spikish cymes, which are densely covered with minute scaly bracts. The flowers are pure white, and the corolla tube is up to 1cm long and swollen near the throat. The fruits are pairs of trigonal follicles up to 1.5cm long (Figure 34.15).

34.11.2   Ethnopharmacology

A paste of roots is applied to the ulcerated nose of syphilitic sufferers in Malaysia. An important contribution to the chemistry of this plant and Apocynaceae in general has been provided by the work of Kam et al.[15,16] They isolated indole alkaloids of the eburnan type such

Figure 34.15 *Kopsia larutensis* King & Gamble. [From: Harvard University Herbarium, Flora of Malay Peninsula. Forest Department. Geographical localization: Kledang Saiong Perak. March 23, 1931. No: 25654. Field collector: C. F. Symington. Botanical identification: V. D. Sleesen, September 1959.]

as (+)-eburnamonine, (+)-eburnamonine *N*(4)-oxide, -eburnamine, (+)-isoeburnamine, and larutenine from the leaves. To date the pharmacological properties of *Kopsia larutensis* King & Gamble have not been explored. An interesting development would be to test the above-mentioned alkaloids against *Treponema pallidum*.

34.12 *WILLUGHBEIA EDULIS* RIDL.

[From: Latin *edulis* = edible.]

34.12.1   Botany

Willughbeia edulis Ridl. (*Ancylocladus cochinchinensis* Pierre, *Willughbeia cochinchinensis* (Pierre) K. Schum., and *Willughbeia martabanica* Wall.) is a woody climber which grows in the primary rain forests of Southeast Asia and is particularly native to the Philippines. The stems are reddish, lenticelled, and terete, showing some tendrils. After excision, they exude a milky latex which produces a rubber called *chittagong*. The leaves are simple, opposite, and exstipulate. The blade is dark green, 10cm × 3.5cm – 9.6cm × 4.5cm – 10.5cm × 6cm, broadly elliptical, notched at the apex, and showing 22 pairs of nerves below. The fruits are indehiscent, globose-

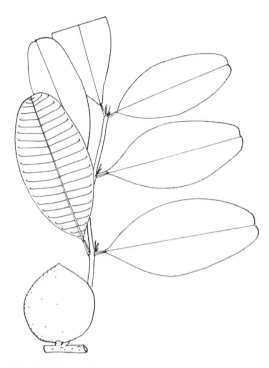

Figure 34.16 *Willughbeia edulis* Ridl. [From: Federated Malay States. Malay Peninsula. Geographical localization: Ulu Rumpin, Pahang. April 24, 1919. Field collector: A. R. Yeob. Botanical identification: D.
Middleton, November 1992.]

like dwarfy watermelons, which are 5cm in diameter, green, smooth, and thick-walled like a
mangosteen (Figure 34.16).

34.12.2   Ethnopharmacology

In the Philippines the plant is called *tabu*, and in Indo-Malaya it is called *jintam, gitah jetan
obat purut*. In Cambodia, Laos, and Vietnam, the latex is used as a plaster for sores and the stems
are used to treat yaws, dysentery, and liver discomfort. In Malaysia, the latex is applied to yaws
and the roots are used internally to treat jaundice, heartburn, and diarrhea. The pharmacological
properties of this plant have not yet been explored. It will be interesting to learn whether further
pharmacological study on this plant discloses any agents against *Treponema pertenue*.

34.13 *WRIGHTIA PUBESCENS* R. BR.

[After William Wright (1735–1819), a Scottish naval surgeon in Jamaica, and from Latin
pubescens = downy.]

34.13.1   Botany

Wrightia pubescens R. Br. (*Anasser lanitii* Blanco, *Wrightia annamensis* Eberh. & Dubard,
Wrightia calycina A. DC., *Wrightia candollei* S. Vidal, *Wrightia javanica* A. DC., *Wrightia lanitii*
(Blanco) Merr., *Wrightia spanogheana* Miq., and *Wrightia tomentosa* var. *cochinchinensis* Pierre
ex Pierre) is a tree that grows at an altitude of 400m in the primary rain forests of Cambodia, India,
Indonesia, Malaysia, the Philippines, Thailand, Vietnam, Australia, and China.

The tree grows to a height of 35m with a girth of 55cm, producing wood of commercial value. The bark is smooth and yellowish-brown. The stems are yellowish, pubescent at the apex, lenticelled, terete, and exuding a milky latex after incision. The leaves are simple, opposite, and exstipulate. The petiole is 1cm long and channeled above. The blade is oblong, 4cm × 4.5cm – 8cm × 3.3cm – 11.4cm × 4cm, papery, velvety below, and showing 8–15 pairs of secondary nerves. The inflorescences are 5cm-long cymes which are pubescent. The flowers are smelly and white–pink. The calyx consists of five sepals which are ovate and 2–5mm long. The corolla tube is 5–6.5mm long and produce five contorted lobes which are oblong and 1–2cm long. The mouth of the corolla tube shows a fringed corona. The fruits are pairs of 15–30cm × 1–2cm follicles which are connate, sublinear, and contain several fusiform hairy seeds (Figure 34.17).

Figure 34.17 *Wrightia pubescens* R. Br. [From: Flora of the Malay Peninsula. Forest Department. Geographical localization: Merbok, Kuala Kunda, Kedah. September 1925. No: 9012. Botanical identification: T. C. Whitmore.]

34.13.2 Ethnopharmacology

The Javanese ingest a few drops from the seeds to stop dysentery. The vernacular names of the plant include *mentoh*, *benteli lalaki* (Indonesian), *thung muc long* (Vietnamese), *dao diao bi* (Chinese), and *mentah* (Malay). A remarkable advance in the pharmacology of *Wrightia pubescens* R. Br. has been provided by the work of Kawamoto et al.[17] They isolated from the plant wrightiamine A, a pregnane alkaloid, which inhibited vincristine-resistant murine leukemia P-388 cells cultured *in vitro*. Pregnanes abound in the similar Asclepiadaceae Family, which is described in the next chapter.

REFERENCES

1. Kam, T. S. and Choo, Y. M. 2004. Alkaloids from *Alstonia angustifolia*. *Phytochemistry*, 65.
2. Chattopadhyay, D., Maiti, K., Kundu, A. P., Chakraborty, M. S., Bhadra, R., Mandal, S. C., and Mandal, A. B. 2001. Antimicrobial activity of *Alstonia macrophylla*: a folklore of bay islands. *J. Ethnopharmacol.*, 77, 49.
3. Arunachalam, G., Chattopadhyay, D., Chatterjee, S., Mandal, A. B., Sur, T. K., and Mandal, S. C. 2002. Evaluation of anti-inflammatory activity of *Alstonia macrophylla* Wall ex A. DC. leaf extract. *Phytomedicine*, 9, 632.
4. Pal, R., Kulshreshtha, D. K., and Rastogi, R. P. 1975. A new lignan from *Carissa carandas*. *Phytochemistry*, 14, 2302.
5. Vohra, M. M. and De, N. N. 1963. Comparative cardiotonic activity of *Carissa carandas* L. and *Carissa spinarum* A. DC. *J. Med. Res.*, 51, 937.
6. Lindsay, E. A., Berry, Y., Jamie, J. F., and Bremner, J. B. 2000. Antibacterial compounds from *Carissa lanceolata* R. Br. *Phytochemistry*, 55, 403.
7. Taylor, R. S. L., Hudson, J. B., Manandhar, N. P., and Towers, G. H. N. 1996. Antiviral activities of medicinal plants of southern Nepal. *J. Ethnopharmacol.*, 53, 105.

8. El-Fiky, F. K., Abou-Karam, M. A., and Afify, E. A. 1996. Effect of *Luffa aegyptiaca* (seeds) and *Carissa edulis* (leaves) extracts on blood glucose level of normal and streptozotocin diabetic rats. *J. Ethnopharmacol.*, 50, 43.

9. Cao, J. X., Pan, Y. J., Lu, Y., Wang, C., Zheng, Q. T., and Luo, S. D. 2005. Three novel pregnane glycosides from *Epigynum auritum*. *Tetrahedron*, 61, 6630.

10. Einhorn, J., Monneret, J. C., and Khuong-Huu, Q. 1972. Alcaloïdes des feuilles *de l'Holarrhena crassifolia*. *Phytochemistry*, 11, 769.

11. Voravuthikunchai, S., Lortheeranuwat, A., Jeeju, W., Sririrak, T., Phongpaichit, S., and Supawita, T. 2004. Effective medicinal plants against enterohemorrhagic *Escherichia coli* O157:H7. *J. Ethnopharmacol.*, 94, 49.

12. Chakraborty, A. and Brantner, A. H. 1999. Antibacterial steroid alkaloids from the stem bark of *Holarrhena pubescens*. *J. Ethnopharmacol.*, 68, 339.

13. Kavitha, D., Shilpa, P. N., and Devaraj, S. N. 2004. Antibacterial and antidiarrhoeal effects of alkaloids of *Holarrhena antidysenterica* WALL. *Indian J. Exp. Biol.*, 42, 589.

14. Cave, A., Potier, P., and Le Men, J. 1967. Alkaloids of the bark of *Kibatalia gitingensis* (Elm.) Woods (Apocynaceae). Steroid alkaloids of *Apocynaceae*. 14th report. *Ann. Pharm. Fr.*, 25, 107.

15. Kam, T. S., Tan, P. S., and Chen, W. 1993. Absolute configuration of C-16 of eburnane alkaloids from *Kopsia larutensis*. *Phytochemistry*, 33, 921.

16. Kam, T. S., Tan, P. S., and Chuah, C. H. 1992. Alkaloids from leaves of *Kopsia larutensis*. *Phytochemistry*, 31, 2936.

17. Kawamoto, S., Koyano, T., Kowithayakorn, T., Fujimoto, H., Okuyama, E., Hayashi, M., Komiyama, K., and Ishibashi, M. 2003. Wrightiamines A and B, two new cytotoxic pregnane alkaloids from *Wrightia javanica*. *Chem. Pharm. Bull. (Tokyo)*, 51, 737.

CHAPTER 35

Medicinal Plants Classified in the Family Asclepiadaceae

35.1 GENERAL CONCEPT

The family Asclepiadaceae (R. Brown, 1810 nom. conserv., the Milkweed Family) consists of approximately 250 genera and 2000 species of tropical climbers, herbs, and shrubs. The botanical signature includes the production of a white latex and simple, opposite, exstipulate, and somewhat fleshy leaves. The flowers are tubular and 5-lobed, which are contorted and include very characteristic binocular-like organelles called pollinia which are inserted in polliniaria along a pentagonal, tabular stigma (Figure 35.1). The fruits of Asclepiadaceae are very much like the fruits of Apocynaceae. Several plant species are ornamental, especially the *Hoya* species.

Well-known examples of medicinal Asclepiadaceae can be found in India and Europe. In India, *Calotropis gigantea* (Willd.) Dry. ex WT. Ait. (*yercum* or *madar* fiber), *Marsdenia tenacissima* W. and A. (*bahjmahal* hemp) have been

Figure 35.1 Botanical hallmarks of Asclepiadaceae. **(See color insert following page 168.)**

used medicinally for a very long time. The former is still being used to eliminate illegally unwanted newborn girls in Tamil Nadu. In Europe, the dried roots of *Hemidesmus indicus* (*Hemidesmus, British Pharmaceutical Codex*, 1934) have been used to treat syphilis, rheumatism, psoriasis, and eczema.

With regard to the chemical constituents found in the family, the evidence currently available demonstrates the existence of two main classes of natural products, pregnanes and phenanthroindolizidine alkaloids. *Asclepias, Calotropis, Carissa, Cryptostegia, Gomphocarpus, Menabea, Xysmalobium*, and *Periploca* species owe their toxic properties to a series of pegnane glycosides. One such glycoside is periplocin (*British Pharmaceutical Codex*, 1967), a cardiac glycoside from the bark of *Periploca graeca*, that has been used in Russia instead of digitalin (1mL ampoule of 0.25mg) (Figure 35.2). Note that pregnanes from *Cryptostegia grandiflora* (Rubber Vine and Pink Alla-

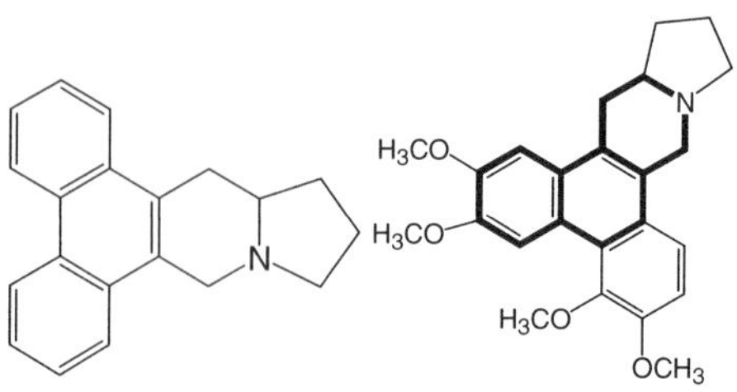

Phenanthroindolizidine

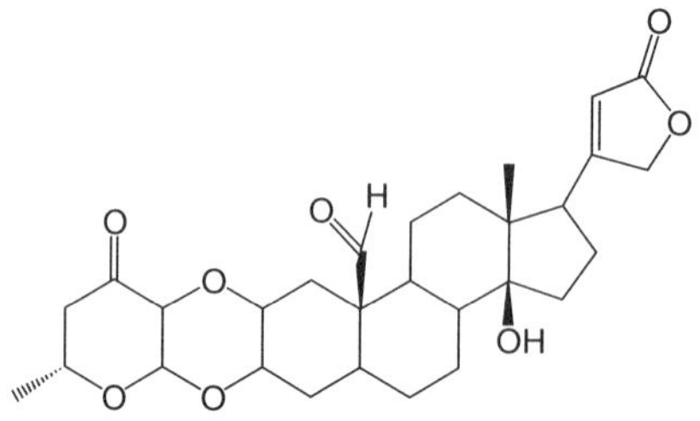

Oxypregnane aglycone

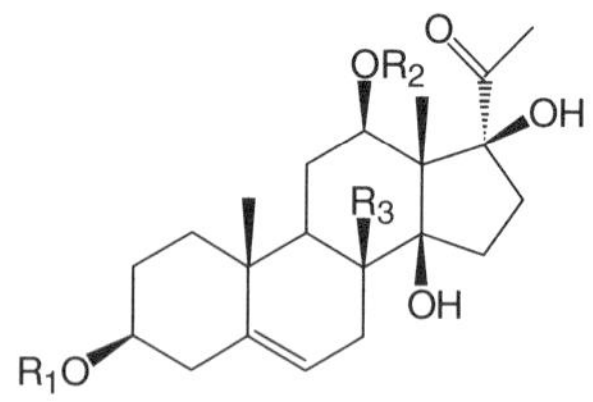

Pregnane aglycone

Figure 35.2 Bioactive natural products from the family Asclepiadaceae.

manda) and *Parquetina nigrescens* have chemotherapeutic potential. What the precise molecular mechanism of action of such compounds might be is a key question with this first group.

The second main group of pharmacologically active products found in Asclepiadaceae consists of planar, glucocorticoid-like, phenanthroindolizidine alkaloids such as tylocrebine, characteristic of *Tylophora crebiflora*, which might hold some potential as a source of chemotherapeutic agents. Note, however, that unmanageable central nervous side effects are common in this group of products.

In the Pacific Rim, about 50 species of plants classified within Asclepiadaceae are used for medicinal purposes, but are virtually untapped in terms of pharmacological potential. Note that the latex and the leaves which abound with pregnanes are often used to make arrow poison, to counteract putrefaction, to mitigate pain, to reduce fever, to induce vomiting, and to relieve the bowels from costiveness. It will be interesting to learn whether a more intensive study on this family discloses any molecules of therapeutic interest. Among the most exciting potential candidates to be studied are *Hoya coriacea* Bl., *Hoya coronaria* Bl., *Hoya diversifolia* Bl., *Streptocaulon cumingii* (Turcz.) F.-Vill., and *Telosma cordata* (Burm. f.) Merr.

35.2 *HOYA CORIACEA* BL.

[After Thomas Hoy, head gardener of the Duke of Northumberland and known to Robert Brown (1773–1858), and from Latin *coriacea* = thick, coriaceous.]

35.2.1 Botany

Hoya coriacea Bl. (*Centrostemma coriaceum* [Bl.] Meisn.) is a climber that grows in the rain forests of Thailand, Malaysia, and Sumatra. The plant is grown as an ornamental climber on account of its magnificence. The stems are terete, smooth, and exude a milky latex after incision. The leaves are simple, opposite, and exstipulate. The petiole is 2–4.5cm. The blade is thick, lanceolate, apiculate at the apex, and about 5–8cm × 15cm × 20cm. The inflorescences are umbel-like, extra-axillary heads of starry, whitish flowers. The corolla is fleshy, rotate, and reflexed. The corona has five lobes, and there are two pollinia per pollinarium that are oblong and erect with a raised, translucent margin. The fruits are pairs of follicles which are 30cm × 2.5cm (Figure 35.3).

35.2.2 Ethnopharmacology

A decoction of leaves is used as a drink to promote expectoration and to treat asthma. To date, the pharmacological properties of *Hoya coriacea* Bl. are unexplored. Saponins are most likely responsible for the properties mentioned above. Tylocrebine and congeners might be involved in the antiasthma property.

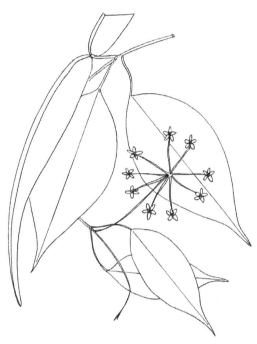

Figure 35.3 *Hoya coriacea* Bl. [From: Phytochemical Survey of the Federation of Malaysia. KL No: 2231. Geographical localization: Ulu Langat, Selangor, Mampil, Sempadan Looi. Field collector: G. A. Umbai for A. H. Millard, Nov. 11, 1960. Botanical identification: R. E. Rintz, Aug. 2, 1976.]

35.3 *HOYA CORONARIA* BL.

[After Thomas Hoy, head gardener of the Duke of Northumberland and known to Robert Brown (1773–1858), and from Latin *corona* = crown.]

35.3.1 Botany

Hoya coronaria Bl. is a climber that grows in the rain forests of Indonesia and Malaysia. The plant is grown as an ornamental. The stems are terete, smooth, and exude a milky latex after incision. The leaves are simple, opposite, and exstipulate. The petiole is 1.3cm long. The blade is broadly oblong, glossy above, velvety below, and shows about 10 pairs of well spaces and secondary nerves looping at the margin. The inflorescences are umbel-like, extra-axillary heads of starry pink flowers. The corolla is fleshy, rotate, reflexed, with five lobes. There are two pollinias per pollinarium, which are oblong and erect, with a raised, translucent margin. The fruits are pairs of follicles (Figure 35.4).

Figure 35.4 *Hoya coronaria* Bl. [From: Flora of Malaya, Kepong FRI No: 3707. Geographical localization: East Pahang, Kuantan, Teloh Chempedak, sandy beach at waterline. May 15, 1967. Field collector: T. C. Whitmore. Botanical identification: R. E. Rintz.]

35.3.2 Ethnopharmacology

In Indonesia, the latex is used to induce vomiting. The pharmacological properties are as of yet unexplored. Note that the acridity of saponins might trigger emesis, hence the traditional use of the plant.

35.4 *HOYA DIVERSIFOLIA* BL.

[After Thomas Hoy, head gardener of the Duke of Northumberland and known to Robert Brown (1773–1858), and from Latin *diversifolia* = leaves variously shaped.]

35.4.1 Botany

Hoya diversifolia Bl. is a climber that grows in the rain forests of Southeast Asia. The stems are terete and smooth, and exude a milky latex after incision. The leaves are simple, opposite, and exstipulate. The petiole is 7mm long. The blade is elliptical, 7cm × 3cm – 6.5cm × 3.5cm. The inflorescences are umbel-like, with extra-axillary heads of about 15 starry flowers on 5.7cm-long pedicels. The flowers are light pink on the petals, brighter towards the calyx, and 4mm long. The corolla is fleshy, rotate, and reflexed. The corona has five lobes. There are two pollinia per pollinarium. The fruits are pairs of follicles (Figure 35.5).

35.4.2 Ethnopharmacology

In Cambodia, Laos, Vietnam, and Malaysia, a decoction of the leaves is mixed with hot water. This mixture is applied externally to ease the pain of rheumatism. Methanolic extracts of the plant have exhibited some levels of antinematodal activity *in vitro* against *Bursaphelenchus xylophilus*.[1]

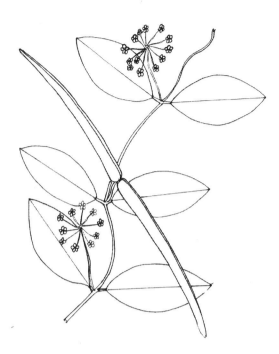

Figure 35.5 *Hoya diversifolia* Bl. [From: Flora of Malaya. Distributed from the Forest Research Institute. Kep. Field No: 99662. Geographical localization: Batu Feringgi, Penang, on seashore.]

35.5 *STREPTOCAULON CUMINGII* (TURCZ.) F.-VILL.

[From: Latin *strepto* = twisted and *cauli* = stem, and after Hugh Cuming (1791–1865), conchologist and botanist, who explored the Philippines.]

35.5.1 Botany

Streptocaulon cumingii (Turcz.) F.-Vill. (*Triplolepis cumingii* Turcz.) is a climber that grows in the primary rain forests of the Philippines. The stems are terete, smooth, glabrous, and exude a milky latex after incision. The leaves are simple, opposite, and exstipulate. The petiole is 2.5–3cm long. The blade is broadly lanceolate, and shows six pairs of secondary nerves. The inflorescences are 7cm long and terminal. The flower pedicel is 1.5cm long. The fruits are oblong follicles filled with hairy seeds (Figure 35.6).

35.5.2 Ethnopharmacology

In the Philippines, the juice squeezed from the stems is applied externally to insect bites. The pharmacological properties of *Streptocaulon cumingii* are to date unexplored. Ueda et al.[2] showed that a crude polar extract of *Streptocaulon juventas* inhibits the proliferation of cancer cells on account of a series of cardenolides, including digitoxigenin gentiobioside, digitoxigenin 3-*O*-[*O*-beta-glucopyranosyl-(1–>6)-*O*-beta-glucopyranosyl-(1–>4)-3-*O*-acetyl-beta-digitoxopyranoside], digitoxigenin 3-*O*-[*O*-beta-glucopyranosyl-(1–>6)-*O*-beta-glucopyranosyl-(1–>4)-*O*-beta-digitalopyranosyl-(1–>4)-beta-cymaropyranoside], and (17α)-*H*-periplogenin-3-*O*-β-glucopyranosyl-(1-4)-2-*O*-acetyl-3-*O*-methyl-β-fucopyranoside via the induction of apoptosis (Figure 35.7).[3–5]

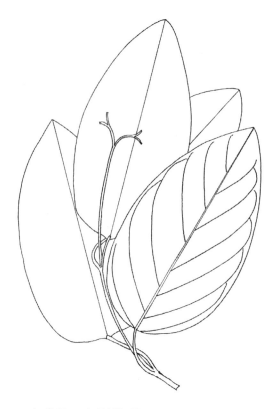

Figure 35.6 *Streptocaulon cumingii* (Turcz.) F.-Vill. [From: Flora of the Philippines, Herbarium Bureau of Sciences. Manila. Plants of Luzon. Collected and presented by A. Loher. Geographical localization: Rizal Province. January 1913. No: 13911.]

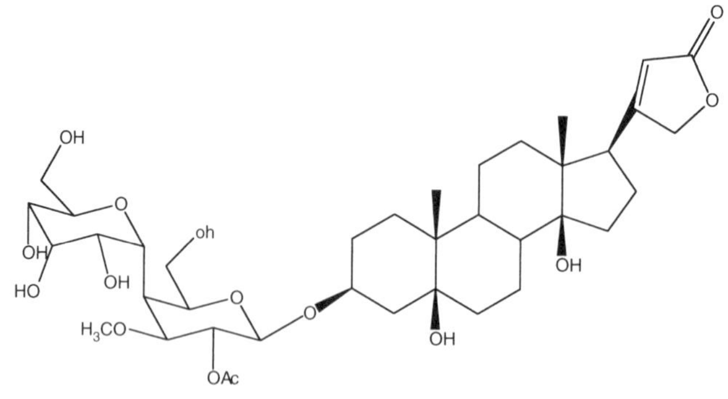

$(17\alpha)-H$-periplogenin-3-O-β-glucopyranosyl-(1-4)-2-O-acetyl-3-O-methyl-β-fucopyranoside

Figure 35.7 A cardenolide of *Streptocaulon cumingii*.

35.6 *TELOSMA CORDATA* (BURM. F.) MERR.

[From: Latin *cordata* = heart-shaped, cordate, referring to the leaves.]

35.6.1 Botany

Telosma cordata (Burm. f.) Merr. (*Cynanchum odoratissimum* Lour. and *Pergularia minor* Andr.) is a woody climber that grows to a length of 10m in the rain forests of China, India, Kashmir, Burma, Pakistan, and Vietnam. The plant is ornamental. The stems are terete, 3mm in diameter, lenticelled, and exude a white latex after incision. The leaves are simple, opposite, and exstipulate. The petiole is 1.5–5cm long. The blade is cordate and lanceolate, 11cm × 4.3cm – 11cm × 5.5cm, and shows 8–9 pairs of secondary nerves and a few tertiary ones. The inflorescences are 4.4cm-long cymes which have 15–30 flowers. The flower pedicels are a peduncle, 5mm – 1.5cm long. The calyx is 7mm long and puberulent. The sepals are oblong–lanceolate and puberulent. The corolla is yellowish-green to red. The corolla tube is about 1cm long, puberulent outside, pilose or glabrous with a pilose throat inside. The tube produces five lobes which are oblong, 6–12mm × 3–6mm, and ciliate. The corona lobes are slightly fleshy. The pollinia are oblong or reniform. The fruits are pairs of follicles which are 7–13cm × 2–3.5cm, glabrous, somewhat obtusely 4-angled, and contain several hairy seeds (Figure 35.8).

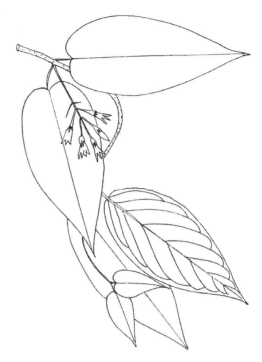

Figure 35.8 *Telosma cordata* (Burm. f.) Merr. [From: Flora of North Borneo. Distributed from The Herbarium of The Forest Department, Sandakan, Borneo. District: Temburong, Kuala Belalong. Alt.: 200ft. March 18, 1957. Field collectors: B. E. Smythies, G. H. S. Wood, and P. Ashton. Botanical identification: G. H. S. Wood.]

35.6.2 Ethnopharmacology

The plant is known as Chinese Violet, Fragrant Telosma, Tonkin Creeper, or *ye lai xiang* (Chinese). In Cambodia, Laos, and Vietnam, the plant is used internally to promote urination. In China, the flowers are very fragrant and yield perfumed oil. They are used in cooking and medicinally to treat conjunctivitis. Huan et al.[6] isolated a series of polyoxypregnane glycosides from *Telosma procumbens* (Blco.) Merr. Are these antiinflammatory or diuretics?

REFERENCES

1. Alen, Y., Nakajima, S., Nitoda, T., Baba, N., Kanzaki, H., and Kawazu, K. 2000. Antinematodal activity of some tropical rain forest plants against the pinewood nematode *Bursaphelenchus xylophilus*. *Z. Naturforsch.*, 55, 295.
2. Ueda, J. Y., Tezuka, Y., Banskota, A. H., Le Tran, Q., Tran, Q. K., Harimaya, Y., Saiki, I., and Kadota, S. 2002. Antiproliferative activity of Vietnamese medicinal plants. *Biol. Pharm. Bull.*, 25, 753.

3. Ueda, J. Y., Tezuka, Y., Banskota, A. H., Tran, Q. L., Tran, Q. K., Saiki, I., and Kadota, S. 2003. Antiproliferative activity of cardenolides isolated from *Streptocaulon juventas, Biol. Pharm. Bull.*, 26, 1431.

4. Ueda, J. Y., Tezuka, Y., Banskota, A. H., Tran, Q. L., Tran, Q. K., Saiki, I., and Kadota, S. 1975. Constituents of the Vietnamese medicinal plant *Streptocaulon juventas* and their antiproliferative activity against the human HT-1080 fibrosarcoma cell line. *J. Nat. Prod.*, 66, 1427.

5. Khine, M. M., Franke, K., Arnold, N., Porzel, A., Schmidt, J., and Wessjohann, L. A. 2004. A new cardenolide from the roots of *Streptocaulon tomentosum. Fitoterapia*, 75, 779.

6. Huan, V. D., Ohtani, K., Kasai, R., Yamasaki, K., and Tuu, N. V. 2001. Sweet pregnane glycosides from *Telosma procumbens. Chem. Pharm. Bull. (Tokyo)*, 49, 453.

CHAPTER 36

Medicinal Plants Classified in the Family Solanaceae

36.1 GENERAL CONCEPT

The family Solanaceae (A. L. de Jussieu, 1789 nom. Conserv., the Potato Family) consists of about 85 genera and 2800 species of prickly herbs, shrubs, climbers, and small trees, and is well represented in South America, and known to produce tropane alkaloids derived from ornithine, pyridine, and steroidal alkaloids. The leaves are alternate, simple, often soft and dull, somewhat untidy, green, and without stipules. The flowers are tubular, funnel-shaped to starry, and 5-lobed. The lobes are folded, contortate, or vulvate. The androecium consists of five stamens, the anthers of which often fuse into a conical body which is bright and yellow. The fruits are 2-celled berries or capsules (Figure 36.1).

Figure 36.1 Botanical hallmarks of Solanaceae. **(See color insert following page 168.)**

Solanaceae are commercially important. They are native to South America and were brought to Europe by early Spanish conquistadors: *Solanum tuberosum* L. (potato), *Lycopersicum esculentum* Mill. (tomato), *Nicotiana tabacum* L. (tobacco), and *Capsicum frutescens* L. (chilies). *Solanum tuberosum* L. (potato) was initially used in Europe to feed pigs and later humans, thanks to Parmentier. Tobacco smoking is the cause of millions of deaths annually and, despite its harmfulness, its consumption remains legal in most parts of the world.

An historically interesting example of medicinal Solanaceae used in the West is *Mandragora officinarum* L., or Mandragora, Mandrake, or Satan's Apple, the use of which can be recorded from the time of the Kings of Thebes, 1800 years before Christ. The plant has always excited curiosity because of its human-shaped roots which were known to Theophrastus, Dioskurides, and Hippocrates. Later in the middle ages, witches during the Sabbath used to smear themselves with the pastes of *Atropa belladona* L. (Deadly Nightshade), with other Solanaceae including *Datura stramonium* L., *Hyoscyamus niger* L. (Black Henbane), and *Mandragora officinalis* L. to enter into ecstasies, rapture, and extreme exaltation.

Mandragora was collected using dogs tethered to the stems as it was believed that pulling the roots would "make a scream that would make the unfortunate collector insane." The same Solanaceae have since been incorporated in several European pharmacopoeias. The dried leaves, or leaves and other aerial parts of *Atropa belladonna* L., collected when the plants are in flower and containing not less than 0.3% of alkaloids calculated as hyoscyamine (Belladona Herb, *British Pharmacopoeia* 1963); the dried leaves and flowering tops of *Datura stramonium* L. containing not less than 0.25% of alkaloids calculated as hyoscyamine (Stramonium, *British Pharmacopoeia*, 1963); and the dried leaves and flowering tops of *Hyoscyamus niger* L. containing not less than 0.05% of alkaloids calculated as hyoscyamine (Hyoscyamus, *British Pharmacopoeia*, 1963) are strongly antispasmodic and used for intestinal colic, gastric ulcer, spasmodic asthma, whooping cough, and bladder and urethral spasms, on account of hyoscyamine (Figure 36.2).

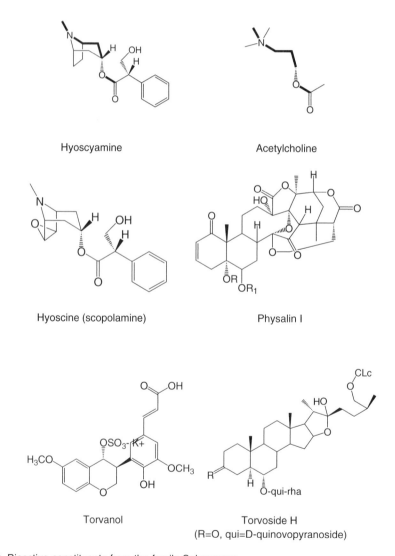

Figure 36.2 Bioactive constituents from the family Solanaceae.

Hyoscyamine is a parasympatholytic tropane alkaloid which exerts a selective blocking action on muscarinic receptors, hence mydriasis and tachycardia, decreased production of saliva, sweat, gastric juice, constipation, and inability to urinate. The traditional systems of medicine of the Asia–Pacific have been using about 50 species of Solanaceae, a number of which are of American origin, such as *Capsicum minimum* Roxb., *Solanum ferox* var. *laniocarpum, Solanum mammosum* L., *Solanum nigrum* L., and *Solanum verbascifolium* L.

36.2 *CAPSICUM MINIMUM* ROXB.

[From: Latin *capsa* = case and *minimus* = of diminutive size.]

36.2.1 Botany

Capsicum minimum Roxb. (*Capsicum frutescens* L. and *Capsicum fastigiatum* Bl.) is a herb that grows to a height of 1.2m and is 2.5cm in diameter. The plant is native to Central America and widespread in the tropical world as a source of chilies. The stems are glabrous, terete, pitted, and 2mm in diameter. The leaves are simple, spiral, and exstipulate in groups of 2–3. The petiole is 6mm – 1.5cm long. The blade is asymmetrical at the base, 4.7cm × 2.3cm – 6cm × 12.7cm – 4cm × 2cm – 2.6cm × 1.9cm, membranaceous, light green, and shows about five pairs of secondary nerves below. The apex is tailed. The flowers are white and minute. The calyx is 4mm long in fruits. The fruits are fusiform, fleshy berries which are green to red, glossy, and edible. The fruit pedicels are about 2–5cm long and 2mm in diameter (Figure 36.3).

36.2.2 Ethnopharmacology

Capsicum minimum Roxb. is also known as African Chilies, Chilies, Red Pepper, Bird Pepper, Capsicum, Hot Pepper, and Tabasco Pepper. *Capsicum* (*British Pharmaceutical Codex*, 1963) or the dried fruits of the plant containing about 0.5%–0.9% of capsaicin have been used internally in the form of a liquid extract (Capsicum Liquid Extract, *British Pharmaceutical*

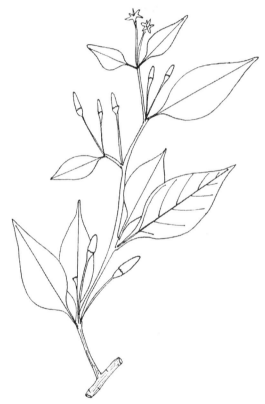

Figure 36.3 *Capsicum minimum* Roxb. [From: Sarawak Forest Department. No. S: 64202. March 16, 1992. Geographical localization: Long Busang, Btg. Balui, Kapit. Planted.]

Codex, 1923) and tincture (*British Pharmaceutical Codex*, 1963), to stimulate digestion. Externally Capsicum Liniment (*British Pharmaceutical Codex*, 1934) has been used to counteract the pain of lumbago, neuralgia, and rheumatism. In Cambodia, Laos, and Vietnam, the fruits are used externally as a rubefacient and eaten raw to stimulate digestion, and to treat jaundice and liver diseases. In Malaysia, the fruits are used to promote digestion and to treat diarrhea and vomiting. In the Philippines, the plant is used externally as a counterirritant. In North Borneo, the plant is used for cuts and wounds. It is also believed there that it "drives away evil spirits." For this latter purpose,

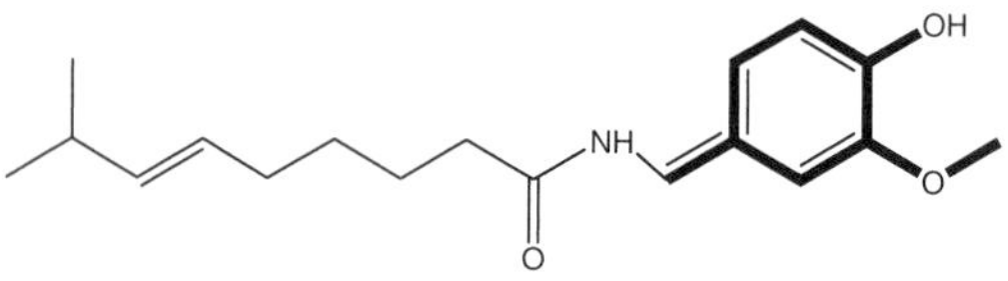

Capsaicin

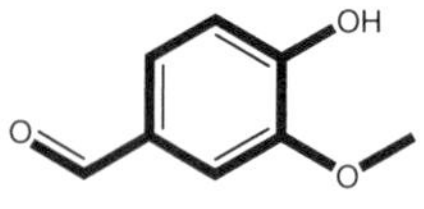

Vanillin

Figure 36.4

the fruits are crushed with other ingredients, wrapped in a cloth, and burned. The smoke will "chase away evil beings." The vernacular names for *Capsicum minimum* Roxb. in Borneo are *lia keli* (Kenya) and *lada padi*, *lada kecil* (Malay).

The first pharmacological study on the gastric effects of chilies is the work of Toriola and Solanke.[1] They studied the effect of *Capsicum minimum* Roxb. on gastric acid secretion in patients with duodenal ulcer and demonstrated that red pepper increases gastric acid production. The active principles involved here are capsaicin (trans-8-methyl-*N*-vanillyl-6-nonenamide) and congeners. Perhaps no other natural product has aroused more interest in the field of pharmacology than capsaicin.[2]

An interesting property of capsaicin is that it binds to the vanilloid receptor 1 expressed in small-diameter primary sensory afferent neurons, especially in nociceptive sensory nerves. In normal physiological conditions, the vanilloid receptor 1 responds to noxious stimuli including heat, acidification, and capsaicin. At high doses, it has been found that capsaicin inhibits sensory nerves, hence its potential in topical and injectable analgesic drugs.[3–6]

However, one should know that capsaicin induces apoptosis in glioma cells, hepatocarcinoma, thymocytes, and human B cells via caspase cascades.[7–11] One might wonder if capsaicin could be harmful for the nervous system (Figure 36.4).

36.3 *SOLANUM FEROX* VAR. *LANIOCARPUM*

[From: Latin *Solanum* = quieting, in reference to the narcotic properties of some species, and from *ferox* = ferocious.]

36.3.1 Botany

Solanum ferox var. *laniocarpum* is an herb that grows to a height of 2m in villages of Indonesia, Thailand, the Philippines, and Malaysia. The stems are spiny, velvety, pitted, fleshy, and the spines are 3mm long. The leaves are simple, spiral, and exstipulate. The petiole is 8–11cm long and spiny. The blade is palmately lobed, membranaceous, spiny on secondary nerves, densely velvety below, 25cm × 11.6cm – 11cm × 10.5cm, and shows five pairs of secondary nerves. The inflorescences are axillary cymes. The fruits are 2cm-diameter berries seated on a 5-lobed calyx. The sepals are velvety and 8mm × 7mm (Figure 36.5).

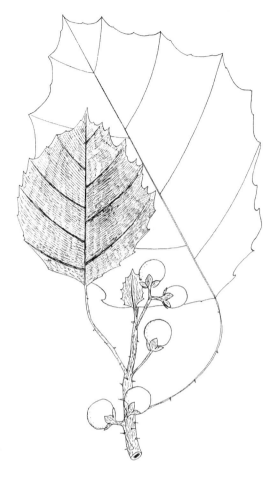

Figure 36.5 *Solanum ferox* var. *laniocarpum*. [From: Sarawak Forest Department. Field collectors: P. Ilias et al. No. S: 51930. Aug. 3, 1986. Geographical localization, Tamawan Tingtian Hill, Mujang, Skerang, Sri Amar Simanggan. In secondary forest.]

36.3.2 Ethnopharmacology

In North Borneo, the plant provides a remedy for toothaches. The seeds are put onto a watchglass and roasted until hot, and then covered with a coconut shell which has holes. The smoke emerging from the holes is inhaled through the mouth for 30 minutes to an hour "so that the worm in the teeth will come out to die." The Bornean name for this plant is *terong gigi*. In Malaysia, the roots are reduced to a paste and boiled to make a drink taken to assuage pains and to treat syphilis. The pharmacological properties of this plant are unexplored as of yet. An interesting development would be to assess the activity of the plant against Treponema.

36.4 *SOLANUM MAMMOSUM* L.

[From: Latin *Solanum* = quieting, in reference to the narcotic properties of some species, and from *mammosum* = with breasts or nipples.]

36.4.1 Botany

Solanum mammosum L. is an herb that is native to Central America which grows to a height of 1m throughout the Pacific Rim and is often cultivated for its very unusually shaped fruits. The

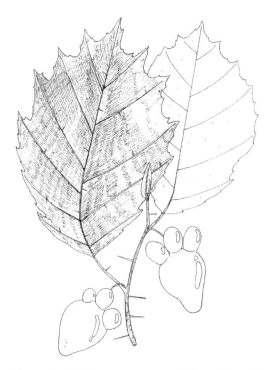

Figure 36.6 *Solanum mammosum* L. [From: Sarawak Forest Department. Field Collectors: Rantai Jawa et al. No. S: 67362. Aug. 7, 1994. Botanical identification: R. Jawa, Jan. 16, 1995. Geographical localization: Rh. Belawan, Ng. Apo, Ulu Sg. Kanowit, Julau. Planted near longhouse compound.]

stems are velvety and spiny at the apex. The thorns are 1.5cm long and 1.5mm in diameter. The leaves are simple, spiral, and exstipulate. The blade is hairy, 22cm × 12cm, and spiny, incised with about five pairs of secondary nerves. The flowers are purplish-yellow. The fruits are yellowish-orange, 5.7cm × 4.4cm, with five ovoid basal appendages which are 1.9cm × 1.6cm, glossy, and curiously shaped like some kind of plastic toy. The fruit pedicles are 1.5cm long (Figure 36.6).

36.4.2 Ethnopharmacology

In Borneo, the vernacular names of the plant are *terong semangat* (Bajan), *teron susu* (Brunei), and *terung tujang* (Iban). In Sarawak, the plant is used for treating sore eyes on chickens by slicing a piece of the fruit and squeezing a drop of the juice into their eyes. In Cambodia, Laos, and Vietnam, the plant is used to induce narcosis. The plant is also known as the Apple of Sodom, and Nipple Fruit. To date, there is little pharmacological evidence for the plant. Note, however, that an extract displayed some levels of activity against *Plasmodium*.[12] Are glycoalkaloids involved here?[13,14] Is the plant antiviral against bird-flu viruses?

36.5 *SOLANUM NIGRUM* L.

[From: Latin *Solanum* = quietening, in reference to the narcotic properties of some species, and from *nigrum* = black.]

36.5.1 Botany

Solanum nigrum L. (*Solanum rhumphii* Dun.) is an herb that grows to a height of 1.5m throughout most of the world as a weed. The stems are fleshy. The young shoots are purplish. The leaves are simple, spiral, and exstipulate, and grouped into groups of two to three. The blade is membranaceous, 5.5cm × 3.5cm – 4cm × 2.5cm – 4.8cm × 3.3cm – 7cm × 3.4cm, incised, and showing six pairs of secondary nerves. The base of the blade is tapering. The inflorescences consist of 3cm-long axillary or cauliflorous clusters of six flowers which are 5mm in diameter. The petals are white, 5-lobed, and yellowish at the middle part. The fruits are globose berries, which are reddish-black and glossy when matured, and 8mm in diameter (Figure 36.7).

36.5.2 Ethnopharmacology

Black Nightshade, Petty Morel, Poison Berry, Garden Nightshade, have been used medicinally for a very long time in Europe and Asia. The leaves and flowering tops of *Solanum nigrum* (Black

Nightshade, Morelle Noire, *French Pharmaco-poeia*, 1965), have been used in liniments, poultices, and decoctions for external applications. In China, the plant is known as *lung k'uei, t'ien ch'ien tzu, t'ien p'ao ts'ao*, and *lao ya yen ching ts'ao*. It is a remedy found in the Pentsao. In China, young leaves are used as a vegetable to invigorate men's sexuality and to regulate menses to normal. Externally, the plant offers a remedy for cancerous sores and to heal wounds. In Cambodia, Laos, and Vietnam, the roots are used to promote expectoration, and the fruits are eaten to relieve the bowels from costiveness. Indonesians use the juice squeezed from ripe fruits to clear pus from the eyes of hens. The vernacular names of the plant in Borneo include *nangka, beiwan* (Indonesian), *tutan* (Sabah), and *ladah* (Borneo). In the Philippines, the leaves are used to treat skin diseases, including cancers, and to mitigate pain.

The plant contains a series of steroidal alkaloids, including solanine and solasodine glycosides, solasonine, and solamargine. An ethanol extract from ripe fruits inhibited the proliferation of breast carcinoma cell-line (MCF)-7 human breast cancer cells cultured *in vitro* via apoptosis by a glycoprotein.[15,16] Perez et al.[17] made the interesting observation that an ethanol extract of the fruit given intraperitoneally prolonged pentobarbital-induced sleeping time, produced an alteration in the general behavior pattern, reduced exploratory behavior pattern, suppressed aggressive behavior, affected locomotor activity, and reduced spontaneous motility of rodents.

Figure 36.7 *Solanum nigrum* L. [From: Philippine Plants Inventory. PPI. Flora of the Philippines Joint Project of the Philippine National Museum, Manila and B.P. Bishop Museum, Honolulu. Supported by NSF/USAID. PPI No: 1565. Geographical localization: Batan Island, Mount Iraya, found around trail to Mount Iraya. Soil clay loam, in a regenerating forest.]

36.6 *SOLANUM VERBASCIFOLIUM* L.

[From: Latin *Solanum* = quieting, in reference to the narcotic properties of some species, and from *verbascifolium* = *Verbascum*-like leaves.]

36.6.1 Botany

Solanum verbascifolium L. is a treelet that grows up to 6m high and has branches midway up. The bole is 8cm in diameter. The stems, petioles, blades, and inflorescences are velvety and woolly. The stems are hollowed and thorny, the thorns are 5mm long. The leaves are simple and spiral. The petiole is channeled above and 5–7.4cm long. The blade is 22.3cm × 11.6cm – 16.7cm × 7.9cm – 14cm × 6.1cm, and shows 7–9 pairs of secondary nerves. The nerves are sunken above and raised below. The base is wedge-shaped, the apex is tailed, and the whole blade is velvety. The inflorescences are axillary cymes that are up to 5cm long. The calyx is glandular and tomentose. The

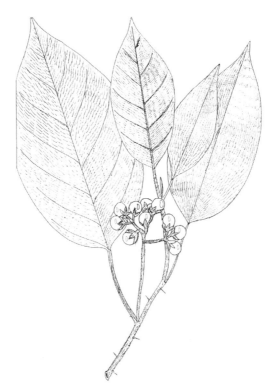

Figure 36.8 *Solanum verbascifolium* L. [From: Herbarium Bogoriense, Harvard University Herbaria. Plants of Indonesia, Bali Timur, Tabanan, southwest corner of Lake Tambligan. Alt.: 1300m. 8°14′ S, 115°6′ E, in primary forest, lakeside. Canopy 15–20m. Dominants include Laportes, Ficus, Ehretia, Homalanthus, Schefflera. Field collectors: J. A. McDonald and R. Ismail. No: 4931, July 30, 1994.

corolla is cream and the five lobes are inflexed. The anthers are yellow. The fruits are globose berries, olive green and glossy at first, then turning black, and 1.2cm in diameter (Figure 36.8).

36.6.2 Ethnopharmacology

The vernacular names of the plant include: *jarong limbang* (Banjar), *ambuggaib* (Dusun), *mansak–mansak*, *kamug–kamu*, and *limbasak* (Borneo). The plant provides a remedy for dysentery in Taiwan and the Philippines, and is used to mitigate intestinal pain in Taiwan and Malaysia. In the Solomon Islands, the leaves are used externally to counteract poisoning and to heal mouth sores. To date, the pharmacological properties of this plant have not been explored. Note that the plant produces cinnamides, steroidal saponins, and vanillic acid.[18–21] What is the activity of vanillic acid towards vanilloid receptors?

REFERENCES

1. Toriola, F. and Solanke, T. F. 1973. The effect of red pepper (*Capsicum frutescens*) on gastric acid secretion. *J. Surgical Res.*, 15, 385.
2. Szolcsányi, J. 2004. Forty years in capsaicin research for sensory pharmacology and physiology. *Neuropeptides*, 38, 377.
3. Bley, K. R. 2004. Recent development in transient receptor potential vanilloid receptor agonist – based therapies. *Expert. Opin. Investing. Drugs*, 13, 1445.

4. Chanda, S., Mould, A., Esmail, A., and Bley, K. 2005. Toxicity studies with pure trans-capsaicin delivered to dogs via intravenous administration. *Regul. Toxicol. Pharmacol.*, 43, 66.

5. Dembiński, A., Warzecha, Z., Ceranowicz, P., Brzozowski, T., Dembiński, M., Konturek, S. J., and Pawlik, W. 2005. Role of capsaicin-sensitive nerves and histamine H_1, H_2, and H_3 receptors in the gastroprotective effect of histamine against stress ulcers in rats. *Eur. J. Pharmacol.*, 508, 211.

6. Brahim, J. S. 2005. Capsaicin as a preventive analgesic in the oral surgery model. *J. Oral Maxillofacial Surg.*, 63, 77.

7. Amantini, C., Mosca, M., Lucciarinin, R., Perfumi, M., Morrone, S., Piccoli, M., and Santoni, G. 2004. Distinct thymocyte substrates express the vanilloid receptor VR1 that mediates capsaicin-induced apoptotic cell death. *Cell Death Differ.*, 11, 1342.

8. Jeftinija, S., Liu, F., Jeftinija, L., and Urban, L. 1992. Effects of capsaicin and resiniferatoxin on peptidinergic neurons in cultured dorsal root ganglion. *Regul. Pept.*, 39, 123.

9. Qiao, S., Li, W., Tsubouchi, R., Haneda, M., Murakami, K., and Yoshino, M. 2005. Involvement of peroxynitrite in capsaicin-induced apoptosis of C6 glioma cells. *Neurosci. Res.*, 51, 175.

10. Wolvetang, E. J., Larm, J. A., Moutsoulas, P., and Lawen, A. 1996. Apoptosis induced by inhibitors of the plasma membrane NADH-oxidase involves Bel-2 and calcineurin. *Cell Growth Differ.*, 7, 315.

11. Jin, H. W., Ichikawa, H., Fujita, M., Yamaai, T., Mukae, K., Nomura, K., and Sugimoto, T. 2005. Involvement of caspase cascade in capsaicin-induced apoptosis of dorsal root ganglion neurons. *Brain Res.*, 1056,139.

12. Muñoz, V., Sauvain, M., Bourdy, G., Callapa, J., Rojas, I., Vargas, L., Tae, A., and Deharo, E. 2000. The search for natural bioactive compounds through a multidisciplinary approach in Bolivia. Part II. Antimalarial activity of some plants used by Mosetene Indians. *J. Ethnopharmacol.*, 69, 139.

13. Alzerreca, A. and Hart, G. 1982. Molluscicidal steroid glycoalkaloids possessing stereoisomeric spirosolane structures. *Toxicol. Lett.*, 12, 151.

14. Seelkopf, C. 1968. Alkaloid glycosides of the fruit from *Solanum mammosum* L. *Arch. Pharm. Ber. Dtsch. Pharm. Ges.*, 301, 111.

15. Son, Y. O., Kim, J., Lim, J. C., Chung, Y., Chung, G. H., and Lee, J. C. 2003. Ripe fruits of *Solanum nigrum* L. inhibits cell growth and induces apoptosis in MCF-7 cells. *Food Chem. Toxicol.*, 41, 1421.

16. Heo, K. S., Lee, S. J., and Lim, K. T. 2004. Cytotoxic effect of glycoprotein isolated from *Solanum nigrum* L. through the inhibition of hydroxyl radical-induced DNA-binding activities of NF-kappa B in HT-29 cells. *Environ. Toxicol. Pharmacol.*, 17, 45.

17. Perez, G. R. M., Perez, L. J. A., Garcia, D. L. M., and Sossa, M. H. 1998. Neuropharmacological activity of *Solanum nigrum* fruit. *J. Ethnopharmacol.*, 62, 43.

18. Adam, G. and Khoi, N. H. 1980. Solaverbascine — a new 22,26-epiminocholestane alkaloid from *Solanum verbascifolium*. *Phytochemistry*, 19, 1002.

19. Zhou, L. X. and Ding, Y. 2002. A cinnamide derivative from *Solanum verbascifolium* L. *J. Asian Nat. Prod. Res.*, 4, 185.

20. Dopke, W., Mola, I. L., and Hess, U. 1976. Alkaloid and steroid sapogenin content of *Solanum verbascifolium* L. *Pharmazie*, 31, 656.

21. Zhou, Q., Zhu, Y., Chiang, H., Yagiz, K., Morre, D. J., Morre, D. M., Janle, E., and Kissinger, P. T. 2004. Identification of the major vanilloid component in *Capsicum* extract by HPLC-EC and HPLC-MS. *Phytochem. Anal.*, 15, 117.

CHAPTER 37

Medicinal Plants Classified in the Family Verbenaceae

37.1 GENERAL CONCEPT

The family Verbenaceae (Jaume Saint-Hilaire, 1805 nom. conserv., the Verbena Family) consists of approximately 100 genera and 2600 species of herbs, climbers, shrubs, and trees producing mainly diterpenes, iridoids, and flavonoids. In the field, Verbenaceae are recognized by their young stems which are quadrangular; the leaves which are compound, exstipulate, and decussate; and by terminal panicles of pinkish or blue bilabiate flowers or berries.

The dried leaves of *Aloysia triphylla* (L'Hérit.) Britt. and *Lippia citriodorata* Kunth or Lemon Verbena, are used to treat digestive and nervous ailments. *Verbena officinalis* L. (*French Pharmacopoeia*, 1965) is traditionally used to promote urination and to soothe inflamed skin. It was known at the time of the Roman Emperor Theodosius (4th Century A.D.) to remove tumors. *Vitex agnus-castus* L. (chaste tree) has been used medicinally since Greek times, and is still used for the treatment of premenstrual syndrome and menopause. About 50 species of plants classified within the family Verbenaceae are of medicinal value in the Pacific Rim. To date the pharmacological potential of this large family remains to be evaluated. Among the most exciting potential candidates to be studied are *Callicarpa arborea* Roxb., *Clerodendrum deflexum* Wall., *Clerodendrum inerme* (L.) Gaertn., *Duranta plumieri* Jacq., *Gmelina elliptica* Sm., *Peronema canescens* Jack, *Sphenodesme pentandra* Jack, *Sphenodesme trifolia* Wight, and *Teijmanniodendron pteropodium* (Miq.) Bakh. which are described next.

37.2 *CALLICARPA ARBOREA* ROXB.

[From: Greek *Kallos* = beauty, karpos = fruits, and from Latin *arborea* = tree-like.]

37.2.1 Botany

Callicarpa arborea Roxb. (*Callicarpa arborea* Roxb. var. *villosa* Gamble and *Callicarpa tomentosa* [L.] Murr.) is a tree that grows to a height of 18m in the rain forests of India, Thailand, Malaysia, and Sumatra. The whole stems, leaves, and inflorescences are densely covered with whitish hairs. Fewer hairs are seen above the blade except for the midrib. The petiole is 2.5–5cm long. The blade is elliptic–lanceolate, 14cm × 6.5cm – 21cm × 4cm, wavy at the margin, and shows

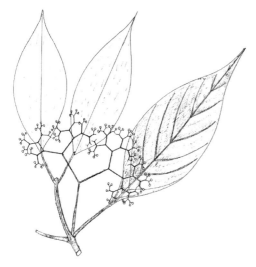

Figure 37.1 *Callicarpa arborea* Roxb. [From: Flora of Malaya. Kep. Field No: 0506. Geographical localization: Gerik–Kroh Road near Kliang Intang Upper Perak, roadside. June 7, 1966. T. C. Whitmore.]

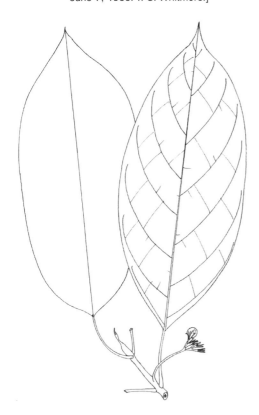

Figure 37.2 *Clerodendrum deflexum* Wall. [From: FRIM No: 76490. Botanical identification: J. Sinclair. Geographical localization: Rembia Island, Sembiland Island. Hillside. Alt.: 200ft.]

10–13 pairs of secondary nerves, only visible below. The base of the blade is cuneate and the apex is tailed. The blade is velvety below. The inflorescences are axillary dichotomous cymes up to 7cm in diameter on an 11cm-long pedicel. The flowers comprise four bluish to purplish petals and four stamens. The fruits are green to dull red, globose little berries (Figure 37.1).

37.2.2 Ethnopharmacology

In Burma, the bark is used to treat skin diseases. The Malays use the leaves to heal sores and to assuage stomachaches, and call the plant *tampang besi*. In China, the plant is used to treat influenza. To date, the pharmacological property of *Callicarpa arborea* Roxb. is unexplored. Note that diterpenes are reported from the genus.[1–3] Some of them should be antiinflammatory.

37.3 *CLERODENDRUM DEFLEXUM* WALL.

[From: Greek *kleros* = casting lots and *dendron* = tree, and from Latin *deflexum* = bent, or turned abruptly downward at a sharp angle.]

37.3.1 Botany

Clerodendrum deflexum Wall. is a shrub that grows to a height of 3m in lowland areas and mountain forests up to 3000m in altitude in Sumatra and Malaysia. The stems are terete and hollowed. The petiole is slender and 4.5–6cm long. The blade is broadly elliptical, 24cm × 12cm, cuneate at the base, acuminate at the apex, and shows seven pairs of secondary nerves and a few tertiary nerves below. The margin is wavy. The inflorescences are cymose, axillary, and 2.5–3cm long. The fruits are bluish-black, glossy, and globose on a persistent calyx, which is red (Figure 37.2). Bioactive diterpenes are probably present here.

37.3.2 Ethnopharmacology

The Malays use the roots to treat fever and intestinal discomfort. The pharmacological properties of this plant are unknown.

37.4 *CLERODENDRUM INERME* (L.) GAERTN.

[From: Greek *kleros* = casting lots, *dendron* = tree, and from Latin *inerme* = unarmed, without prickles.]

37.4.1 Botany

Clerodendrum inerme (L.) Gaertn. (*Volkameria inermis* L., *Clerodendrum neriifolium* [Roxb.] Sch., and *Volkameria neriifolia* Roxb.) is a shrub that grows in estuaries and along the coasts of South China, Australia, the Pacific Islands, Southeast Asia, and India. The plant is grown as an ornamental. The stems are quadrangular and glabrous, and show 5.5–4–3.5cm internodes. The leaves are simple, exstipulate, and decussate. The petiole is 9mm long. The blade is papery, elliptical–lanceolate, 6.5cm × 2.8cm – 7.5cm × 3cm, and show 6–9 pairs of secondary nerves. The flowers are pure white. The corolla tube is 2–3cm long and 2mm in diameter at the throat. The corolla lobes are elliptical and 7mm long. The inflorescences are 3-flowered cymes. The fruits are cordate, 1.6cm in diameter on a 2cm-long pedicel. The axillary cymes are gray–yellow, obovoid to subglobose, and 6–11mm (Figure 37.3).

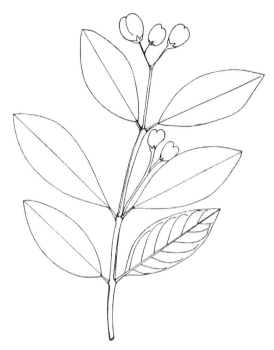

Figure 37.3 *Clerodendrum inerme* (L.) Gaertn. [From: Federated Malay States, Malay Peninsula. Geographical localization: Pasir Hitam Island, Perak in mangrove swamps. April 16, 1919. Field collector: V. P. Borger. C.F. No: 0264.]

37.4.2 Ethnopharmacology

In Burma, the plant is used to counteract the putrefaction of genitals. In Taiwan and China, the leaves are used externally to treat skin diseases. In Cambodia, Laos, and Vietnam, the leaves are roasted and boiled in water to make a drink that is taken to treat beriberi, and the plant is used to reduce fever. Indonesians use the seeds to counteract venom poisoning. The plant is called *setulang* and *chera puteh* in Indonesia and Malaysia. In the Solomon Islands, the steam from the water boiled with the leaves is used to clear vision. An interesting feature of the plant is that it elaborates clerodane diterpenes among which are some very unusual dimers, the pharmacological potential of which would be worth assessing (Figure 37.4).

Examples of such diterpenes are inermes A and B, 14,15-dihydro-15-methoxy-3-epicaryoptin, and 14,15-dihydro-15-hydroxy-3-epicaryoptin.[4,5] Also, the genus contains ethylsterols, such as 4α-methylsterol, 4-methyl-24-ethyl-5-cholesta-14,25-dien-3-ol.[6–11]

Bashwira et al. have made the interesting observation that *Clerodendrum myricoides* contains a series of cyclopeptides including the hexapeptide cleromyrin.[12,13] An interesting development would be to investigate *Clerodendrum inerme* (L.) Gaertn. for the pharmacological properties of its terpenes and cyclopeptide. Note that there is a slowly growing interest in cyclopeptides from flowering plants, perhaps based on the fact that cyclopeptides of some marine organisms, including ascidians from the genus *Lissoclinum,* such as haterumalide B, exert potent antitumor properties.[14]

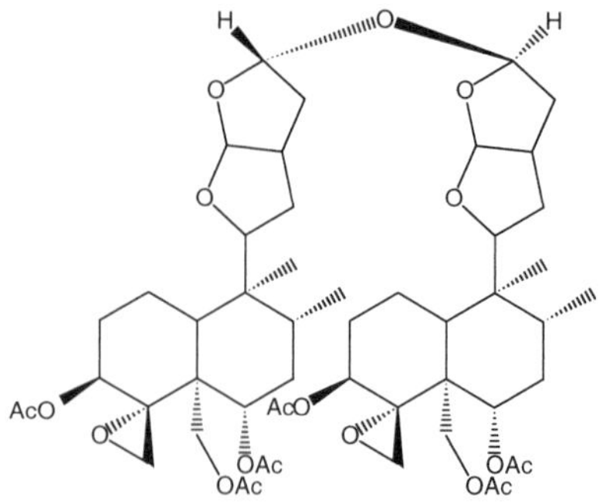

Inerme A

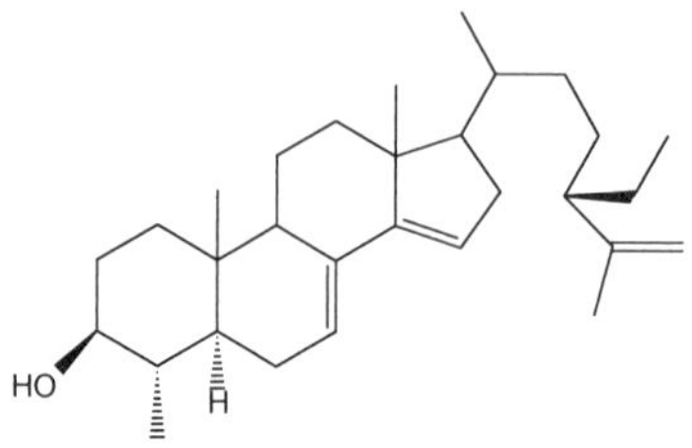

4-methyl-24-ethyl-5-cholesta-14, 25-dien-3-ol

Figure 37.4 Diterpenes from the *Clerodendrum* species.

37.5 *DURANTA PLUMIERI* JACQ.

[After Charles Plumier (1646–1707), French botanist.]

37.5.1 Botany

Duranta plumieri Jacq. (*Duranta repens* L.) is a shrub native to Central America, introduced into the Asia–Pacific for its ornamental value. It is in fact one of the most common ornamental plants of the tropics and can be easily recognized with its pendulous spikes of bluish flowers and pea-sized bright orange fruits. The stems are lenticelled, squarish, and pubescent. The leaves are simple, spiral in groups of three, and exstipulate. The blade is 4cm × 1.7cm and often light green. The inflorescences are slender, pendulous, terminal spikes of bluish flowers which are tubular and 5-lobed, and the tube is 7mm long. The fruits are orange berries which are 5mm–1.7cm in diameter (Figure 37.5).

37.5.2 Ethnopharmacology

Sky Flower, Golden Dew Drop, and Pigeon Berry are its vernacular names. In China, the fruits are used to treat malaria. In Cambodia, Laos, and Vietnam, the plant is used to induce urination. The Chinese name for the plant is *jia lian qiao*. Note that flavonoids occur in the plant as well as a series of clerodanes and saponins.[15]

Of particular interest are C-alkylated flavonoids 7-O-α-D-glucopyranosyl-3,5-dihydroxy-3′-(4″-acetoxyl-3″-methylbutyl)-6,4′-dimethoxyflavone, 7-O-α-D-glucopyranosyl-3,4′-dihydroxy-3′-(4″-acetoxyl-3″-methylbutyl)-5,6-dimethoxyflavone, 3,7,4′-trihydroxy-3′-(8″-acetoxy-7″-methyloctyl)-

Figure 37.5 *Duranta plumieri* Jacq. [From: The Botanic Gardens, Singapore. Malay Peninsula. Geographical localization: Singapore, cultivated in the Botanic Gardens. July 20, 1938. Field collector: C. X. Furtado.]

5,6-dimethoxyflavone and a trans-clerodane type diterpenoids (−)-6-hydroxy-5,8,9,10β-cleroda-3,13-dien-16,15-olid-18-oic acid, (+)-hardwickiic acid, and (+)-3,13-clerodadien-16,15-olid-18-oic acid, which exhibited some levels of activity against β-glucosidase *in vitro*.[16]

Castro et al.[17] made the interesting observation that an extract of the fruit inhibits the survival of *Plasmodium berghei in vitro*.

37.6 *GMELINA ELLIPTICA* SM.

[After J. G. Gmelin (1709–1799), German botanist, and from Latin *elliptica* = elliptical.]

37.6.1 Botany

Gmelina elliptica Sm. (*Gmelina villosa* Roxb. and *Gmelina asiatica* L. var. *villosa* [Roxb.] Bakh.) is a tree that grows to a height of 9m. Its stem is 3mm in diameter and minutely velvety and lenticelled. Leaves are simple, exstipulate, and decussate. The petiole is 2.9cm long, slender, and channeled above. The blade is broadly elliptical, papery, 8cm × 5cm – 9cm × 5.4cm – 5.3cm × 2cm, and shows four pairs of secondary nerves. The inflorescences are axillary and 4.5cm long. The fruit is globose, 8mm in diameter, that starts out green ripening to yellow (Figure 37.6).

37.6.2 Ethnopharmacology

In Indonesia, the juice squeezed from the fresh leaves and fruits is instilled into the ears to assuage earaches. The fruits are used to calm itchiness. In Malaysia, a paste of the plant is applied to the head to assuage headaches and to prevent alopecia. The analgesic property of the plant is

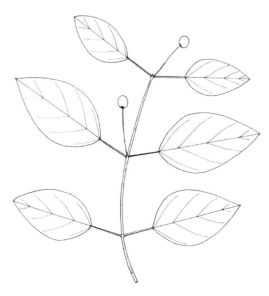

Figure 37.6 *Gmelina elliptica* Sm. [From: Flora of Malaya. FRI Field No: 27167. Patani River, Kedah, hedge of rain forest. Feb. 22, 1978. Botanical identification: F. S. P. Ng, July 6, 1979.]

not confirmed yet, but an extract of *Gmelina asiatica* L. given orally has protected rodents against yeast-induced fever.[18] Another observation is that an alcoholic extract of the root lowered dose-dependently the glycemia of both normal and diabetic rodents.[19]

The accompanying antipyretic and hypoglycemic effects, plus the use of the plant as an analgesic, adds strength to the hypothesis that a mechanism involving some steroidal hormone mechanism could be involved. Another trail to explore would be to look into the iridoid content of the plant.

It is a matter of fact that iridoids which occur in the *Gmelina* species and in general in the Verbenaceae, Lamiales, and Scrofulariales are known for exerting both antidiabetic and antiinflammatory properties.[20] One such compound is harpagoside-B from *Harpagophytum procumbens* DC. or Devil's Claw (family Pedaliaceae, order Scrofulariales), a South African plant involved in a business worth multimillions in U.S. dollars of annual benefits.

37.7 *PERONEMA CANESCENS* JACK

[From: Greek *pero* = disabled and *nema* = a thread, referring to the two missing stamens, and from Latin *canescens* = gray-downy.]

37.7.1 Botany

Peronema canescens Jack (*Peronema heterophyllum* Miq.) is a timber tree that grows to a height to 15m with a girth of 60cm in the rain forests of Indonesia and Malaysia. The bole is scaly and soft. The bark is gray. The wood is yellow. The stems are quadrangular and 8mm in diameter. The leaves are compound, exstipulate, and decussate. The rachis is 30cm long and winged at the internodes, which are about 5cm long. The folioles are subopposite, 20cm × 6cm – 17cm × 5.5cm – 13.5cm × 4cm – 12cm × 4cm – 10cm × 3cm, lanceolate, and membranaceous. The blade shows about 21 pairs of secondary nerves that are conspicuous below. The inflorescences are axillary cymes on 7.5cm-long pedicels. The fruits are tiny spiny capsules that are 5mm in diameter (Figure 37.7).

37.7.2 Ethnopharmacology

In Indonesia, the leaves are used to mitigate toothache and to reduce fever. In Malaysia, it is used for the same and is also used to remove ringworm infections. The Malays and Indonesians call the plant *sungkai*. A significant advance in the pharmacology of this plant has been provided by the work of Kitagawa et al.[21] They isolated from the plant a series of clerodane diterpenes

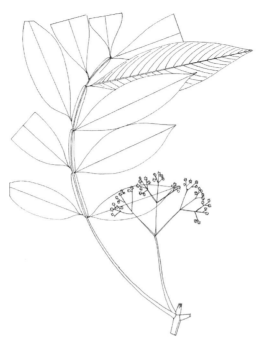

Figure 37.7 *Peronema canescens* Jack. [From: Flora of Malaya. Kep. Field No: 16296. Geographical localization: Compt. 12, Sungkap Forest Reserve, Kedah, flat land. Nov. 3, 1971. Field collector and botanical identification: K. M. Kochummen.]

including seven new peronemins B2, A2, B1, C1, B3, A3, and D1 which are perhaps involved in the antibabesial activity measured by Subeki et al.[22]

37.8 *SPHENODESME PENTANDRA* JACK

[From: Greek *spheno* = wedges, *penta* = 5, and *andros* = male, referring to the androecium.]

37.8.1 Botany

Sphenodesme pentandra Jack is a climber that grows to a length of 4.5m in the rain forests of Laos, Cambodia, Thailand, Malaysia, and Burma. The stems are squarish or terete, 2mm in diameter, lenticelled, and glabrous. The leaves are simple, decussate, exstipulate, and are medium green above, and shining pale below. The base of the blade is cordate, and the apex is acuminate. The blade is asymmetrical and measures 6cm × 2.4cm – 10.5cm × 5.2cm – 8.3cm – 4.2 cm – 2.7 cm × 9mm. The blade shows seven pairs of secondary nerves and tertiary nerves which are scalariform. The petiole is 7mm – 1cm long and channeled above. The petals are blue with numerous blue hairs. There are five stamens. The bract is green. The fruits are 3-winged (Figure 37.8).

37.8.2 Ethnopharmacology

The leaves and roots are boiled in water to make a liquid which is applied externally to mitigate rheumatic pains. The stems are used to tether buffalos. The pharmacological potential of this plant are to date unexplored. Does the plant contain harpagoside-like principles?

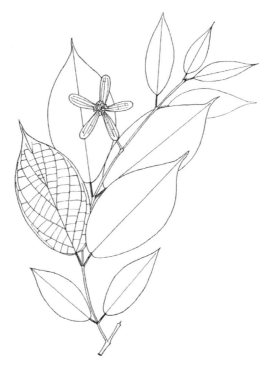

Figure 37.8 *Sphenodesme pentandra* Jack. [From: Flora of Johor, distributed by The Herbarium Botanic Gardens, Singapore. S.F. No: 40690. July 30, 1955. Field collector and botanical identification: J. Sinclair. Geographical localization: 2.5 miles from Kota Tinggi-Mawai Road, Dusun of J. A. le Doux.]

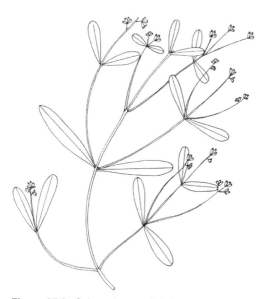

Figure 37.9 *Sphenodesme trifolia* Wight. [From: Flora of Malaya. Geographical localization: 16 miles Genting-Dempah Road. May 6, 1980. Field collector and botanical identification: K. M. Kochummen.]

37.9 *SPHENODESME TRIFOLIA* WIGHT

[From: Greek *spheno* = wedges, and from Latin *trifolia* = 3 leaves.]

37.9.1 Botany

Sphenodesme trifolia Wight is a woody climber that grows to a length of 15m in Malaysia and Singapore. The stems are smooth, angled, and arranged dichotomously. The leaves are simple, spathulate, sessile, and exstipulate. The blade is 7.5cm × 1cm – 6.8cm × 2.5cm. The flowers have pink heads in racemous clusters (Figure 37.9).

37.9.2 Ethnopharmacology

In Singapore, the plant is used to bring fever down, but is known to be stupefying. The pharmacological potential of this plant is to date unexplored, and one could investigate this species for central nervous system properties.

37.10 *TEIJMANNIODENDRON PTEROPODIUM* (MIQ.) BAKH.

[After Johannes Elias Teysmann (1808–1882), Curator, Botanic Gardens, Bogor, and from Greek *pteron* = winged and *podo* = foot, referring to the winged petiole.]

37.10.1 Botany

Teijmanniodendron pteropodium (Miq.) Bakh. (*Vitex pteropoda* Miq. and *Vitex perelata* King) is a tree that grows to a height of 15m with a girth of 1.5m in the swampy spots of the rain forests of Malaysia, Indonesia, the Philippines, and Papua New Guinea. The crown is dense, green, and cylindrical. The bark is lenticelled, smooth, and grayish-white. The inner bark is yellow and the sapwood is yellow and acrid. The stems are squarish, lenticelled, and 1.2cm in diameter. The leaves are curious, whorled in sixes, purple–black when young, and somehow araliaceous. The petiole is 16cm long and winged, the wings are nerved and 4cm transversally, and tapering to the apex. The blade is lanceolate to elliptical or obovate, thick, 30cm × 14cm – 20cm × 10cm – 10cm × 6cm, and shows 5–10 pairs of secondary nerves. The inflorescences are 17cm-long racemes of pale violet flowers. The fruits are ovoid and fleshy, 3.5cm × 2.7cm, green and glossy, pointed at the apex, and seated on vestigial calyces (Figure 37.10).

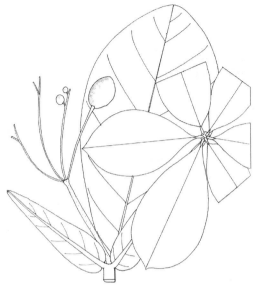

Figure 37.10 *Teijmanniodendron pteropodium* (Miq.) Bakh. [From: Forest Department Malaysia, Kep. No: 98412. K. M. Kochummen. June 1966. Geographical localization: Tapah, Palawan River State, in swamps, inhabited by ants at nodes.]

37.10.2 Ethnopharmacology

Indonesians use the seeds both internally and externally to normalize defecation.

REFERENCES

1. Hu, Y., Shen, Y., Gan, F., and Hao, X. 2002. Four diterpenes from *Callicarpa pedunculata*. *Biochem. System. Ecol.*, 30, 999.
2. Singh, A. K. and Agpawal, P. K. 1994. 16,17-Isopropylideno-3-oxo-phyllocladane, a diterpenoid from *Callicarpa macrophylla*. *Phytochemistry*, 37, 587.
3. Fujita, E., Ochiai, M., Ichida, I., Chatterjee, A., and Desmukh, S. K. 1975. Confirmation of the structure of calliterpenone, a diterpene from *Callicarpa macrophylla*. *Phytochemistry*, 14, 2249.
4. Barton, D. H. R., Cheung, H. T., Cross, A. D., Jackman, L. M., and Smith, M. M. 1961, Diterpenoid bitter priciples: Part III. The constitution of clerodin. *J. Chem. Soc.*, 5061–5073.
5. Pandey, R., Verma, R. K., and Gupta, M. M. 2005. Neo-clerodane diterpenoids from *Clerodendrum inerme*. *Phytochemistry*, 66, 643.
6. Akihisa, T., Ghosh, P., Thakur, S., Nagata, H., Tamura, T., and Matsumoto, T. 1990. 24,24-Dimethyl-25-dehydrolophenol, a 4α-methylsterol from *Clerodendrum inerme*. *Phytochemistry*, 29, 1639.
7. Akihisa, T., Ghosh, P., Thakur, S., Oshikiri, S., Tamura, T., and Matsumoto, T. 1988. 24-Methyl-cholesta-5, 22E, 25-trien-3β-ol and 24α-ethyl-5a-cholest-22E-en-3β-ol from *Clerodendrum fragrans*. *Phytochemistry*, 27, 241.

8. Akihisa, T., Matsubara, Y., Ghosh, P., Thakur, S., Shimizu, N., Tamura, T., and Matsumoto, T. 1988. The 24α and 24β-epimers of 24-ethylcholesta-5, 22-dien-3β-ol in two *Clerodendrum* species. *Phytochemistry*, 27, 1169.

9. Akihisa, T., Tamura, T., Matsumoto, T., Kokke, W. C. M. C., Ghosh, P., and Thakur, S. 1990. (22Z, 24S)–Stigmasta-5,22,25-trien-3-ol and other novel sterols from *Clerodendrum scandens*: first report of the isolation of *cis*-D22-unsaturated sterol from a higher plant. *J. Chem. Soc. Perkin Trans. India*, 2213.

10. Pandey, R., Verma, R. K., Subhash, C., Singh, C., and Gupta, M. 2003. 4-Methyl-24-ethyl-5-cholesta-14,25-dien-3-ol and 24-ethylcholesta-5, 9(11), 22E-trien-3-ol, sterols from *Clerodendrum inerme*. *Phytochemistry*, 63, 415.

11. Akihisa, T., Matsubara, Y., Ghosh, P., Thakur, S., Tamura, T., and Matsumoto, T. 1989. Sterols of some *Clerodendrum* species (Verbenaceae): occurrence of the 24- and 24-epimers of 24-ethylsterols lacking a 25-bond. *Steroids*, 53, 625.

12. Bashwira, S. and Hoostelé, C. 1988. Myricoidine and dihydromyricoidine, two new macrocyclic spermidine alkaloids from *Clerodendrum myricoides*. *Tetrahedron*, 44, 4521.

13. Bashwira, S., Hoostelé, C., Tourwé, D., Pepermans, H., Laus, G., and Van, G. 1989. Cleromyrine I, a new cyclohexapeptide from *Clerodendrum myricoides*. *Tetrahedron*, 45, 5845.

14. Ueda, K. and Hu, Y. 1999. Haterumalide B: a new cytotoxic macrolide from an Okinawan ascidian *Lissoclinum* sp. *Tetra. Lett.*, 40, 6305.

15. Hiradate, S., Yada, H., Ishii, T., Nakajima, N., Ohnishi-Kameyama, M., Sugie, H., Zungsontiporn, S., and Fujii, Y. 1999. Three plant growth inhibiting saponins from *Duranta repens*. *Phytochemistry*, 52, 1223.

16. Iqbal, K., Malik, A., Mukhtar, N., Anis, I., Khan, S. N., and Choudhary, M. I. 2004. Alpha-glucosidase inhibitory constituents from *Duranta repens*. *Chem. Pharm. Bull. (Tokyo)*, 52, 785.

17. Castro, O., Barrios, M., Chinchilla, M., and Guerrero, O. 1996. Chemical and biological evaluation of the effect of plant extracts against *Plasmodium berghei*. *Rev. Biol. Trop.*, 44, 361.

18. Ikram, M., Khattak, S. G., and Gilani, S. N. 1987. Antipyretic studies on some indigenous Pakistani medicinal plants: II. *J. Ethnopharmacol.*, 19, 185.

19. Kasiviswanath, R., Ramesh, A., and Kumar, K. E. 2005. Hypoglycemic and antihyperglycemic effect of *Gmelina asiatica* LINN. in normal and in alloxan induced diabetic rats. *Biol. Pharm. Bull.*, 28, 729.

20. Ahmed, B., Al-Rehaily, A. J., Al-Howiriny, T. A., El-Sayed, K. A., and Ahmad, M. S. 2003. Scropolioside-D2 and harpagoside-B: two new iridoid glycosides from *Scrophularia deserti* and their antidiabetic and antiinflammatory activity. *Biol. Pharm. Bull.*, 26, 462.

21. Kitagawa, I., Simanjuntak, P., Hori, K., Nagami, N., Mahmud, T., Shibuya, H., and Kobayashi, M. 1994. Indonesian medicinal plants. VII. Seven new clerodane-type diterpenoids, peronemins A2, A3, B1, B2, B3, C1, and D1, from the leaves of *Peronema canescens* (Verbenaceae). *Chem. Pharm. Bull. (Tokyo)*, 42, 1050.

22. Subeki, M. H., Matsuura, H., Yamasaki, M., Yamato, O., Maede, Y., Katakura, K., Suzuki, M., Trimurningsih, C., and Yoshihara, T. 2004. Effects of Central Kalimantan plant extracts on intraerythrocytic *Babesia gibsoni* in culture. *J. Vet. Med. Sci.*, 66, 871.

Plant Index

A

Abuta, 41
Acokanthera, 247
Acokanthera ouabaio, 247
Acokanthera schimperi, 247
Acronychia baueri Scott, 212
Adenia cordifolia Engl., 102–104
Adenium, 247
Aegiceras corniculatum, Blco., 54–55
Aegle marmelos Correa, 212–215
Aeschrion excelsa, 185
Aglaia aphanamixis Pellegr., 195
Aglaia beddomei (Kosterm.) Jain & Gaur, 196
Aglaia odorata Lour., 191, 193, 195
Aglaia polystachya Wall. ex Roxb., 196
Ailanthus, 186
Aloysia triphylla (L'Hérit.) Britt., 279
Alstonia angustifolia Wall. ex A. DC., 247
Alstonia macrophylla Wall. ex G. Don., 248–249
Alstonia pangkorensis King & Gamble, 248
Alstonia spatulata Bl., 250–251
Alstonia spectabilis R. Br., 249–250
Alstonia villosa Bl., 249
Amoora aphanamixis Roem. & Schult., 191
Amoora grandifolia Bl., 195
Amoora rohituka (Roxb.) Wight & Arn., 196
Amygdalus communis var. *dulcis*, 124
Anamirta cocculus, 41
Anamirta paniculata, 41
Anasser lanitii Blanco, 258
Ancyllocladus cochinchinensis Pierre, 257
Anisophyllea disticha Hook. f., 121–122
Anplectrum barbatum Wall. ex C.B. Clarke, 135
Anplectrum cyanocarpum (Bl.) Triana, 135
Anplectrum divaricatum (Willd.) Triana, 135
Anplectrum glaucum (Jack) Triana, 135
Anplectrum patens Geddes, 135
Anplectrum stellulatum Geddes, 135
Antidesma frutescens Jack, 166
Antidesma ghaemsembilla Gaertn., 166
Antidesma paniculatum Bl., 166
Antidesma pubescens Roxb., 166
Aphanamixis cumingiana C. DC., 196
Aphanamixis grandifolia Bl., 191, 195–196
Aphanamixis polystachya (Wall.) Parker, 196

Aphanamixis rohituka (Roxb.) Pierre, 191, 196
Aphanamixis sinensis How & Chen, 196
Arcangelica flava (L.) Merr., 42–44
Ardisia boissieri A. DC., 63
Ardisia corolata, Roxb., 55–56
Ardisia elliptica Thunb., 56
Ardisia fuliginosa Bl., 56–57
Ardisia hainanensis Mez., 58
Ardisia humilis Vahl, 57–58
Ardisia lanceolata Roxb., 58–59
Ardisia littoralis Andr., 56
Ardisia odontophylla Wall., 59
Ardisia oxyphylla Wall., 59–60
Ardisia pyramidalis (Cav.) Pers., 61–62
Ardisia pyrgina Saint Lager, 58
Ardisia pyrgus Roemer & Schultes, 58
Ardisia ridleyi King & Gamble, 62
Ardisia squamulosa, Presl., 63
Arduina carandas (Linnaeus) K. Schumann, 251
Aristolochia clematis, 31
Aristolochia contorta, 31
Aristolochia kaempferi, 31
Aristolochia philippinensis Warb., 31–32
Aristolochia recurvilabra, 31
Aristolochia reticulata, 31
Aristolochia serpentaria, 31
Asarum europeaum, 31
Asclepias, 261
Atalantia kwangtungensis Merr., 216
Atalantia monophylla DC., 215
Atalantia roxburghiana Hook. f., 216
Atalantia spinosa Tanaka, 215

B

Backeria barbata (Wall. ex C.B. Clarke) Raizada, 135
Baeobotrys ramentacea Roxb., 68
Beilschmiedia pahangensis Gamb., 16–17
Beilschmiedia tonkinensis Ridl., 17–18
Blastus cogniauxii Stapf., 134–135
Brucea, 186
Brucea javanica (L.) Merr., 186
Bruguiera eripetala W. & A. ex Arn., 148
Bruguiera sexangula (Lour.) Poir., 148
Bryonia cochinchinensis Lour., 110

Chemical Compounds Index

Subject Index

A

Adenia cordifolia Engl., 102–104
Aegiceras corniculatum Blco., 54–55
Aegle marmelos Correa, 212–215
Ageing
    antioxidant
        extracts, 35, 95
        lignans, 12
        naphthoquinones, 73
        tannins, 144, 168, 203
    free radicals, 73, 124
        scavenging by
            DPPH, 13, 56
            flavonoids, 133
            lignans, 12
            principles, 73
            tannins, 168
Aglaia odorata Lour., 191
Alstonia angustifolia Wall. ex A. DC., 247
Alstonia macrophylla Wall. ex G. Don., 248–249
Alstonia spatulata Bl., 250–251
Alstonia spectabilis R. Br., 249–250
Analgesic
    alkaloids, 226, 272
    extracts, 172
    lignans, 235
Anisophyllea disticha Hook. f., 121–122
Antibacterial
    extracts, 142, 177, 208, 219, 221, 249
    lignans, 12
    phenolics, 9, 23
    principles, 68, 92, 147, 201, 249, 252
    quinones, 77, 208, 219, 221, 249
Antidesma ghaemsembilla Gaertn., 166
Antidiarrheal
    alkaloids, 256
    extracts, 206, 208, 214
Antifeedant, 198, 199, 202, 249, 252
Antifungal, 249
    alkaloids, 219
    extracts, 214
    limonoids, 197, 198
    naphthoquinones, 77
Antimycobacterial
    alkaloids, 219, 223

lactone, 224
    limonoids, 219
Antiparasites
    extracts, 203, 264, 283
    flavonoids, 179
Antiplasmodial
    alkaloids, 44
    amide, 28
    atovaquone, 73
    coumarin, 226
    flavonoids, 130
    principles, 45, 274
    quassinoids, 185, 188
    quinine, 17
Antiprotozoal
    alkaloids, 44, 256
    principles, 26, 44
Antipyretics, 172, 214
Antivenom, 97, 105
Antiviral
    alkaloids, 34, 159, 161, 215, 228
    coumarins, 219
    diterpenes, 95, 161, 163, 228
    extracts, 124, 159, 172, 174
    lignans, 21
    limonoids, 219
    naphthoquinones, 77, 172, 174
    phenolics, 95, 97, 179
    polyacetylenic fatty acids, 157
    polysaccharides, 147, 153
    principles, 33, 63, 73, 77, 87, 122, 130, 147, 154, 219,
        239, 252
    protein, 112
    tannins, 133, 168
    xanthones, 239
Aphanamixis grandifolia Bl., 195–196
Aphanamixis rohituka (Roxb.) Pierre, 196
Apoptosis induction, 84, 113, 221, 224, 265, 272, 275
Arcangelina flava (L.) Merr., 42–44
Ardisia corolata Roxb., 55–56
Ardisia fuliginosa Bl., 56–57
Ardisia humilis Vahl, 57–58
Ardisia lanceolata Roxb., 58–59
Ardisia odeontophylla Wall., 59
Ardisia oxyphylla Wall., 59–60
Ardisia pyramidalis (Cav.) Pers., 61–62
Ardisia ridleyi King & Gamble, 62